Gerhard Pfaff (Ed.)

Encyclopedia of Color, Dyes, Pigments

Also of Interest

Encyclopedia of Color, Dyes, Pigments.
Volume 1: Antraquinonoid Pigments - Color Fundamentals
Gerhard Pfaff (Ed.), 2021
ISBN 978-3-11-058588-9, e-ISBN (PDF) 978-3-11-058807-1,
e-ISBN (EPUB) 978-3-11-058663-3

Encyclopedia of Color, Dyes, Pigments.
Volume 2: Color Measurement - Metal Effect Pigments
Gerhard Pfaff (Ed.), 2021
ISBN 978-3-11-058684-8, e-ISBN (PDF) 978-3-11-058710-4,
e-ISBN (EPUB) 978-3-11-058694-7

Inorganic Pigments
Gerhard Pfaff, to be published in 2024
ISBN 978-3-11-074391-3, e-ISBN (PDF) 978-3-11-074392-0,
e-ISBN (EPUB) 978-3-11-074399-9

Applied Inorganic Chemistry.
Volume 1: From Construction to Photofunctional Materials
Rainer Pöttgen, Thomas Jüstel, Cristian A. Strassert (Eds.),
to be published in 2022
ISBN 978-3-11-073814-8, e-ISBN (PDF) 978-3-11-073314-3,
e-ISBN (EPUB) 978-3-11-073332-7

Applied Inorganic Chemistry.
Volume 2: From Magnetic to Bioactive Materials
Rainer Pöttgen, Thomas Jüstel, Cristian A. Strassert (Eds.),
to be published in 2022
ISBN 978-3-11-073837-7, e-ISBN (PDF) 978-3-11-073347-1,
e-ISBN (EPUB) 978-3-11-073357-0

Encyclopedia of Color, Dyes, Pigments

Volume 3: Mixed Metal Oxide Pigments – Zinc Sulfide Pigments

Edited by
Gerhard Pfaff

DE GRUYTER

Editor
Prof. Dr. Gerhard Pfaff
Allee der Kosmonauten 69
12681 Berlin
Germany
pfaff.pigmente@gmx.de

ISBN 978-3-11-058686-2
e-ISBN (PDF) 978-3-11-058712-8
e-ISBN (EPUB) 978-3-11-058688-6

Library of Congress Number: 2021948066

Bibliographic information published by the Deutsche Nationalbibliothek
The Deutsche Nationalbibliothek lists this publication in the Deutsche Nationalbibliografie; detailed bibliographic data are available on the Internet at http://dnb.dnb.de.

Cover image: Gettyimages/Sagittarius Pro
Typesetting: Integra Software Services Pvt. Ltd.
Printing and binding: CPI books GmbH, Leck

www.degruyter.com

FSC www.fsc.org MIX Papier aus verantwortungsvollen Quellen FSC® C083411

Preface

The three-volume "Encyclopedia of Color, Dyes, Pigments" represents this attempt to summarize the current expertise in the fields stated in the title of the book. The main objective is to present the information that is available today encompassed by these three terms in a scientifically and technically correct, high-level and up-to-date manner. All aspects from theory to practical application are covered.

As the title suggests, all major classes of dyes and pigments are covered in detail. Emphasis is given so that the reader obtains an overview of the basic principles, the synthesis possibilities, the production, the chemical and physical properties as well as the technical application of these colorants. Separate chapters are provided for the main areas of application of dyes and pigments, for color fundamentals and color measurement as well as for historical pigments, dyes and binders. The encyclopedia is addressed to color specialists in industry and academia as well as to dye and pigment users in the applications areas of coatings, paints, cosmetics, plastics, printing inks, ceramics, and building materials. In order to make orientation as easy as possible for the interested reader, the topics covered in the book are arranged in alphabetic order.

Color has always played an important role in the lives of humans and animals. Dyes and pigments were therefore important substances early on in the development of mankind for expressing oneself through color, white or black and for shaping life. Colors and their functions have always been fascinating, playing a major role in the human psyche, and are of great importance in the design of a wide variety of surfaces. While natural colorants were initially used by people for thousands of years, the development of modern natural sciences, especially chemistry, led to the introduction of synthetically produced dyes and pigments in the lives of people. These play a predominant role among today's colorants.

The treatment of the individual topics differs in many cases from that in other reference books. New ways of presentation are chosen for different classes of dyes and pigments, but also for the application of colorants in various systems. The encyclopedia is thus up to date, especially since the latest findings from the field have been included. The authors involved are very familiar with the contents of their chapters, most of them having researched and worked on them for many years.

The objective of the Encyclopedia of Color, Dyes and Pigments is to provide a comprehensive overview of the state of knowledge about dyes and pigments, while at the same time identifying the connections to the relevant coloristic and application fundamentals. Special attention was paid to developing a clear structure of the approximately 70 chapters to provide access to the desired information quickly and without a long search. Figures and tables inserted illustrate fundamental aspects, dye and pigment structures, manufacturing details as well as application examples. At the end of each chapter, references to further reading are given. These include a wide range of journal articles, reference books and patents.

https://doi.org/10.1515/9783110587128-202

The chapters to follow are written by some of the most knowledgeable authors in the subject areas covered in the encyclopedia. Most of them were and are involved in innovative developments and practical applications of colorants. They are authors of a variety of publications, patents, presentations and lectures in the respective areas. Their insights will certainly prove to be valuable to the reader and their contributions to the Encyclopedia of Color, Dyes, Pigments are greatly appreciated.

At this point, I would like to take the opportunity to express my sincere thanks to all the authors involved in this book project. Without their support and cooperation, the completion of the encyclopedia would not have been possible. Finally, a big thank you goes to Karin Sora from de Gruyter, who has supported and accompanied the project from the early beginning and to Vivien Schubert and Esther Markus from the same publisher, who have done an extraordinarily good job in the completion of the encyclopedia.

Berlin, December 2021 — Gerhard Pfaff

Contents

Volume 2

Volume 3

List of contributors

Adrian Abel
DCC Europe
Rossendale
Lancashire
United Kingdom
adrianfixby@gmail.com

Michael R. Bartelt
Business Unit Effect Pigments
Schlenk Metallic Pigments GmbH
Barnsdorfer
Hauptstr. 5, Roth
91154 Germany
michael.bartelt@schlenk.de

Robert Christie
School of Textiles & Design
Heriot-Watt University
Scottish Borders Campus
Galashiels TD1 3HF, United Kingdom
rmchristie66@gmail.com

Magdalene Gärtner
Maltechnik
Institut für Maltechnik
Muenstergasse
3, Schwaebisch Gmuend
73525 Germany
magdalene.gaertner@yahoo.de

Uwe Hempelmann
Technical Operations, Lanxess Deutschland GmbH
Rheinuferstraße 7-9, Krefeld
Nordrhein-Westfalen
47812 Germany
uwe.hempelmann@-lanxess.com

Heinz Langhals
Department of Chemistry
Ludwig-Maximilians-Universität
Butenandtstr. 13,
Munich D-81377
Germany
langhals@lrz.uni-muenchen.de

Ghita Lanzendörfer-Yu
Owner
DejaYu Kosmetikblog
Fuchsgrube 36
Mülheim
45478 Germany
info@dejayu.de

Frank Maile
Business Unit Effect Pigments
Schlenk Metallic Pigments GmbH
Barnsdorfer
Hauptstr. 5, Roth
91154 Germany
frank.maile@schlenk.de

Heinz Mustroph
Former FEW Chemicals GmbH
Technikumstraße 1,
Bitterfeld-Wolfen
06756 Germany
mustroph@few.de

Gerhard Pfaff
Former Merck KGaA
Allee der Kosmonauten 69, Berlin
12681 Germany
pfaff.pigmente@gmx.de

https://doi.org/10.1515/9783110587128-204

Thomas Rathschlag
Weilburger Graphics GmbH
Am Rosenbühl 5,
Gerhardshofen 91466
Germany
t.rathschlag@weilburger-graphics.de

Werner Rudolf Cramer
Optical research
Cramer Lackdesign
Hafenweg
22, Muenster 48155
Germany
info@wrcramer.de

Andrew Town
Lambson Ltd., Clifford House
York Road, Wetherby
United Kingdom of Great Britain and Northern Ireland
andrew.towns@lambson.com

Gerhard Pfaff

45 Mixed metal oxide pigments

Abstract: Mixed metal oxide (MMO) pigments, also knowns as complex inorganic color pigments (CICPs), are an important class of inorganic pigments. They are representatives of the blue, green, yellow, brown, and black pigments. These pigments are called mixed complex because they contain two or more different metals in the composition. Their main advantages are high opacity, heat stability, useful infrared properties, light- and weather fastness, and chemical resistance. MMOs are solid solutions of metal oxides, which are homogenously distributed in their crystal lattices. They are based on different types of crystal structures, such as rutile, spinel, inverse spinel, hematite, priderite, and pseudobrookite. Other products belonging to the CICPs besides the rutile and spinel type MMOs are pigments based on zirconium silicate (zircon, $ZrSiO_4$). Such compositions belong to the class of ceramic pigments (color stains for glazes).

Keywords: mixed metal oxide pigments, rutile type pigments, spinel type pigments, zirconium silicate color stains

45.1 Fundamentals and properties

Mixed metal oxide (MMO) pigments, also knowns as complex inorganic color pigments (CICP), are an important class of inorganic pigments. These pigments are called mixed complex because they contain two or more different metals in the composition. The combination of two or more metals in suitable structures allows the generation of a wide range of colors for MMO pigments.

Most of the MMO pigments are characterized by outstanding fastness properties, which are used in paints, plastics, building materials, glass coatings and ceramics [1–5].

A multitude of metals can be incorporated in the crystal lattice of MMO pigments as metal ions besides of oxygen ions. Most relevant for colored pigments are V, Cr, Mn, Fe, Co, Ni, and Cu. Some other metals in the pigments do not really contribute to the color, but they are needed for the stability of the crystal structure. They are therefore added as modifiers to balance the charge of the crystal lattice, or to slightly modify the shade produced by the coloring metal ions. Metals playing a role in this sense are Al, Si, Ti, Zn, Nb, Mo, Sb, and W.

This article has previously been published in the journal Physical Sciences Reviews. Please cite as: G. Pfaff, Mixed Metal Oxide Pigments *Physical Sciences Reviews* [Online] 2021, 6. DOI: 10.1515/psr-2020-0183

https://doi.org/10.1515/9783110587128-045

MMOs are solid solutions of metal oxides, which are homogenously distributed in their crystal lattices. They are based on different types of crystal structures, such as rutile, spinel, inverse spinel, hematite, priderite, and pseudobrookite (Table 45.1). The most decisive factor for the structure of MMOs is the metal/oxygen (M/O) ratio. It is for rutile MO_2 2.00, for spinel M_3O_4 1.33 and for hematite or corundum M_2O_3 1.50.

Table 45.1: Mixed metal oxide pigments with their chemical constituents, structures, and colors [3].

Color index	Chemical constituents	Structure	Colors
Pigment Yellow 53	Ni(II), Sb(V), Ti(IV)	Rutile	Green shade Yellow
Pigment Brown 24	Cr(III), Sb(V), Ti(IV)	Rutile	Ocher shade
Pigment Yellow 162	Cr(III), Nb(V), Ti(IV)	Rutile	Ocher shade
Pigment Yellow 164	Mn(II), Sb(V), Ti(IV)	Rutile	Brown
Pigment Yellow 119	Zn(II), Fe(II,III)	Spinel	Ocher-brown
Pigment Blue 28	Co(II), Al(III)	Spinel	Red shade blue
Pigment Blue 36	Co(II), Cr(III), Al(III)	Spinel	Green shade blue
Pigment Green 26	Co(II), Cr(III)	Spinel	Dark green
Pigment Green 50	Co(II), Ti(IV)	Inverse spinel	Green
Pigment Brown 29	Fe(II), Cr(III)	Hematite	Brown
Pigment Brown 35	Fe(II,III), Cr(III)	Spinel	Dark brown
Pigment Black 30	Ni(II), Fe(II,III), Cr(III)	Spinel	Black
Pigment Black 26	Mn(II), Fe(II,III)	Spinel	Black
Pigment Black 22	Cu(II), Cr(III)	Spinel	Black
Pigment Black 28	Mn(II), Cu(II), Cr(III)	Spinel	Black
Pigment Black 27	Co(II), Cr(III), Fe(II)	Spinel	Bluish black

Compositions with an M/O ratios of about 2.00 and metal ions of a radius similar to Ti(IV) crystallize preferred in the rutile structure. These compositions contain a main fraction of TiO_2 as the base oxide acting as the host lattice. Rutile MMOs have, not unexpected, several physical properties comparable to titanium dioxide powders in the rutile modification. The titanium ions in the rutile structure are coordinated octahedrally by six oxygen ions. In MMO pigments, where titanium is partly substituted by other metal ions, these ions are evenly arranged on Ti positions.

The most important rutile-type MMO pigments are nickel rutile yellow and chromium rutile yellow. Nickel ions, also known as chromium ions, located at titanium positions are responsible for the color of these pigments. The valence states of Ni(II) and Cr(III) differ from that of Ti(IV). The charge equalization is achieved by simultaneous incorporation of Sb(V) in the crystal lattices. The compositions of the pigments can therefore be described with $(Ti,Ni,Sb)O_2$ for nickel rutile yellow and $(Ti,Cr,Sb)O_2$ for chromium rutile yellow. In case of manganese rutile brown $(Ti,Mn,Sb)O_2$, titanium is substituted by manganese and antimony. It is likewise possible to replace antimony by tungsten or niobium. Pigments with such substitutions are the yellows $(Ti,Ni,Nb)O_2$, $(Ti,Ni,W)O_2$, $(Ti,Cr,Nb)O_2$ and $(Ti,Cr,W)O_2$ as well as the browns (Ti,Mn,Nb)O_2 and (Ti,Mn,W)O_2 [3, 6]. The amount of TiO_2 as the base oxide is typically in the range of 70 to 90 wt % TiO_2 for all MMO pigments of the rutile type.

The spinel structure AB_2O_4 (M_3O_4) occurs in many compositions containing metal ions of a main group element (A) together with ions of a transition metal element from the row Ti to Zn (B) and oxide ions. A number of transition metal oxides adopt this structure with the M/O ratio of 1.33 when the radii are fitting to the spinel lattice. The structure is energetically so favored that also stoichiometries different from M_3O_4 can crystallize in the spinel lattice. The crystals contain typically metal ion vacancies in such cases.

Metal ions in the spinel lattice are generally divalent (+ 2) and trivalent (+ 3), although differently charged ions can be accommodated. The structural coordination of the metal ions in the spinel lattice occurs in two different ways. The B ions in the AB_2O_4 structure are octahedrally coordinated by the oxide ions, the A ions are located in a tetrahedral environment. Normal spinel compositions are characterized by divalent metal ions in tetrahedral coordination and trivalent metal ions in the octahedrally surrounded positions. Inverse spinels, on the other hand, contain also divalent metal ions in sixfold coordination and some trivalent metal ions in positions with the coordination number four. Table 45.1 contains those spinel and inverse spinel compounds that are of most interest for industrial pigments.

Cobalt blue with the formula $CoAl_2O_4$ belongs to the most important inorganic blue pigments. It is a red-shaded blue pigment that crystallizes in the normal spinel structure. The partial replacement of cobalt by chromium leads to green-shaded blue pigments. These pigments crystallize also in the normal spinel structure. Typical cobalt blue pigment compositions contain up to 25 wt % Co and up to 27 wt % Cr [2].

An important MMO pigment based on the inverse spinel structure is cobalt titanate with the formula Co_2TiO_4. The pigment has a green color and is therefore called cobalt green. Typical cobalt green compositions contain up to 15 wt % Co. A partial replacement of cobalt by zinc and nickel is possible and leads to pigments with a specific green color shade.

Zinc ferrite with the formula $ZnFe_2O_4$ is characterized by an ocher-brown color and can also be used for pigment purposes. It crystallizes in the spinel structure. The variation in the Zn: Fe ratio allows a broad variety of shades from brighter to medium browns. Typical zinc values are in the range of about 40 wt %.

The color black is also covered by pigments based on the spinel structure. An example is spinel black, a copper chromite with the formula $CuCr_2O_4$. Chromium can be substituted partially in this composition by manganese and/or iron. Other compositions for black MMO pigments are iron chromium spinels. Cobalt or copper can additionally be introduced in the spinel lattice in this case.

Dark brown pigments are obtained when iron(III) in the iron oxide spinel Fe_3O_4 is replaced by chromium(III). Such pigments are referred as chromium iron brown and contain up to 17 wt % Cr.

Other products belonging to the CICPs besides the rutile and spinel type MMOs are pigments based on zirconium silicate (zircon, $ZrSiO_4$). Such compositions belong to the class of ceramic pigments (color stains for glazes).

Zirconium silicate ($ZrSiO_4$) stains are used almost exclusively in enamels and ceramics and can therefore be regarded as high-temperature pigments. Their advantage is the combination of sufficient color properties with extreme thermal and chemical stability. The pigments cover the color areas blue, yellow and red. The generation of many intermediate color tones is likewise possible [7–9]. All these pigments are based on the ZrO_2 crystal lattice, in which transition metal ions are incorporated. Most important representatives of the zircon-based pigments are zircon vanadium blue and zircon praseodymium yellow. The content of vanadium and praseodymium in the pigment compositions is only a few percent.

Zirconium silicate color stains cover also the so-called inclusion pigments [10]. There is no incorporation of coloring ions in the zircon crystal lattice in this case. The basic principle is the envelopment of colorants by the thermally and chemically stable $ZrSiO_4$ structure. Color-providing components are included in this crystal lattice. They are therewith protected and can be exposed to the extreme conditions of the ceramic glaze firing. Examples for zirconium silicate inclusion pigments are zircon iron pink (with up to 15 wt % Fe_2O_3) and products containing cadmium sulfoselenide included in $ZrSiO_4$. Other compounds usable as enveloping substances besides zirconium silicate are tin dioxide and silicon dioxide.

45.2 Production of mixed metal oxide pigments

Rutile and spinel MMO pigments are produced using solid state reactions of metal oxides, hydroxides or carbonates at temperatures in the range of 800 to 1300 °C. The amounts of the starting components are calculated corresponding to the compositions of the desired final pigments. The starting compounds are used in form of fine-grained powders. Intimate blending is necessary before the high-temperature reaction starts. The solid-state reaction is a high-temperature process, where the mixture of the metal compounds is converted into a compound with preliminary pigment properties. Finishing procedures such as micronizing, washing and drying are necessary to obtain powders with the desired final pigment characteristics.

45.2.1 Rutile mixed metal oxide pigments

MMO pigments crystallizing in the rutile lattice are formed by calcination of homogeneous mixtures of titanium dioxide as the later host component and the color giving guest components at temperatures above 1000 °C in air. These can be introduced as oxides but also as non-heat resistant, decomposable compounds of the necessary coloring metal. Metal hydroxides, oxide hydrates, carbonates, acetates, and nitrates are deployed. TiO_2 is mostly used in the more reactive anatase modification. An alternative to titanium dioxide is dried titanium dioxide hydrate. After the high-temperature reaction, the pigment brick is ground wet or dry to get the desired grain size for the pigment. The surface of so-obtained pigment particles can be equipped with an additional inorganic or organic treatment to enhance the weather resistance and the behavior in the intended binder systems.

The synthesis of nickel rutile yellow and chromium rutile yellow can be described exemplary by the following equations:

$$0.85\,TiO_2 + 0.05\ NiO + 0.05\ Sb_2O_5 \rightarrow (Ti_{0.85}Ni_{0.05}Sb_{0.10})O_2 \quad (45.1)$$

$$0.90\,TiO_2 + 0.025\ Cr_2O_3 + 0.025\ Sb_2O_5 \rightarrow (Ti_{0.90}Cr_{0.05}Sb_{0.05})O_2 \quad (45.2)$$

Sb_2O_3 is typically used as antimony-delivering raw material. Sb_2O_5 is formed during the reaction at high temperatures by oxidation of Sb_2O_3 with oxygen. The Sb_2O_5 formation is important in order to achieve electroneutrality in the final pigment composition. Sb_2O_4 can be detected as an intermediate, which reacts further with anatase to $(Ti,Sb)O_2$. This composition crystallizes already in the rutile lattice. The reaction of the intermediary formed $(Ti,Sb)O_2$ with the nickel or chromium oxide and oxygen to $(Ti_{0.85}Ni_{0.05}Sb_{0.10})O_2$ or $(Ti_{0.90}Cr_{0.05}Sb_{0.05})O_2$ proceeds in the final phase of the solid-state reaction. Other intermediates that could be detected during the formation of nickel rutile yellow are $NiSb_2O_6$ and $NiTiO_3$ [6].

45.2.2 Spinel mixed metal oxide pigments

MMO pigments with spinel structure are synthesized starting from intense mixtures of the fine-particular powdered raw materials. The reaction to the spinel pigments takes place in a calcination step at temperatures of 800 to 1300 °C. As in the case of the rutile MMO pigments, oxides or suitable decomposable metal compounds are used for the manufacture. Homogeneous starting mixtures can also be obtained by simultaneous precipitation of the hydroxides or carbonates of the involved metals. The calcination takes place in mineral crucibles in directly heated bogie hearth furnaces or tunnel kilns. Indirectly heated metallic rotary kilns are likewise possible. The products obtained after calcination can be regarded as raw pigments. They are ground as suspension in ball mills, washed, filtrated, dried, and dry ground.

The synthesis of cobalt blue and cobalt green can be described by the following equations:

$$CoO + Al_2O_3 \rightarrow CoAl_2O_4 \quad (45.3)$$

$$2\,CoO + TiO_2 \rightarrow Co_2TiO_4 \quad (45.4)$$

The synthesis of MMO pigments with hematite, priderite or pseudobrookite structure takes place similar to the described manufacture of the spinel type pigments.

45.2.3 Zirconium silicate pigments (ceramic colors)

Zirconium silicate pigments are produced in a two-step process. The first step includes the formation of the zircon host lattice. In a second step, the inclusion of the color-providing component in the previously formed zircon structure takes place. Basically, similar process steps are used as described for the synthesis of rutile or spinel MMO pigments.

Alkali metal halides, mainly sodium fluoride, in the starting mixture promote the color formation during the solid-state reaction. The color of the final pigments depends on various factors, such as the proportions of the color giving component and sodium fluoride, the purity of the raw materials, the particle sizes of the reactants and their initial degree of compaction. In particular, the atmosphere used for the high-temperature reaction plays a decisive role.

Some examples for zirconium silicate pigments are provided below (percent values correspond to wt %):
Zircon vanadium blue $(Zr,V)O_2$ [7]:
Starting mixture: 62% ZrO_2, 30% SiO_2 (host lattice former), 5% NH_4VO_3, 3% NaF (mineralizer)
Calcination conditions: 850–1000 °C, 2–3 hrs

Zircon praseodymium yellow $(Zr,Pr)O_2$ [18]:
Starting mixture: 62% ZrO_2, 30% SiO_2, 5% Pr_6O_{11}, 3% Na(F,Cl)
Calcination conditions: 1250 °C, 2–3 hrs

Zircon iron pink $(Fe_2O_3/ZrSiO_4)$ [7]:
Starting mixture: 45% ZrO_2, 22% SiO_2 (envelop former), 21% $FeSO_4 \cdot 7\,H_2O$, 12% Na(F,Cl)
Calcination conditions: 880–1100 °C, 1 h

Zircon-based colorants have to be produced using very pure chemicals. An expensive grade of ZrO_2 is required for the manufacture of ceramic colors with highest chroma. In order to save costs, a new process has been developed for the synthesis of zircon-based colorants directly from zircon sand, which is comparably cheap and readily available. Equimolar proportions of ZrO_2 and SiO_2 are used to form a specific $ZrSiO_4$ host lattice, which is the basis for intense colorants at the end of the process. The so-formed zircon is calcined with an alkali compound to form an alkaline silicozirconate. This intermediate is decomposed by an acid treatment and an intense zirconia-silica mixture is formed. A next calcination step is necessary to form a zircon-based ceramic colorant. This process can be used for the production of all zirconium silicate pigments.

45.3 Pigment properties and uses

MMOs are representatives of the inorganic blue, green, yellow, brown, and black pigments. Their main advantages are high opacity, heat stability, useful infrared properties, light- and weather fastness, and chemical resistance. The excellent heat stability of the MMO pigments is due to the high temperatures used for the production.

The mean particle sizes of MMO pigments are in the range from 0.5 to 2.5 µm. The high-temperature synthesis of the pigments often leads to the formation of larger and disturbing agglomerates. Longer grinding times are necessary to obtain pigment particles with a suitable size distribution. It is to be noted that MMOs, particularly nickel rutile yellow and chromium rutile yellow, exhibit a strong color dependency on the grinding time and intensity. Grinding of the MMO particles leads in many cases to more spherical particles with favorable color properties. Darker and more saturated pigments are thus possible [3].

MMO pigments are characterized by low surface porosities. Specific surface areas are mostly in the range of 3 to 6 m^2/g for rutile type MMOs and 3 to 10 m^2/g for spinel type MMOs. The pigments show in most cases a favorable dispersion behavior in their application media [3].

Some of the MMO pigments with spinel and hematite structure find interest because of their infrared reflective properties. They can be used for camouflage applications as well as for applications with a defined minimum total solar reflectance [11, 12]. MMOs developed specifically for infrared reflectance are the brownish to black pigments Pigment Brown 29, Pigment Brown 35 and Pigment Black 30 (Table 45.1).

Rutile MMO pigments have purer color shades in comparison with iron oxide yellow pigments. The application possibilities in the yellow color range are therefore broader. Reflectance spectra of nickel rutile yellow and chromium rutile yellow are seen in Figure 45.1.

Rutile MMO pigments are often combined with high performance organic pigments in their application in order to achieve brilliant colors. Compared with lead chromate, lead molybdate or cadmium pigments, such combinations have the advantage of nontoxicity. The addition of rutile yellows to formulations with other inorganic pigments leads mostly to an improvement of the weather stability of the system due to the UV absorbing properties of the inorganic MMO pigments.

Spinel MMO pigments are characterized by outstanding fastness properties. Chemical stability and weather fastness are excellent, also in brightening blends with titanium dioxide. The reflectance spectra of cobalt blue and cobalt green are shown in Figure 45.2.

Cobalt blue pigments represent the color area from red shaded to greenish turquoise shaded blue. The substitution of aluminum by chromium leads to a bathochromic shift in the maximum absorption of the spectra. Development work has led to specific cobalt blues, which enable an improvement of the UV opacity and the weathering properties. Pigment Green 26 is of interest for camouflage applications [3].

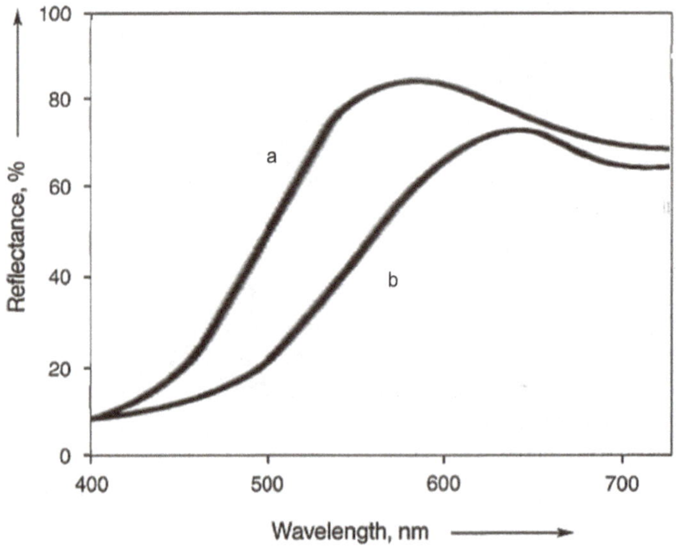

Figure 45.1: Reflectance spectra of (a) nickel rutile yellow and (b) chromium rutile yellow [5].

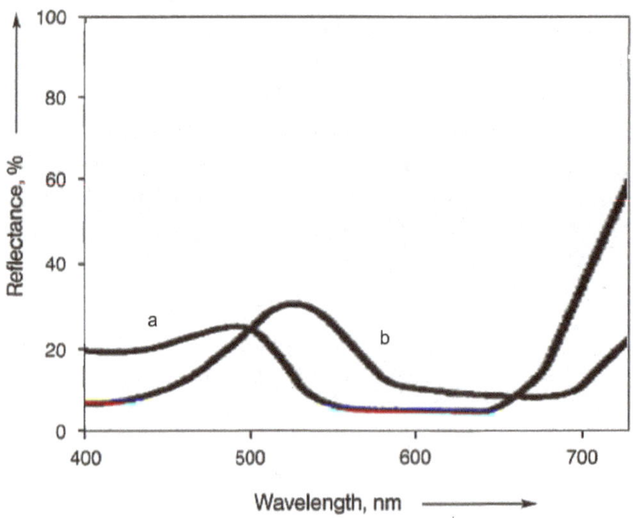

Figure 45.2: Reflectance spectra of (a) cobalt blue and (b) cobalt green [5].

Zinc ferrite pigments are interesting for the yellowish-brown color area. The pigments offer higher heat stability than iron oxide yellow pigments and can be used in plastics and coatings at processing temperatures above 120 °C. Compared with rutile yellows, zinc ferrite pigments offer less clean color shades.

Manganese titanate brown pigments have excellent weathering properties. They are often used in plastics, particularly in rigid-PVC (vinyl siding), where they replace iron containing pigments, which can degrade PVC catalytically under the influence of UV radiation [3].

Some of the spinel-type brown and black pigments, e. g., Pigment Brown 35 and Pigment Black 30, but also Pigment Brown 29 with its hematite structure are used for infrared reflecting materials. In contrast to other inorganic brown and black pigments like iron oxides, copper chromite or carbon black, these pigments are able to reflect wavelengths in the near infrared region [13].

Copper chromite black pigments like Pigment Black 22 are used because of their extreme durability and temperature resistance. They do not exhibit infrared reflectance. They are mainly applied in coatings for black and gray colors. Cobalt chromites like Pigment Black 27 are preferably used in applications, where higher heat stability is required. Manganese ferrites like Pigment Black 26 are seen as cost-effective black alternatives to other inorganic black pigments [3].

The zircon-based ceramic colorants vanadium zircon blue, praseodymium zircon yellow and iron zircon coral fulfill the demands for high-temperature applications in a very effect manner. These colorants are stable up to temperatures of 1350 °C in all types of glazes. Cross-mixing of the pigments is the basis for a wide range of high-temperature stable colors.

Vanadium zircon blue is estimated to be the best ceramic colorant. The amount of vanadium in the zircon lattice controls the color strength of this pigment. The incorporation of the vanadium in the lattice depends to a large extent on the mineralizer component used in the reaction mixture.

Praseodymium zircon yellow is a bright yellow pigment for high-temperature applications. The addition of cerium oxide allows the production of orange shades. Lead compounds as mineralizers instead of alkali metal halides lead to more intense colors. Praseodymium zircon yellow is brighter and cleaner in comparison with other temperature stable yellow colorants.

MMO pigments with rutile and spinel structure do not exhibit acute toxicity (LD_{50} value rat oral: >5000 mg/kg). They are not irritating to skin or mucous membranes. Most of the MMOs are highly inert chemical compounds, containing no relevant bioavailable or degradable components. The pigments are considered nontoxic and comply with food contact as well as toy safety regulations [5].

The incorporated metal oxides exist in the pigments no longer as chemical individuals. They are fixed components of the rutile or spinel lattice. Investigations with the focus on antimony, chromium, cobalt, copper, iron, manganese, and zinc have shown no indication for a relevant bioavailability of these metals. The rutile MMOs as well as the spinel MMOs are no subjects to labelling as hazardous materials [1].

Toxicology and occupational health of the zirconium silicate inclusion pigments containing cadmium sulfoselenide have to be considered separately and discussed together with the pure cadmium containing pigments.

References

1. Endriss H. Aktuelle anorganische Buntpigmente, Zorll U, editor. Hannover: Vincentz Verlag, 1997:94.
2. Etzrodt G. In Kittel - Lehrbuch der Lacke und Beschichtungen. Spille J, editor. vol. 5. 2nd ed. Stuttgart/Leipzig: S. Hirzel Verlag, 2003:96.
3. Brussaard H. In industrial inorganic pigments. Buxbaum G, Pfaff. G, editors. 3rd ed. Weinheim: Wiley-VCH Verlag, 2005:116.
4. White J. In high performance pigments. Faulkner EB, Schwartz RJ, editors. 2nd ed. Weinheim: Wiley-VCH Verlag, 2009:41.
5. Pfaff G. Inorganic pigments. Berlin/Boston: Walter de Gruyter GmbH, 2017:115.
6. Maloney J. In high performance pigments. Faulkner EB, Schwartz R, editors. 2nd ed. Weinheim: Wiley-VCH Verlag, 2009:53.
7. Batchelor RW. Modern inorganic pigments. Trans Br Ceram Soc. 1974;73:297.
8. Kleinschmidt P. Ein Kapitel angewandte Festkörperchemie: Zirkonsilicat-Farbkörper Chem Unserer Zeit. 1986;20:182.
9. Pfaff G. in Winnacker – Küchler, Chemische Technik, Prozesse und Produkte. Dittmeyer R, Keim W, Kreysa G, Oberholz A, editor. vol. 7. 5th ed. Weinheim: Wiley-VCH Verlag, 2004:334.
10. Beyer HH, Broll A., Kleinschmidt P. Einschlußpigmente, Farbkörper nach einem neuen Prinzip. Chem Ztg. 1977;101:319.
11. White JP. Complex inorganic color pigments: Durable pigments for demanding applications. Paint Coat Ind. 2000;16:54.
12. Bendiganavale AK, Malshe VC. Infrared reflecting inorganic pigments. Recent PaT Chem Eng. 2008;1:67.
13. Burkhart G, Detrie T, Swiler D. When black is white. Paint Coat Ind. 2001;17:30.

Robert Christie and Adrian Abel

46 Monoazo (Monohydrazone) pigments based on acetoacetanilides

Abstract: The monoazoacetoacetanilide series of pigments, traditionally known as Hansa Yellows, are long-established products that entered the market in the early twentieth century. They are mostly inexpensive products offering bright colors of moderate intensity covering the entire yellow area of the spectrum, good lightfastness, but inferior solvent resistance. The technical properties of the pigments may be explained by their molecular structures, which adopt the ketohydrazone tautomeric form, and their crystal structures. Their good lightfastness is attributed mainly to intramolecular hydrogen-bonding, while their generally inferior fastness to organic solvents is explained by the relatively weak intermolecular interactions in the crystal structure. The monoazoacetanilide pigments are synthesized by the traditional two-stage process of diazotization of a primary aromatic amine, followed by an azo coupling reaction of the resulting diazonium salt with an acetoacetanilide coupling component. Their main use is in decorative paints, although a few products are suitable for printing inks.

Keywords: monoazoacetoacetanilide, monohydrazone, Hansa yellow, decorative paints, intramolecular hydrogen-bonding, phenylsulfonamide group, calcium salt pigments, diazotization, azo coupling, CI Pigment Yellow 1, CI Pigment Yellow 3, CI Pigment Yellow 6, CI Pigment Yellow 65, CI Pigment Yellow 73, CI Pigment Yellow 74, CI Pigment Yellow 97, CI Pigment Yellow 98, CI Pigment Yellow 111

46.1 Fundamentals

The most important classical yellow azo pigments are structurally described as the azoacetoacetanilides, which are derived from azo coupling reactions with acetoacetanilide derivatives as the coupling components. This chapter focusses on the long-established monoazoacetanilide pigments, generally referred to as Hansa Yellows, which exist structurally as monohydrazones, while the structurally-related disazoacetoacetanilides are covered in a separate chapter.

This article has previously been published in the journal Physical Sciences Reviews. Please cite as: R. Christie, A. Abel, Monoazo (Monohydrazone) Pigments Based on Acetoacetanilides *Physical Sciences Reviews* [Online] 2021, 6. DOI: 10.1515/psr-2020-0185

https://doi.org/10.1515/9783110587128-046

46.2 History

The monoazoacetoacetanilide group of pigments includes some of the earliest insoluble colors that did not contain salt-forming acid groups (SO_3H and/or CO_2H) reacted with appropriate metal salts to make them insoluble, a process often referred to as laking. The first patent for this class of yellow pigments was granted to AG Farbwerke vorm Meister Lucius & Brüning (later to become Farbwerke Hoechst, and then Clariant) following their discovery by Wagner in 1909 [1]. They entered the market in 1910 and were given the trade name Hansa Yellow, which is still a registered trade name to Clariant in Europe, but which lost its protection in the US in 1910. The first Hansa yellow pigment introduced into the market was CI Pigment Yellow 1, generally known as Hansa Yellow G, and became the forerunner of a series of structurally related monoazoacetanilide pigments. They remain an important group of classical organic yellow pigments, although some of the original products no longer occupy as an important position as they did up to the twenty-first century. Many are still in use for specific applications, while others have disappeared from the market.

46.3 Structures and properties

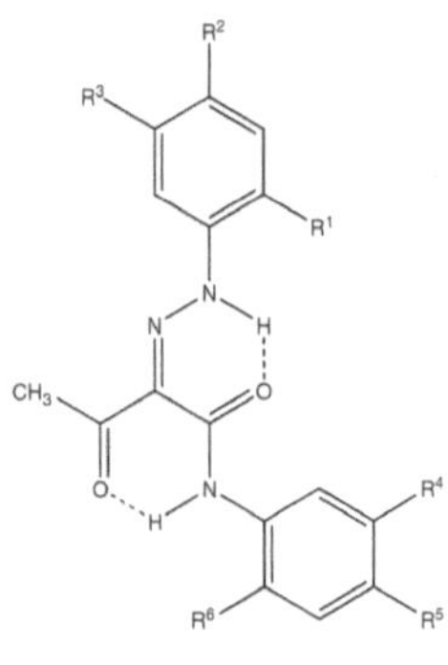

1

Figure 46.1: General structure of monoazoacetoacetanilide pigments (Hansa Yellows) (**1**).

Compounds (**1a**) – (**1j**) (Figure 46.1 and Table 46.1) represent the more important commercial examples of the monoazoacetoacetanilides. They are, in general, pigments of relatively low cost that are characterized by bright yellow colors of moderate intensity, and generally good lightfastness, but inferior resistance towards organic solvents. They tend therefore to be used in applications where this deficiency is not critical, such as in water-based systems and in solvent-based alkyd resins for decorative (architectural) paints [2–5].

This group of azo pigments was the focus of many of the earliest crystal structure determinations of industrial organic pigments using single crystal X-ray techniques, including compounds (**1a**) [6], (**1b**) [7], (**1c**) [8], (**1d**) [9], (**1f**) [10], (**1h**) [11], and

Table 46.1: Substituent pattern in monoazoacetoacetanilide pigments.

Compound	CI Pigment Yellow	R^1	R^2	R^3	R^4	R^5	R^6
1a	1	NO_2	CH_3	H	H	H	H
1b	3	NO_2	Cl	H	Cl	H	H
1c	6	NO_2	Cl	H	H	H	H
1d	65	NO_2	OCH_3	H	OCH_3	H	H
1e	73	NO_2	Cl	H	OCH_3	H	H
1f	74	OCH_3	NO_2	H	OCH_3	H	H
1g	75	NO_2	Cl	H	H	OC_2H_5	H
1h	97	OCH_3	SO_2NHPh	OCH_3	OCH_3	Cl	OCH_3
1i	98	NO_2	Cl	H	CH_3	Cl	H
1j	111	OCH_3	NO_2	H	OCH_3	H	Cl

(**1i**) [12]. The results of these studies have been reviewed [13]. The molecules all exist in the ketohydrazone tautomeric form and most adopt an essentially planar molecular configuration, although slight twisting of the amide aryl ring is observed in some cases. The molecular structure of CI Pigment Yellow 1 (**1a**), Hansa Yellow G (Figure 46.2) serves as a useful illustration of the features that may be used to account for the properties of this series in application. The good lightfastness of this product may be attributed mainly to the stability resulting from intramolecular hydrogen-bonding in 6-membered rings, in which the *ortho*-nitro and amide groups participate as illustrated in Figure 46.2. Intramolecular hydrogen bonding, a feature commonly encountered throughout organic pigment classes, is proposed to enhance lightfastness by offering electronic and steric protection of the azo (hydrazone) chromophore towards photochemical degradation. The electronic protection arises from reduction in electron density at the chromophore, thus reducing susceptibility towards photo-oxidation. The inferior fastness to organic solvents arises from the relatively weak intermolecular interactions in the crystal structure which mainly involve only van der Waals' forces, although close molecular packing, facilitated by molecular planarity, maximizes such interactions. In the case of compounds (**1b**), (**1d**) and (**1f**), weak intermolecular H-bonding has been suggested [13]. CI Pigment Yellow 74 (**1f**) is an especially important member of the monoazoacetoacetanilide series because of its higher color intensity and generally improved technical properties. In its crystal structure, essentially planar molecules are stacked on top of one another in columns. Between the molecules in the columns, it is proposed that there is weak (C-H ... O) hydrogen bonding between a methyl group in one molecule and a carbonyl oxygen atom in another, a factor that may contribute towards enhance technical performance. The enhancement in color strength has been demonstrated as be due to the nitro group *para* to the azo (hydrazone) linkage, rather than in the *ortho* position as is the case with most other members of the series [14]. CI Pigment Yellow 97 (**1 h**) is a product of some importance, since it offers excellent color strength and lightfastness, and

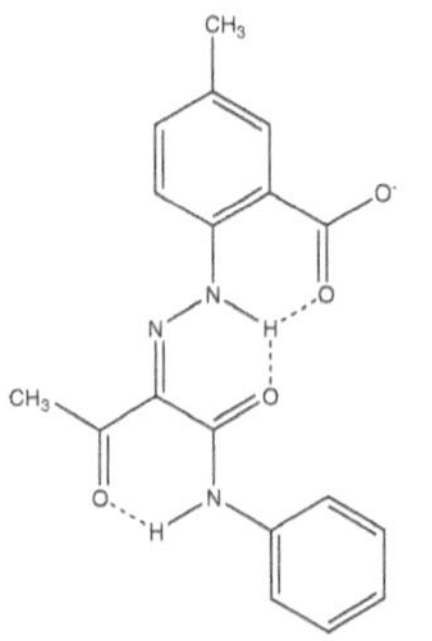

1a

Figure 46.2: Structure of CI Pigment Yellow 1 (Hansa Yellow G) (**1a**).

significantly improved solvent resistance compared with the rest of the Hansa Yellow series. The X-ray single crystal structure of CI Pigment Yellow 97 illustrates that it exists as a ketohydrazone with strong intramolecular hydrogen bonding [11]. The phenylsulfonamide group, the presence of which is relatively unusual in the range of commercial organic pigments, appears to play a vital role in determining superior technical performance. The phenyl ring of the phenylsulfonamide group is positioned like an open lid over the dimethoxy aromatic ring with an angle between the planes of 52°. It is conceivable that the phenylsulfonamide ring may be acting as a UV filter leading to improved lightfastness. The sulfonamide group also participates in intermolecular hydrogen bonding in which two groups form a dimeric arrangement, thus enhancing solvent resistance. CI Pigment Yellow 111 (**1j**) shows superior fastness properties, attributed to the presence of two additional amide groups, although its industrial importance is restricted because it lacks color intensity.

There are a few commercially produced yellow monoazoacetoacetanilide derivatives that are calcium salt pigments, including CI Pigments Yellow 61 (**2a**) and 62 (**2b**), as illustrated in Figure 46.3. These products are of modest importance in plastics applications.

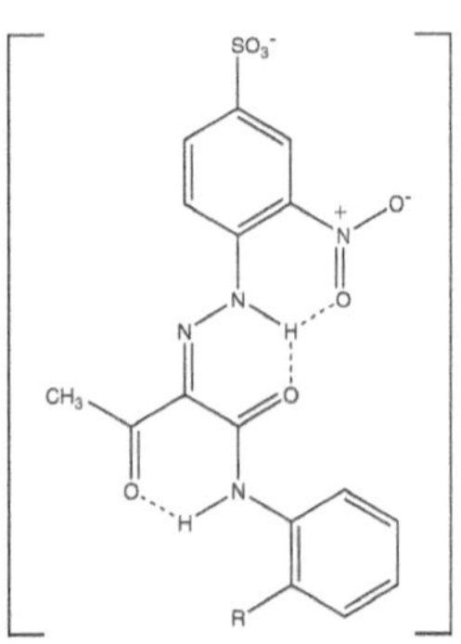

Ca^{2+}

2a: R=H

2b: R=CH_3

Figure 46.3: Metal salt monoazoacetoacetanilide pigments.

46.4 Synthesis and manufacture

The monoazoacetanilide pigments are synthesized by the traditional two-stage process used for essentially all azo dyes and pigments, as discussed in detail in the separate chapter Azo Pigments: General Principles. This route involves the diazotization of appropriately substituted aromatic amines to form the diazonium salt, followed by azo coupling. In the case of monoazoacetanilide pigments, diazotized anilines are coupled to acetoacetanilide derivatives. Azo coupling is conducted within the pH range 4–6 and takes place at the reactive methylene group carbon, which reacts via the deprotonated anionic form. The overall process is illustrated, as an example, for CI Pigment Yellow 1 (**1a**) in Figure 46.4. In this case, the diazo component is *o*-nitro-*p*-toluidine and the coupling component is the unsubstituted acetoacetanilide. In the manufacture of CI Pigment Yellow 97 (**1 h**), control of the particle size distribution is required to ensure its superior fastness properties. This can be achieved by conditioning the pigment with a solvent treatment, thereby eliminating very fine particles.

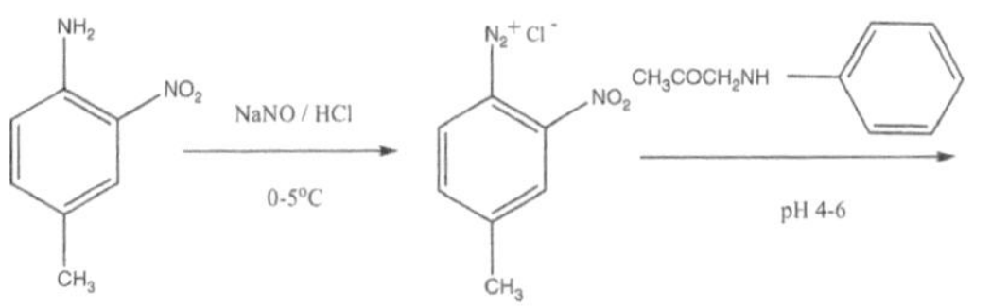

Figure 46.4: Synthesis of CI Pigment Yellow 1 (**1a**).

46.5 Applications

The monoazoacetoacetanilide group of pigments covers the entire yellow area of the spectrum, from primrose yellow to red shade yellow. They serve as useful examples of how the shade, strength, and fastness properties can be influenced by the various substituents that can be added to the basic chromophoric structure. They are characterized by moderate through to very good lightfastness, but many suffer from solubility in more powerful organic solvents, especially when exposed to elevated temperatures in application. Their main use is in decorative paints. They mostly provide moderate color strength, although pigments in the series that are the strongest tinctorially may find application in printing inks.

CI Pigment Yellow 1 (1a)

This was the first pigment in the series to be introduced. It is often referred to as Yellow G, the letter G recognizing that its shade is just on the green side of yellow [15]. The pigment provides reasonably good lightfastness but its inferior fastness to solvents limits its use to decorative (architectural) paints, especially water-based products. Other uses are in artists colors and textile printing if fastness to dry cleaning fastness is not required. Most grades provide moderate opacity, with smaller particle size grades showing increased color strength, a shift towards greener shades, and increased transparency. Although thermally stable to 140°C, it is not generally considered suitable for stoving paints, as the pigment shows some solubility in the solvents used. This eventually leads to reprecipitated pigment on the surface of the paint, a phenomenon known as blooming. A similar phenomenon also occurs when pigments dissolve in molten polymer then migrate to the surface. In plastics it is generally known as migration. CI Pigment Yellow 1 is little used in printing inks due to its low color strength but is occasionally used if better lightfastness is require than can normally be achieved from disazoacetoacetanilide (diarylide) yellow pigments. It is normally quite easily dispersed into most media and there are special easy-dispersible grades, which require only a high- speed stirrer to achieve optimum dispersion. This is a useful feature because the dispersion process, especially in bead mills, generates heat. As the temperature rises, the pigment may dissolve, initially leading to high strength and purity of color. However, as the system subsequently cools, the pigment starts to recrystallize, resulting in reduced color strength and a shift to a redder shade. Since the recrystallized particles can act as seeds for crystal growth, these color shifts can take place over a prolonged period, even weeks. This feature is especially troublesome when the yellow pigment is used in combination with other pigments, such as copper phthalocyanine blue. The blue pigment remains insoluble throughout, whereas the yellow pigment dissolves, then recrystallizes and loses its strength, resulting in shifting final shade towards the bluer side. The popularity of CI Pigment Yellow 1 has declined in recent years in favor of CI Pigment Yellow 74 (**1f**).

CI Pigment Yellow 3 (1b)

This pigment is often known as Yellow 10 G, recognizing its very green shade. The chloro substituents appear to contribute to a greener shade and reduced color strength compared with CI Pigment Yellow 1 (**1a**). The pigment offers very good to excellent lightfastness but slightly inferior solvent fastness compared with CI Pigment Yellow 1, making it of little use in stoving paints. While never achieving the popularity of CI Pigment Yellow 1, it has been widely used in combination with copper phthalocyanine blue to produce a wide range of pure green colors. However, the decline in the price of phthalocyanine green pigments has made this combination less attractive economically. It can be used in several special applications, such as artists colors, textile printing, and applications where its poor solvent resistance can be tolerated. Easily dispersible grades are widely available that can be fully dispersed using a high-speed stirrer.

CI Pigment Yellow 6 (1c)

This pigment is slightly greener than CI Pigment Yellow 1 (**1a**) and was one of the original pigments discovered, then designated as Hansa Yellow 3 G. It is currently of little significance.

CI Pigment Yellow 65 (1d)

This pigment was discovered by IG Farben. It was originally named Hansa Yellow 3RN, denoting its reddish shade. Its use in Europe diminished many years ago, but it has retained a place in the US market, partly on account of its use in road lining paints.

CI Pigment Yellow 73 (1e)

This pigment is slightly greener in shade than CI Pigment Yellow 1 (**1a**), has similar tinctorial strength, and slightly lower lightfastness, while usually a little more expensive. Its major advantage was realized with the introduction of modern bead mills. It shows much less color instability due to recrystallization arising from the heat generated in these mills, compared with most other monoazoacetoacetanilide pigments. This feature makes it a useful pigment for packaging inks, where higher lightfastness is required than can be obtained from disazoacetoacetanilide (diarylide) yellow pigments.

CI Pigment Yellow 74 (1f)

This pigment was developed in the 1960s and has replaced CI Pigment Yellow 1 (**1a**) for many applications. In its original form, it possesses about two and a half times the tinctorial strength of traditional CI Pigment Yellow 1, but with similar lightfastness, at the same depth of shade. It is often known as Yellow 5GX, signifying its greener shade, with X denoting higher color strength. It is not considered suitable for stoving paints as its fastness to solvents is slightly inferior to that of CI Pigment Yellow 1. Because of its higher color strength, it finds favor in some printing ink applications, especially water-based inks, having a similar color strength to CI Pigment Yellow 12, a disazoacetoacetanilide (diarylide) yellow, in both offset and letterpress applications. Its shade meets some yellow standards for three- and four-color printing processes. It is a little too green for the European standard, although this can be corrected with additions of redder shade pigments. Its lightfastness is good, decreasing to moderate in paler white reductions, although it fades at less than a quarter the rate of CI Pigment Yellow 13, an equivalent shade disazoacetoacetanilide (diarylide) yellow pigment. However, inferior heat stability makes it less useful for inks that are calendered or require to be sterilized. Special grades have been developed for the ink industry, which are even stronger and somewhat greener than the pigment as originally developed. This extra strength is achieved by careful control of the aqueous heat treatment finishing process after completion of coupling, leading to smaller particle size with a uniform size distribution compared with the conventional pigment. The paint industry has turned to an alternative grade of CI Pigment Yellow 74. For this product, processes were developed to grow the particles and provide a narrow particle size distribution that maximizes the opacity of the commercial pigment. The conventional grade has a surface area of around 34 m^2g^{-1} and an average particle size of around 135 nm, while the opaque grade has a surface area of around 16 m^2g^{-1} and an average particle size of around 340 nm. The opaque pigment grade provides a shade and color strength close to CI Pigment Yellow 1, lower oil absorption, which means that it can be used at higher concentrations, and improved levels of gloss in application. The opaque grade has very good lightfastness at least double that of the conventional pigment. The generally inferior fastness to recrystallization of CI Pigment Yellow 74 means that it has limited its value in tinting systems, where color stability is a pre-requisite although the opaque grades offer much lower coloration costs compared with the high-performance pigments that are often used in such systems.

CI Pigment Yellow 97 (1h)

Introduced by Hoechst in the 1960s, this pigment has more of the characteristics of a high-performance pigment than a Hansa yellow. The structural reasons proposed for the significantly enhanced insolubility and very good lightfastness of this pigment have been discussed in a previous section of this chapter. However, it has been observed that some commercial grades give lower lightfastness and solvent fastness than is normally associated with this pigment, due to differences in the conditioning processes in the manufacture of the pigment. Therefore, users must test any product in their systems to ensure that their requirements are fully met, not assuming that all pigments with this designation are identical. The pigment has similar color strength shade compared to CI Pigment Yellow 1 (**1a**). It is widely used in stoving paints where it is usually stable to 180°C. For a period, it was even used for automotive paints, although there are now many products with superior lightfastness and better resistance to bleeding, so that it no longer meets modern requirements for automotive finishes but is still marginally acceptable for automotive refinishes. In decorative paints, CI Pigment Yellow 97 held a special place in tinting systems that were used to color alkyd paints with a high aromatic solvent content, where its stability to recrystallization was a key feature. The current trend towards solvents with a lower aromatic content or ideally with no volatile organic compounds (VOCs) has reduced this demand. Although not widely used in printing inks, where its moderate color strength and higher cost limit its appeal, it occasionally finds use where its high lightfastness meets the demands for packaging and posters that may be exposed to significant levels of sunlight. When introduced, the pigment was recommended for the coloration of many polymers used in plastics, especially rigid PVC, but it tends to bleed when used at low concentrations in plasticized PVC. Above 200°C, the pigment dissolves in many polymers, giving it some of the characteristics of solvent soluble dyes. This can be tolerated in crystalline polymers but is unacceptable for the more amorphous polymers such as polyolefins, in which the pigment can migrate. In the search for organic alternatives to the use of lead chromate pigments, which are discouraged on environmental grounds, CI Pigment Yellow 97 was used as a replacement for lead chromate in the coloration of decorative laminate paper. Disazoacetoacetanilide (diarylide) yellow pigments generally had too low lightfastness, whereas most monoazoacetanilide pigments had inferior stability to the resins. In this case, high-performance organic pigments proved economically unacceptable. CI Pigment Yellow 97 proved to be a highly acceptable compromise and is still widely used, for example as the background to pine wood effect bases.

Table 46.2 presents a summary of some color and fastness properties of the more important monoazoacetoacetanilide yellow pigments, as assessed in decorative paints based on an alkyd resin binder. The data are compiled from pattern cards and literature sources, but should not be regarded as absolute, as different commercial products, even from the same company, can vary in performance. As far as possible the data refer to a standard product. The data are presented to allow a broad comparison between the products, but only as a guide to selection as there are so many variables that can influence the properties in application. It is essential that potential users test the actual pigments they intend to use in their own system based on their own test procedures. In the table, qualitative lightfastness ratings are given: E- excellent, VG – very good, G – good, R – reasonable, M – moderate, F – fair, P – poor. For solvent fastness, the ratings refer to the staining of the solvent, from 5 (no staining) to 1 (severe staining), when the paint sample is immersed in the solvent. The heat stability assessments refer to the maximum temperature to which the pigment is stable when subjected to that temperature for 30 minutes.

Table 46.2: Summary of some color and fastness properties of the more important monoazoacetoacetanilide yellow pigments.

Color	CI Pigment Yellow	Relative color strength	Lightfastness		Solvent Fastness			Heat stability/
Greenish yellow			Full shade	1/3 std. depth	Xylene	Ethanol	Methyl ethyl ketone	°C for 30 min.
↓	3	0.4	VG-E	VG-E	2	2	2	140
	74 transparent	2.3	VG-E	G	2	3–4	2–3	140
Mid yellow	97	0.7	E	VG	4	4–5	2–3	180
↓	73	1	VG-E	G-VG	2–3	3–4	2–3	160
	1	1	VG-E	R	2–3	3	2–3	140
Reddish yellow	74 opaque	0.9	E	VG	2	3–4	3	140
	65	1	G	R-G	1–2	3	1	140

CI Pigment Yellow 98 (1i)

This pigment was growing in popularity until it was withdrawn when one of the components used in its manufacture was no longer available. It was a very green shade, similar to CI Pigment Yellow 3 (**1b**) and was often known as Yellow 10GX.

CI Pigment Yellow 111 (1j)

This pigment was first introduced as Irgalite Yellow F4G in the early 1970s by Geigy as a greenish shade yellow pigment that mainly found application in printing inks. In most printing inks it offers high tinctorial strength, with good lightfastness. It provides exceptionally good flow properties in application. This pigment has declined in use, although ECHA (European Chemicals Agency) records that between 10–100 tonnes are still manufactured or imported into Europe annually. Only three companies registered it for REACH and one of them, Clariant, withdrew its registration in 2020.

CI Pigment Yellows 2, 4, and 5 are no longer regarded as important products although still registered in the Colour Index (2020).

References

1. Farbwerke vorm Meister Lucius & Brüning. DE 257,488 (17th Oct 1909).
2. Christie RM. Colour chemistry, 2nd ed, London: RSC, Ch 3, 2015.
3. Zollinger H. Color chemistry, 3rd ed, Weinheim: Wiley-VCH Verlag GmbH, Ch 7, 2003.
4. Christie RM. The organic and inorganic chemistry of pigments, surface coatings reviews, London: OCCA, 2002.
5. Hunger K, Schmidt MU. Industrial organic pigments, 4th ed, Weinheim: Wiley-VCH Verlag GmbH, Ch 2, 2019.
6. Paulus EF. Molecular and crystal structure of C.I. Pigment Yellow 1. N-Phenyl-2-(4-methyl-2-nitro-phenyl-hydrazono)-3-oxo-butyramide (Hansagelb G). Z Kristallogr. 1984:167:65.
7. Whitaker A. The crystal structure of C.I. Pigment Yellow 3, α-(1-hydroxyethylidene)-2-chloro-acetanilide-α-azo-(4′-chloro-2′-nitrobenzene). Z Kristallogr. 1983:163:19.
8. Whitaker A. The crystal structure of aceto-acetanilide azo-pigments II. C.I. Pigment Yellow 6, α-(1-hydroxyethylidene)acetanilide-α-azo-(4′-chloro-2′-nitrobenzene). Z Kristallogr. 1983:163;139.
9. Whitaker A. CI Pigment Yellow 65. Acta Crystallogr. 1987:C43;2141.
10. Whitaker A, Walker NP. CI Pigment 74. 2-((2-Methoxy-4-nitrophenyl)hydrazono)-N-(2-methoxyphenyl)-3-oxobutanamide. Acta Crystallogr. 1987:C43;2137.
11. Christie RM, Hill J, Rosair G. The crystal structure of CI Pigment Yellow 97, a superior performance Hansa yellow. Dyes Pigm. 2006:71;194–8.
12. Whitaker A. The crystal structure of acetoacetanilide azo pigments IV. C.I. Pigment Yellow 98, α-(1-hydroxyethylidene)-acet-2-methyl-4-chloroanilido-α-azo(4′-chloro-2′-nitrobenzene). Z Kristallogr. 1984:167;225.
13. Whitaker A. Crystal structures of azo pigments derived from acetoacetanilide. J Soc Dyers Colour. 1988:104;294.
14. Christie RM, Standring PN, Griffiths J. Colour and constitution relationships in organic pigments. Part 1—Monoazoacetoacetanilides. Dyes Pigm. 1988:9;37.
15. Kositze G. In: Patton TC, editor. In pigment handbook. New York: John Wiley & Sons; p.444, 1973.

Robert Christie and Adrian Abel

47 Monoazo (Monohydrazone) pigments based on benzimidazolones

Abstract: A series of azo pigments containing the benzimidazolone ring were introduced in the mid to late twentieth century as high-performance organic pigments in the yellow, orange, red, and brown shade areas. The structures of the commercial benzimidazolone azo pigments are derived from either the monoazoacetoacetanilide or monoazonaphtharylamide classical azo pigments systems and exist in the ketohydrazone tautomeric forms. The high-performance properties of the pigments have been attributed to a network of intermolecular hydrogen bonds involving the benzimidazolone group, and efficient molecular packing, as demonstrated by X-ray crystal structure determinations. The manufacturing processes leading to the pigments involve traditional diazotization and azo coupling reaction procedures, although they require special conditioning aftertreatments to optimize their performance. Although benzimidazolone azo pigments were initially developed for the coloration of plastics, they have probably had a greater impact on the paint industry. The application properties of the benzimidazolone azo pigments are discussed for individual products.

Keywords: azo pigments, benzimidazolone azo pigments, benzimidazolones, monohydrazone, azoacetoacetanilide, azonaphtharylamide, intermolecular hydrogen bonds, tetrahydroquinoxalinedione, CI Pigment Yellow 120, CI Pigment Yellow 151, CI Pigment Yellow 154, CI Pigment Yellow 175, CI Pigment Yellow 180, CI Pigment Yellow 181, CI Pigment Yellow 194, CI Pigment Yellow 213, CI Pigment Orange 36, CI Pigment Orange 62, CI Pigment Orange 64, CI Pigment Red 171, CI Pigment Red 175, CI Pigment Red 176, CI Pigment Red 185, CI Pigment Red 208, CI Pigment Viloet 32, CI Pigment Brown 25

47.1 Fundamentals

As high-performance pigments were developed throughout the twentieth century to fulfil the requirements of demanding applications, such as automotive paints, and to complement copper phthalocyanine blue and green pigments in the other colors, attention turned to whether azo pigments could be designed to meet such demands. Two approaches proved successful. One approach aimed to incorporate amide groups and structurally related heterocyclic ring systems to raise the level of fastness properties,

This article has previously been published in the journal Physical Sciences Reviews. Please cite as: R. Christie, A. Abel, Monoazo (Monohydrazone) Pigments Based on Benzimidazolones *Physical Sciences Reviews* [Online] 2021, 6. DOI: 10.1515/psr-2020-0184

https://doi.org/10.1515/9783110587128-047

utilizing the ability to participate in strong hydrogen bonding, to enhance intermolecular association in the crystalline solid state. This approach was pioneered mainly by Hoechst through a comprehensive investigation of azo pigments containing heterocyclic side-groups [1]. The most successful outcome was the development of azo pigments containing the benzimidazolone ring system (Figure 47.1). The second approach principally involved increasing molecular size, leading to the range of disazo condensation pigments, as described in a separate chapter.

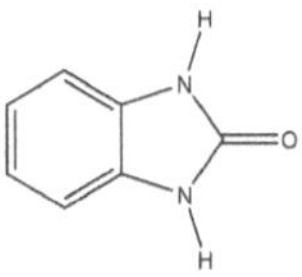

Figure 47.1: The benzimidazolonebenzimidazolone ring system.

47.2 History

The first patents claiming benzimidazolone azo pigments were applied for in 1960 [2, 3]. Their commercial introduction was the ultimate result of intense research carried out with the aim to produce pigments that were sufficiently heat stable to color polymers used in plastics such as polypropylene and styrene co-polymers and to increase the lightfastness and weatherability in coatings to meet the ever-increasing demands of the automotive manufacturers and coil coating producers. Following a wide-ranging investigation of azo pigments containing heterocyclic side-groups [1], Hoechst introduced their range of high-performance pigments containing the benzimidazolone ring system (Figure 47.1). The pigments are of either the azoacetoacetanilide (**1a-e**) or the azonaphtharylamide (**2a-d**) types, as illustrated in Figure 47.2, with the substituent patterns in the individual products listed in Tables 47.1 and 47.2, providing colors from yellow through orange, red and violet to brown [4, 5]. CI Pigment Orange 36 (**1g**) was the first benzimidazolone azo pigment, introduced to the market in 1964, with CI Pigment Yellow 120 (**1a**), the first yellow benzimidazolone azo pigment, launched in 1969. After these original introductions, several others were launched over time and new improved products have continued to be developed, including in more recent decades. The benzimidazolone group has also been used to enhance the performance and extend the color range of other chemical classes of organic pigments, for example the dioxazines.

47.3 Structures and properties

The structural cores of the commercial benzimidazolone azo pigments are derived from either the monazoacetoacetanilide or monoazonaphtharylamide systems. As

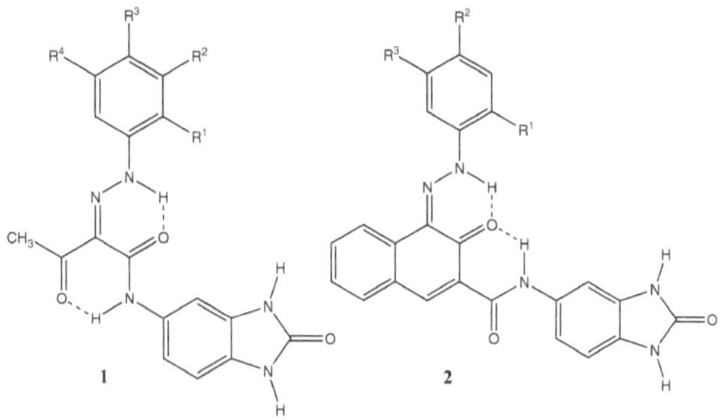

Figure 47.2: The structures of benzimidazolone azo pigments.

Table 47.1: The substituent pattern in benzimidazolone azo pigments based on monoazoacetoacetanilides (**1**).

Compound	CI Pigment	R^1	R^2	R^3	R^4
1a	Yellow 120	H	CO_2CH_3	H	CO_2CH_3
1b	Yellow 151	CO_2H	H	H	H
1c	Yellow 154	CF_3	H	H	H
1d	Yellow 175	CO_2CH_3	H	H	CO_2CH_3
1e	Yellow 181	H	H	*p*-$CONH(C_6H_4)CONH_2$	H
1f	Yellow 194	OCH_3	H	H	H
1g	Orange 36	NO_2	H	Cl	H
1h	Orange 62	H	H	NO_2	H

Table 47.2: The substituent pattern in benzimidazolone azo pigments based on monoazonaphtharylamides (**2**).

Compound	CI Pigment	R^1	R^2	R^3
2a	Red 171	OCH_3	NO_2	H
2b	Red 175	CO_2CH_3	H	H
2c	Red 176	OCH_3	$CONHC_6H_5$	H
2d	Red 185	OCH_3	SO_2NHCH_3	H
2e	Red 208	$CO_2C_4H_9(n)$	H	H
2f	Violet 32	OCH_3	SO_2NHCH_3	OCH_3
2g	Brown 25	Cl	H	Cl

in the case of these two groups of classical azo pigments, the benzimidazolone azo pigments exist in the ketohydrazone rather than the hydroxyazo tautomeric forms, for reasons discussed in the separate chapter – Azo pigments: General Principles. However, the properties are vastly different from those of the traditional monoazo pigments. The high stability of these pigments towards light and heat and their excellent solvent resistance has been attributed to extensive two-dimensional interlocking networks of intermolecular hydrogen bonds involving the benzimidazolone group. The first demonstrations of this feature arose from structural determinations of one compound of each type, established conclusively by single crystal X-ray investigations [6]. Neither of these compounds was a commercial product. Such studies of the industrial products proved problematic because of difficulty in growing suitable single crystals for analysis. Subsequently, crystal structures were obtained for CI Pigments Yellow 151 (**1b**), Yellow 154 (**1c**), Yellow 181 (**1e**), and Orange 36 (**1 g**) based on high quality powder diffraction data, confirming in each case the network of hydrogen bonds due to the benzimidazolone group, frequently involving other substituents, such as the $-CO_2H$ group in CI Pigment Yellow 151 (**1b**) [7]. However, analysis showed that there is no specific hydrogen bonding pattern in the series. Rather, the patterns vary to allow efficient molecular packing and formation of strong intermolecular hydrogen bonds. It was also noted that the pigments give rise to considerable polymorphism.

Figure 47.3 illustrates three pigments that are structurally different from the traditional benzimidazolone azo pigments (**1**) and (**2**). CI Pigment Yellow 180 (**3**) is a greenish yellow disazo pigment. The structure of CI Pigment 213 (**4**) contains the tetrahydroquinoxalinedione group, as opposed to the benzimidazolone group, derived from the coupling component. Thus, although not a benzimidazolone azo pigment, the pigment is structurally related and is often considered in the same series. CI Pigment Orange 64 (**5**) is a yellowish-orange pigment used in plastics. Its structure is rather different from pigments (**1**) and (**2**) in that the benzimidazolone group is derived from the diazo component.

47.4 Synthesis and manufacture

Pigments (**1**) and (**2**) are manufactured using traditional diazotization and azo coupling processes using coupling component (**6**) or (**7**), respectively, containing the benzimidazolone group (Figure 47.4). An exception is CI Pigment Orange 64 (**5**) in which an aminobenzimidazolone is diazotized and undergoes azo coupling with barbituric acid. The procedures commonly lead initially to crude products which xare hard in texture, low in crystallinity and show poor coloristics. Thus, they require a subsequent controlled conditioning process to develop crystal properties that are necessary for their performance in application. In a commonly used method, the aqueous pigment suspension is heated to 100–150 °C under pressure,

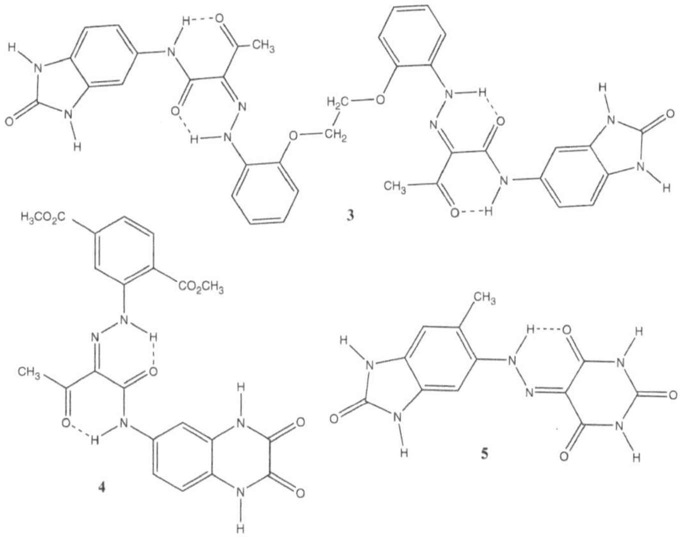

Figure 47.3: The structures of CI Pigment Yellow 180 (**3**), CI Pigment Yellow 213 (**4**), and CI Pigment Orange 64 (**5**).

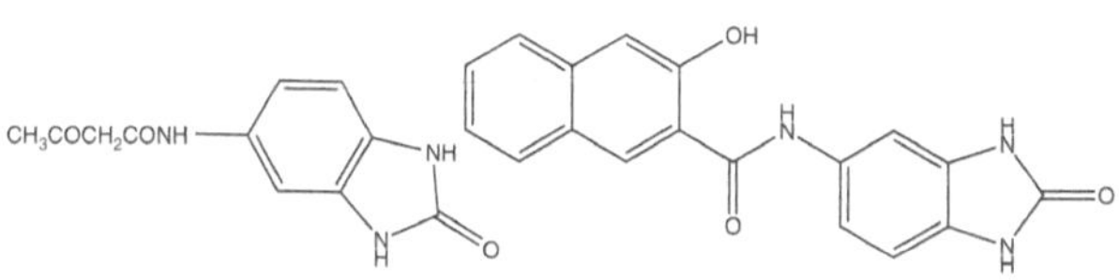

6 7

Figure 47.4: Coupling components used in benzimidazolone azo pigment synthesis.

optionally in the presence of organic solvents or surfactants. Alternatively, the crude product is heat-treated in organic solvents to develop the optimum crystal form and particle size distribution. Specific conditioning processes are especially important with the many benzimidazolone azo pigments that exhibit polymorphism to ensure that the process leads to the desired polymorph.

47.5 Applications

Although benzimidazolone azo pigments were initially developed for the coloration of plastics, they have probably had a greater impact on the paint industry, certainly in the case of the yellows. Most pigments of the red series tend to produce either rather dull shades with very good to excellent lightfastness or brighter pigments with lower lightfastness.

Yellow and orange series

Benzimidazolone azo pigments have had a more profound effect in the yellow and orange parts of the spectrum of colors, both for the coatings and plastic industries, than in the red (including brown) areas. Whereas no new red pigments of this type have been introduced for several decades, new yellows and oranges have been brought to the market in more recent times. Some pigments with the highest level of fastness properties have become alternatives to lead chromate and molybdate pigments. They cover a wide range of shades from very green shade yellows to red shade oranges. While the paint industry has been the main beneficiary of this chemistry, plastics and printing inks have also benefited.

CI Pigment Yellow 120 (1a)

This pigment, introduced in 1969, was the first yellow benzimidazolone azo pigment to be marketed, and is known as Yellow H2G, indicating that it is just on the green side of yellow. It is mainly used for the coloration of plastics, notably PVC. It does not bleed in plasticized PVC and has excellent lightfastness, even in pale reductions. In polyolefins, it is heat stable up to 270–290 °C. It can create some dimensional instability, leading to warpage in some injection moldings. It is suitable for coloring polystyrene, at temperatures up to 240 °C. It can be employed in specialist ink applications where its lightfastness is slightly better than that of CI Pigment Yellow 97, a monoazoacetoacetanilide pigment. It is particularly useful for gravure printing of PVC film, where powerful ketone and ester solvents are used. It has had some use in paints, although more suitable pigments were subsequently developed for coatings.

CI Pigment Yellow 151 (1b)

This was the first of a "next generation" of benzimidazolone azo pigments entering the market in 1970. The pigment was designated Yellow H4G, indicating that it is a little greener than CI Pigment Yellow 120 (**1a**). It has a large particle size, which results in lower tinctorial strength, but high opacity. Therefore, it is used mainly in full shades and has provided a natural alternative to yellow lead chromate, especially in automotive original equipment and vehicle refinish paints. It has excellent lightfastness in deep shades and only slightly lower in paler shades. Data obtained from external exposure tests carried out in Florida assess the pigment as unchanged after one year of exposure in full and strong shades, with a slight change in paler reductions. It is fast to over-coating up to 160 °C, and heat stable up to 200 °C. However, the presence of the carboxylic acid group in its structure means that it is sensitive to strong alkali, so that it has inferior fastness to soap and other alkalis and overall is not recommended for water-based systems. It also has limited suitability in amine-cured epoxy systems. Another grade of the pigment was subsequently introduced with even larger particle size, thus further improving opacity, reducing oil absorption, and allowing even higher covering power. In plastics, CI Pigment Yellow 151 can be used in PVC, polyolefins, and polystyrene. It has excellent resistance to migration in plasticized PVC, with excellent lightfastness even in pale shades. For the highest levels of fastness to weathering, CI Pigment Yellow 154 is preferred. Heat stability in polyolefins is assessed as up to 260–300 °C. It has a low tendency to cause warpage in injection moldings. It can be used to spin color polypropylene fibers. In polystyrene, it is stable to 260 °C–300 °C, and around 20 °C lower in ABS, and is suitable for achieving excellent lightfastness. It has some limited use in printing inks where good opacity and excellent lightfastness are required.

CI Pigment Yellow 154 (1c)

Under the designation Yellow H3G and released a few years later than CI Pigment Yellow 151 (**1b**), this pigment has a slightly redder shade. It is characterized by lightfastness rated as excellent in full shade and all reductions down to 1/25 standard depth in both air-drying alkyd and alkyd-melamine industrial paint systems. In an alkyd resin system, it shows no change in color when exposed to weathering for 12 months, and a just perceptible change after 24 months, even in paler depths. Thus, it is one of the most lightfast yellow organic pigments on the market, allowing it to be used in automotive original equipment paints and all industrial finishes. It has excellent solvent fastness to hydrocarbons and very good in ketones and esters. In contrast to CI Pigment Yellow 151, it is suitable for water-based systems. It does not bleed when over-coated up to 130 °C and can be stoved at temperatures up to 160 °C. It is fast to alkali and recommended for amine-cured epoxy systems. It is described as intermediate between transparent and opaque. There do not appear to be grades specifically aimed at the plastics industry. Nevertheless, CI Pigment Yellow 154 can be useful when high lightfastness is required but where high heat stability is not demanded. Its tinctorial strength in PVC is low, although it is suitable for window profiles in impact-resistant PVC and for coil coatings based on plasticized PVC. Its heat stability in polyolefins, up to 210 °C, is not sufficient to be useful. The pigment is tinctorially too weak to be of much use in printing inks.

CI Pigment Yellow 175 (1d)

This pigment is a green shade yellow as indicated by its designation Yellow H6G. It has a very pure shade and is used mainly for paints, where it meets current requirements for automotive original equipment. It is often blended with copper phthalocyanines to produce very pure green shades. It has fairly low tinctorial strength, although about 10% higher than CI Pigment Yellow 154 (**1c**). Although described in technical literature as opaque, it does not have particularly good covering power. It has excellent lightfastness, but not quite attaining that of CI Pigment Yellow 154. Its lightfastness remains excellent in pale shades, rated only slightly lower than in deeper shades.

Fastness to weathering follows a similar pattern. It has excellent resistance to most solvents used in paints but shows some solubility in ketones. When over-coated, it shows no bleed below 140 °C and remains stable at temperatures up to 180 °C. It is not recommended for water-based systems. Its low tinctorial strength limits its use in plastics, and although slightly stronger than CI Pigment Yellow 154, it does not quite achieve its level of lightfastness in PVC. However, it does have better heat stability than CI Pigment Yellow 154, and so can be used for coloring polyolefins, in which it is stable up to 270 °C. It does not promote dimensional instability. Its low tinctorial strength limits its use in inks to specialist applications where very high lightfastness is demanded, such as certain packaging applications and wallcoverings.

CI Pigment Yellow 180 (3)

Introduced as Yellow HG in 1980, this disazo pigment was developed for the plastics industry. It has a pure greenish yellow shade. It has very good heat stability and was one of the pigments developed to replace disazoacetoacetanilide (diarylide) yellows, which had been found to present toxicological issues when heated at temperatures above 200 °C. It is also widely used as an alternative to inorganic pigments based on cadmium and lead. The pigment is stable up to 290 °C in polyolefins, and while its lightfastness does not meet the highest standards, it is much better than the disazoacetoacetanilide (diarylide) yellows. It can also be used for spin dyeing of polypropylene where it shows high tinctorial strength. The pigment does not cause dimensional instability and so can be used for injection molded articles such as crates, where this feature is critical. It is suitable for both rigid and plasticized PVC, showing very good lightfastness in both full shade and reductions. It does not bleed in plasticized PVC and can be used for cable sheathing. In polystyrene, it is stable to 300 °C, but slightly lower in ABS. It can be used in some technical (engineering) polymers although is not recommended for certain polyamides or polyethylene terephthalate (PET) fibers. Although initially aimed at the coloration of plastics, CI Pigment Yellow 180 is finding application in other systems. In printing inks, it provides an alternative to disazoacetoacetanilide (diarylide) yellows where good heat stability is required, such as in metal decorative finishes that are dried above 200 °C. Special grades have been introduced for inkjet printing application, where it shows high color strength, good lightfastness, good transparency, and stability to UV curing.

CI Pigment Yellow 181 (1e)

This pigment was introduced as Yellow H3R around the same time as CI Pigment Yellow 180 (**3**). As the name suggests, it is a reddish shade yellow that was introduced for coloring plastics, specifically polyolefins, allowing a wide gamut of yellow shades when used in combination with CI Pigment Yellow 180, although it is tinctorially much weaker, having only about 40% of the color strength of CI Pigment Yellow 180. However, it has better fastness properties, with heat stability up to 300 °C and excellent lightfastness. It is considered highly suitable for spin coloration of viscose and polypropylene, where it meets the light and weather fastness levels suitable for automotive fabrics, joining a select group of pigments that satisfy the exacting requirements for such applications. Its low tinctorial strength means that it creates little interest for other applications, such as paint and inks.

CI Pigment Yellow 194 (1f)
First launched in the US as Yellow F2G, this is a strong mid-yellow pigment for industrial paints. Although not highly opaque, it may be used to provide alternatives to lead chromate pigments in reductions. The pigment has good to very good lightfastness and moderate weatherability in paler reductions. The pigment is stable to 200 °C but starts to bleed in alkyd-melamine systems at 160 °C. The pigment offers good heat stability (to 230 °C) and a fairly high tinctorial strength in polyolefins although it is not completely free from bleeding in plasticized PVC. It is one of the less expensive pigments in the series.

CI Pigment Orange 36 (1g)
Initially, this pigment was produced as a rather dull red-shade orange with good lightfastness and resistance to solvents. However, its dullness limited its value as in the orange area of the spectrum, the main demand was for a high-quality inexpensive pigment. In the mid-1970s, a new grade of the pigment was introduced with a larger particle size in a narrow distribution in the range that provides maximum opacity, although with lower tinctorial strength than the original grade. The high opacity grade was slightly redder and significantly brighter in shade than the original pigment, and thus a suitable alternative to molybdate orange pigments. It had excellent fastness properties suited to the automotive original equipment finish sector, where it was quickly adopted to provide Ford's Flame Red shade in combination with quinacridone violet. An added advantage is that unlike most molybdate orange pigments it does not darken in industrial atmospheres. The pigment is stable up to 160 °C in stoving paints, with a high resistance to over-coating. In outdoor weathering exposure tests, it is rated as very good after 12 months, but slightly lower after 24 months. Fastness to solvents is generally close to excellent, even in ketones and esters. Both grades find occasional use in printing inks where high lightfastness and resistance to solvents are required, such as those used for wallpaper, security, and laminate printing. The pigments are fast to sterilization and heat stable to 220 °C so that they are ideal for metal decorative inks. Although CI Pigment Orange 36 has been used in plastics, other products now offer better options. Its limited heat stability of up to around 220 °C precludes its use in most polyolefins and polystyrene, but it can be used to color PVC. It does not bleed in plasticized PVC where it has excellent lightfastness in deeper shades and very good in pale reductions. In rigid PVC, its lightfastness and weatherability may not achieve the same level,

depending on the system used. A non-flocculating grade has been introduced, mainly for coatings applications.

CI Pigment Orange 62 (1h)

Introduced as Orange H5G70 around 1973, this pigment has a pure, very yellowish shade. Its large particle size makes it highly opaque and provides excellent rheology in application while also ensuring high lightfastness. The pigment was developed for the paint industry. It is recommended for automotive finishes, although it is mainly used in vehicle refinishes and for commercial vehicles, rather than on mass produced original equipment finishes. Its lightfastness in full and deep shade is excellent, reducing to an extent in pale reductions. After weathering for 12 months, it is rated as unchanged in full shades. It is heat stable up to 180 °C but shows some bleeding above 140 °C. Its high opacity and good fastness properties make the pigment attractive for some specific printing ink applications, but these are severely limited by its low tinctorial strength. In plastics, it can be used in rigid PVC, but not if excellent weatherability is required.

CI Pigment Orange 64 (3)

This pigment was first marketed by Ciba in the mid-1970s under the designation Cromophtal Orange GP. BASF, who subsequently acquired Ciba, no longer market the pigment, but Clariant (and others) have introduced it as Orange H2GL into their range as a pigment for plastics. It provides a strong, brilliant yellow shade orange with excellent fastness properties and high heat stability. It finds use in most plastics other than polyamides and PET. Its heat stability can reach up to 300 °C for polyolefins, polystyrene, and polycarbonate, and a little lower (270 °C) in ABS. It does not bleed in plasticized PVC but has only limited suitability for electric cable sheathing. Lightfastness is assessed as excellent in full shade and as very good in reduction. It causes a little dimensional instability in polyolefins, which can lead to warping. The main application in printing inks is for metal decorative finishes, in which it is stable up to 200 °C.

CI Pigment Yellow 213 (4)

This greenish yellow semi-transparent pigment, designated Yellow H5G, offers high purity of shade with extremely high tinctorial strength and stability to heat up to 200 °C. It has excellent solvent fastness, even to ketones. Its lightfastness in both alkyd and alkyd-melamine resins is excellent both in full shade and pale reductions. Weatherability tests show that the color remains completely unchanged after exposure for 24 months, except in very pale shades in air-drying alkyds where it shows slight fading. It can be used in both solvent and water-based systems for automotive original equipment and repair systems. It is suitable for solid shades, metallics, and in blends with other pigments in the green, yellow, and orange parts of the spectrum.

Pigments that are no longer being marketed by their original company include CI Pigment Yellow 156 and CI Pigment Orange 60 (both Hoechst, early 1970s), which had been introduced as transparent pigments, where the transparency had been achieved by enlarging the particle size beyond that which gives maximum opacity, but which in turn severely reduced the tinctorial strength and therefore economy.

Red and brown series

The range of colors in this series of pigments ranges from rather dull mid-red, through violet to brown. The pigments show very good to excellent lightfastness, often better than with the disazo condensation pigments. The best performing pigments are used in the automotive industry. In contrast, where heat stability is paramount, the disazo condensation pigments often outperform benzimidazolone azo pigments. However, the technology involved in the synthesis of benzimidazolone azo pigments is inherently less expensive. The tinctorial strength of these benzimidazolone azo pigments varies considerably, influenced by the particle size of the pigment. The heat stability also varies among the different products.

CI Pigment Red 171 (2a)

A rather unusual shade, this pigment was originally marketed as Permanent Maroon HFM. As the commercial name suggests, it is a blue shade red. It is transparent, with high tinctorial strength and has been used in paint and plastics. In paints it has good fastness to bleeding with very good to excellent fastness to light and weathering. Its transparency makes it suitable for foil coatings and metallic finishes. Although it does not reach the standards required for automotive original equipment, it has been used in repair finishes. In plastics, it is stable to up to 180 °C in PVC, and migration fast in plasticized PVC. In polyolefins, it is stable up to 240 °C, but its lightfastness is rated only as good. The original manufacturer has ceased to market the pigment, although in Europe it has been registered under REACH regulations.

CI Pigment Red 175 (2b)

This is also a rather dull pigment, with very good solvent fastness, and is used in paints and plastics. In paints, it is mainly used for industrial finishes, including automotive refinishes. It is transparent and so can be used in metallic and pearlescent finishes. Its lightfastness is very good, and it is stable to 200 °C. In plastics, it can be used in both rigid and plasticized PVC, offering very good to excellent lightfastness, and can be considered for external coil coatings, where it satisfies all but the most demanding applications. It can be used in combination with carbon black for wood effect window profiles. It has better heat stability than CI Pigment Red 171 (**2a**) in polyolefins, in which it is stable up to 270 °C, and thus suitable for HDPE and PP fibers. It slightly affects the dimensional stability of the polymer, producing only minor warpage in injection molded articles. It is used for specialist ink applications, as it is fast to sterilization and has very good heat stability.

CI Pigment Red 176 (2c)

In contrast to most pigments in this group, this pigment is a clean, bluish red, one of two benzimidazolone azo pigments designated as carmine, in this case Carmine HF3C. It is primarily used in plastics but is also used for the coloration of decorative laminate papers. It has very good fastness to migration in plasticized PVC and good to very good lightfastness. It is heat stable to 200 °C. Common uses include cables and synthetic leather cloth. In polyolefins it is stable to 270 °C and shows very good lightfastness. The pigment has only a small tendency to cause warping in injection moldings. It is used for the mass coloration of polypropylene used in spin dyeing. In polystyrene, it is stable in full shade to 300 °C, but the stability decreases sharply as the depth of shade is reduced, down to 220 °C. It is used for both printing on to decorative laminate paper and for pulp mass coloration.

CI Pigment Red 185 (2d)

This pigment also provides a clean, bluish shade, designated Carmine HF4C. However, its main application is printing inks, where its shade is close to the standard magenta process color, and thus presents a higher quality alternative to the azonaphtharylamide CI Pigment Red 184. It has excellent fastness to solvents and is stable to sterilization, with heat stability up to 220 °C, making it a favored pigment for metal decorative inks. Its lightfastness is good to very good. In plastics, it is not as useful as CI Pigment Red 176 (**2c**), as it bleeds slightly in PVC and its heat stability limits its use in polyolefins to LDPE, but it may be used in lower temperature spin coloration of polypropylene fibers. It has a much greater tendency to cause warping than CI Pigment Red 176.

CI Pigment Red 208 (2e)

This is a mid-shade red, similar in shade to the 2B toners, metal salt azo pigments designated as CI Pigment Red 48, which inspired its name, Red HF2B. Its main application is in plastics, with some use in specialist inks, especially packaging gravure. In plasticized PVC it does not bloom but bleeds slightly at low concentrations. Its lightfastness is good to very good and it is stable up to 200 °C. In polyolefins, heat stability is dependent on the concentration, with limits varying between 200 and 250 °C depending on the depth of shade at which it is used. It can be used for transparent spin dyeing of polypropylene. It tends to cause warping in injection moldings. Because of its dielectric constant, it is suitable for electric cables. Its moderate lightfastness limits its use to interior applications. It can be used in combination with red shade yellows (e. g., CI Pigment Yellow 191, a calcium azopyrazolone salt) to produce a wide range of orange and scarlet shades. In inks, prints have excellent solvent fastness, and it offers good lightfastness. It is fast to sterilization and heat stable to 200 °C. It is suitable for use in decorative laminate prints. It has limited application in paints, where it can be used for general industrial finishes if high levels of fastness properties are not required.

CI Pigment Violet 32 (2f)

This pigment provides a bordeaux shade and can be used across a range of applications, although mainly plastics and printing inks. It has a similar shade to the thioindigoid pigment CI Pigment Red 88, which has been withdrawn from the market. In plastics, the pigment is fast to bleeding in plasticized PVC and gives high tinctorial strength, roughly double that of CI Pigment Red 88. In full shade, it has very good to excellent lightfastness, although only rated as good in pale shades. In polyolefins it has limited heat stability, up to 200–250 °C depending on the polymer grades and its concentration. It can change shade above 200 °C, but this does not usually prohibit its use. The pigment can be used for the spin coloration of secondary acetate and viscose, for which special grades are available. In printing inks, it provides a clean color tone, with high transparency. It offers good fastness to calendering. It is fast to most solvents but bleeds slightly in ketones. Other applications include wood stains, where it offers good lightfastness, often in combination with yellow and black to give deep brown shades.

CI Pigment Brown 25 (2g)

This pigment provides a reddish-brown shade and is highly transparent. While not completely fast to ketone and ester solvents, it finds application in general industrial paints, including automotive repair and even original finishes if the film is protected by a clear topcoat. Its transparency lends its use to metallic and pearlescent finishes, used in combination with transparent iron oxides, which provide some protection from UV light. It is stable up to 200 °C. It has excellent lightfastness, although decreasing slightly at low depths of shade. In plastics, it can be used for plasticized PVC, provided that its slight migration can be tolerated. It remains stable to 200 °C even in pale reductions, and its lightfastness is close to excellent. It is one of the few pigments suitable for window profiles based on PVC, and for coil coatings. In polyolefins it is stable to 290 °C and retains its excellent lightfastness. It is known to cause some dimensional instability leading to warpage in certain injection moldings produced at lower processing temperatures. It can be used for the spin coloration of polypropylene. Although there is little demand for brown printing inks, it can be used for specialty applications.

References

1. Dietz E, Fuchs O. Neue echte Monoazopigmente. Farbe Lack. 1973:79;1058–63.
2. Dietz E. Water-insoluble benzimidazolone mono-azo-dyestuffs. US 3,124,565. (10th March 1964).
3. Schilling K, Dietz E. US 3,124,565. (13th October 1964).
4. Christie RM. Colour chemistry, 2nd ed. London: RSC, Ch 9, 2015.
5. Hunger K, Schmidt MU. Industrial organic pigments, 4th ed. Weinheim: Wiley-VCH Verlag GmbH, Ch 2, 2019.
6. Hunger K, Paulus EF, Weber D. Über die Molekül- und Kristallstruktur gelber Mono-"azo"-Pigmente. Farbe Lack. 1982:88;453–8.
7. Van de Streek J, Jurgen B, Ivashevskaya SN, Ermrich M, Paulus EF, Bolte M, et al. Structures of six industrial benzimidazolone pigments from laboratory powder diffraction data. Acta Crystallogr. Sect B Struct Sci. 2009:B65;200–11.

Robert Christie and Adrian Abel

48 Monoazo (Monohydrazone) pigments based on 2-naphthol and derivatives

Abstract: This chapter describes the range of industrial monoazo pigments based on the 2-naphthol (β-naphthol) ring system. This group includes some of the earliest organic pigments introduced commercially and is also numerically the largest group of products currently described in the Colour Index. Most of the pigments within this group are red (with a few oranges), thus complementing the azoacetoacetanilides, which are mostly yellows. Three groups of monoazonaphthol-based pigments may be identified categorized according to the chemical structure of the coupling components used in their synthesis. The first group contains products based on 2-naphthol itself, a second is based on amide derivatives of 3-hydroxy-2-naphthoic acid (naphtharylamides), and the third is a series of metal salt azo pigments. The historical development of these pigments, outlined in an early section of this chapter, originated in the late 19th century with pigments described as 'lakes', derived from water-soluble anionic dyes absorbed on to inert colorless substrates, which were the forerunners of products now referred to as metal salt pigments. The non-ionic 2-naphthol-based pigments were introduced soon after. In the early to mid-20th century, a series of monoazonaphtharylamide (Naphthol AS) pigments were developed and introduced commercially. The pigments of this type that are currently manufactured can be sub-divided into products containing a single amide group (group 1) and higher performance products containing more than one amide or sulfonamide groups. Several group 1 pigments have diminished in importance over the years, while some higher performing group 2 pigments have grown in importance. The molecular and crystal structures of the range of pigments are presented and discussed in relation to their performance characteristics. The manufacture of the pigments involves the reaction of a diazotized aromatic amine with the appropriate 2-naphthol-based coupling component, using synthesis conditions typical of phenolic coupling components, followed by conditioning aftertreatments that are typical for azo pigments in general. Finally, there is an extensive discussion of the wide-ranging applications in which the individual pigments are used. While the pigments are especially well-suited to printing ink applications, many products also find use in paints and a few in plastics.

This article has previously been published in the journal Physical Sciences Reviews. Please cite as: R. Christie, A. Abel, Monoazo (Monohydrazone) Pigments Based on 2-Naphthol and Derivatives *Physical Sciences Reviews* [Online] 2021, 6. DOI: 10.1515/psr-2020-0186

https://doi.org/10.1515/9783110587128-048

Keywords: monoazo pigments, monohydrazone, azonaphthol, monoazonaphthol, azonaptharylamide, naphtharylamide, monoazonaphtharylamide, metal salt azo pigments, Naphthol AS, hydrogen bonding, polymorphism, β-oxynaphthoic acid, Lithol Red, Lithol Rubine, 2-naphthol

48.1 Fundamentals

There is a diverse range of classical monoazo pigments containing the 2-naphthol (β-naphthol) ring system. This series of organic pigments is numerically the largest in the Colour Index and includes some of the first organic pigments to be synthesized and introduced commercially. The pigments may be classified into three groups dependent on the coupling component type used in the synthesis. The first group contains products based on 2-naphthol itself. A second is based on the amide derivatives of 3-hydroxy-2-naphthoic acid (commonly known as β-oxynaphthoic acid or βONA), commonly encompassed by the term naphtharylamides, while a third is a series of metal salt pigments. Most of the pigments in this class are red in color (with a few oranges), thus complementing the azoacetoacetanilides, which primarily occupy the yellow shades in classical organic pigments.

48.2 History

The first metal salt azo pigment to become a commercial product was CI Pigment Red 49, a barium salt of an azo dyestuff derived from 2-naphthol (β-naphthol) as the coupling component. It was discovered and patented by Julius of BASF in 1899. Metal salt monoazo pigments derived from β-oxynaphthoic acid (βONA), including CI Pigment Red 53:1 (Lake Red C) and CI Pigment Red 57:1 (Lithol Rubine), were released into the market in 1902 and 1903, respectively. CI Pigment Red 57:1 currently remains one of the most important classical organic pigments. Monoazo pigments based on 2-naphthol were probably the earliest commercial organic pigments that did not contain anionic groups and made insoluble by metal salt formation. The first insoluble colorants of this type were introduced commercially as a class of textiles dyes rather than as pigments. In the application of the dyes, the cotton fibers were impregnated with an alkaline solution of the 2-naphthol and then the cotton immersed in a solution of the diazotized amine, so that an azo coupling reaction formed the insoluble colorants within the fibers. They were originally known as ice colors, as the process involved formation of a diazonium salt, thus requiring ice cold conditions. The dyeings obtained had extremely good washfastness, due to the insolubility of the colorant formed. They became known as azoic dyes, patented by Robert Holliday in 1880 and introduced to the market by Read Holliday, Huddersfield, England [1]. News of the acceptance of his patent reached Robert while on vacation in France, and so he named the initial product

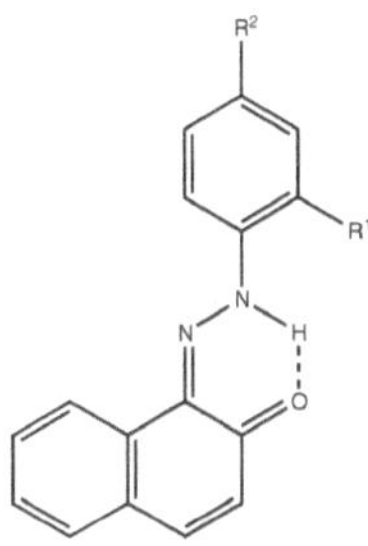

1

Figure 48.1: Structure of monoazonaphthols (1a)–(1e).

Table 48.1: Substituent patterns in monoazonaphthols.

Compound	CI Pigment	R^1	R^2
1a	Red 1	H	NO_2
1b	Orange 5	NO_2	NO_2
1c	Red 3	NO_2	CH_3
1d	Red 4	Cl	NO_2
1e	Red 6	NO_2	Cl

Vacanceine Red, after *vacance*, French for holiday. However, their market lead was soon lost to German companies who developed a whole range of products. Read Holliday ultimately became part of the British Dyestuff Corporation, which in turn formed part of ICI. Over the next few years, it was found that azoic dyeings based on naphtharylamide (Naphthol AS) couplings gave much better substantivity than those based on 2-naphthol. It soon became recognized that many of these azoic dyes could be used as pigments, and indeed many are still in use today. The earliest metal free azo pigment based on 2-naphthol was CI Pigment Red 1 (Para Red), introduced in 1885 by von Gallois and Ullrich of Frankfurt-Höchst. In the same year, its *o*-nitro isomer (CI Pigment Orange 2) followed. Two early products that are still of some importance today are CI Pigment Red 3 (Toluidine Red) and CI Pigment Orange 5 (Dinitroaniline Orange), introduced in 1905 and 1907, respectively. In 1911, Winter, Laska and Zitcher of Griesheim Elektron, Offenbach (later to become part of Hoechst) introduced a series of azo pigments based on the naphtharylamides as coupling components, and patented in the US a year later [2]. These pigments offered a wider range of colors than the 2-naphthol azo pigments and provided better fastness properties, although at higher cost. Further development of the so-called Group 1 azonaphtharylamide pigments continued through to the 1930s, mainly by IG Farben, the conglomerate formed from the large German chemical companies in 1925. After the breakup of IG Farben in the second half of the 1940s, further decades of research both in Germany and the US led to a new group (Group 2) of azonaphtharylamide pigments, aimed at satisfying the demands for the coloration of paints and plastics.

48.3 Structures and properties

The monoazonaphthols, corresponding to the general structure shown in Figure 48.1, are contained in a long-established group of orange and red monoazo pigments with structures that exhibit some similarities to those of the Hansa Yellows [3–6]. Table 48.1 lists the pigments (**1a**) – (**1e**) that are, or have been, of some commercial significance, X-ray crystallographic studies have demonstrated, for example, with compounds (**1c**) [7] and (**1e**) [8], that they exist as ketohydrazones, as illustrated for CI Pigment Red 3 in Figure 48.2.

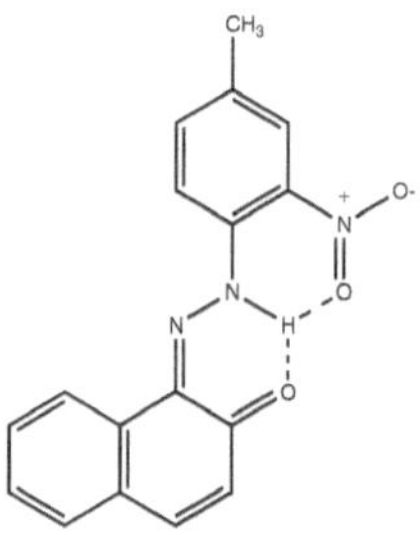

1c

Figure 48.2: Structure of CI Pigment Red 3.

Based on the similarities in their molecular structures and crystal structural arrangements, it is unsurprising that the monoazonaphthols exhibit similar application properties to the monoazoacetoacetanilides (Hansa Yellows). The reasonable lightfastness of the products is dependent on the presence *ortho* to the hydrazone function of a group, especially the nitro group as in CI Pigment Red 3 (Figure 48.2) which participates in strong bifurcated intramolecular hydrogen-bonding with the hydrazone N-H and carbonyl C = O. The monoazonaphthol pigments, like the Hansa Yellows, generally show inferior resistance to organic solvents, because of the weak intermolecular interactions in their crystal lattice structures, and their use is therefore restricted to applications where this issue is not significant, such as in decorative paints.

The azonaphtharylamide pigments are often referred to by the term Naphthol AS, the generic name for amides of 2-hydroxy-3-naphthoic acid (βONA), which are the coupling components used in their synthesis. The term Naphtol AS (adopting the German spelling) is a registered trademark, which currently belongs to Clariant. The pigments may be represented by the general structure (**2**) (Figure 48.3). These pigments show better solvent resistance than the azonaphthols (**1**) because of their increased molecular size and the presence of the amide functionality, and consequently they are suitable for use in a wider range of applications. Additional functional groups in this series of pigments are incorporated into individual products to modify color and enhance fastness properties. The pigments may be classified into two groups. Group 1 products contain only the one amide group. Group 2 pigments are products of superior solvent resistance, obtained by the incorporation of further amide or sulfonamide substituents into the ring systems, the technical performance generally improving with the number of such groups. The amide group especially, by virtue of its ability to participate in strong intra and intermolecular hydrogen bonding, enhances molecular planarity and strengthens intermolecular association, thus further improving the insolubility of the pigments. The additional amide or sulfonamide groups may also link to further aromatic rings, thus building up molecular size. Many Naphthol AS pigments are manufactured, the more important products given in Tables 48.2 (group 1) and 3 (group 2). Over the years, many group 1 pigments have diminished in importance, while some higher performing group 2 pigments have grown in importance.

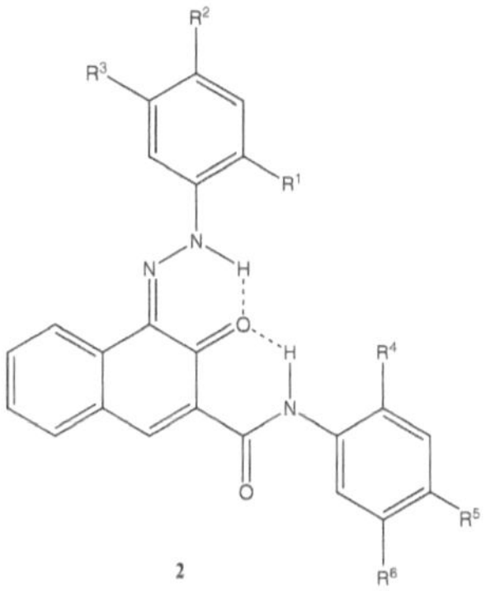

2

Figure 48.3: Azonaphtharylamide (Naphthol AS) Pigments.

Table 48.2: The substituent pattern in azonaphtharylamide (Naphthol AS) Pigments (Group 1).

Compound	CI PigmentRed	R^1	R^2	R^3	R^4	R^5	R^6
2a	2	Cl	H	Cl	H	H	H
2b	8	CH_3	H	NO_2	H	Cl	H
2c	9	Cl	H	Cl	OCH_3	H	H
2d	10	Cl	H	Cl	H	CH_3	H
2e	12	CH_3	NO_2	H	CH_3	H	H
2f	14	NO_2	Cl	H	CH_3	H	H
2g	22	CH_3	H	NO_2	H	H	H
2h	23	OCH_3	H	NO_2	H	H	NO_2
2i	95	OCH_3	H	$SO_2OC_6H_4NO_2(p)$	CH_3	H	H
2j	112	Cl	Cl	Cl	CH_3	H	H

A few single crystal X-ray structures of these pigments have been reported and the literature has been reviewed [9]. As illustrated in Figure 48.4 for CI Pigment Red 2 (**2a**) [10], the pigments exist in the ketohydrazone form, exhibiting extensive intramolecular H-bonding that involves the amide group. Polymorphism is a commonly encountered feature in this series of pigments. CI Pigment Red 170 (**2n**) is a pigment worthy of special mention as a product of some considerable current commercial importance, providing superior technical performance suited to many industrial coating applications. Unusually in organic pigments, CI Pigment Red 170 contains a primary amide (-$CONH_2$) group, together with the much more commonly encountered secondary amide (-CONHR) group. A notable feature of this pigment is its polymorphism, with α-, β-, γ-, and δ- forms well-established. The β- and γ-crystal modifications are both important commercially. The β-form has a bluer shade and is transparent. It is thus suited to high-performance printing ink applications. The γ-form is more yellow and opaque and is commonly used in industrial coatings. The α-form is coloristically uninteresting and is not used commercially. Both β- and γ-forms show extensive intermolecular hydrogen bonding, a feature that is no doubt primarily responsible for their excellent technical performance. In both cases, the primary amide group in one molecule forms a "dimeric" intermolecular hydrogen bonded arrangement with the equivalent group in another molecule [6, 11].

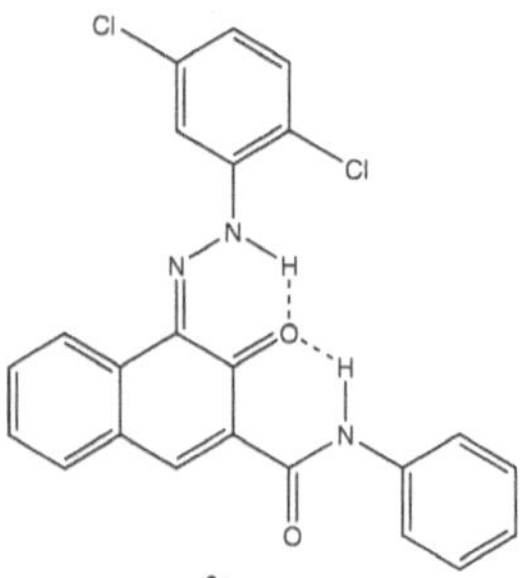

2a

Figure 48.4: Structure of CI Pigment Red 2 (**2a**).

The most important classical red azonaphthol pigments are metal salts [12]. These products are derived from azo compounds containing -$SO_3^-Na^+$ and/or -$CO_2^-Na^+$ groups by replacement of the Na^+ ions with divalent metal ions, notably the alkaline earths, Ca^{2+}, Sr^{2+}, Ba^{2+} and the transition metal Mn^{2+}. The water-soluble sodium salt is often referred to as the "dye". These "dyes" are reacted with various metal salts, such as calcium chloride, producing an insoluble salt. The divalent cations introduce some inorganic character into the pigment structures, a feature that dramatically reduces solubility. Metal salt pigments have evolved from early products referred to as *lakes*, which were essentially anionic azo dyestuffs precipitated on to inorganic substrates such as alumina or barium sulfate. The metal salt azo pigments currently in use no longer contain the inorganic substrate as an integral constituent. However, they are still commonly referred to as lakes (especially in the US) or alternatively as *toners*, for example, in the UK. These products are of prime importance for printing ink applications. Metal salt azo pigments provide colors from orange through to bordeaux. The nature of the metal ion is important as it influences both the shade and the properties of the pigment, including lightfastness and resistance to solvents. The shades generally become progressively bluer in the order (sodium → barium → strontium → calcium → manganese). This order also applies generally to lightfastness and resistance to solvents, which improve from sodium through to manganese. Sodium salts are inexpensive but always have some solubility in water. Barium salts are more expensive, but give rise to toxicological concerns, and so are to be avoided in sensitive applications such as toys, food packaging and children's reading matter. Calcium salts are the most economical and provide little or no toxicological concerns. Strontium salts are expensive but may provide substitutes for barium where there are toxicological issues. Manganese salts give the bluest shade and the highest lightfastness but are more expensive. However, manganese salts can react with certain ink binders causing emission of unpleasant odors due to oxidation of impurities in the resin and can accelerate the drying of inks and paints that dry by air oxidation.

Structurally, metal salt pigments may be grouped into three types according to the type of coupling component used in their synthesis. Of the pigments based on 2-naphthol, the most important are the Lithol Reds (**3b** and **3c**) CI Pigment Red 49 (Figure 48.5), and Lake Red C (**4a**), CI Pigment Red 53:1 (Figure 48.6). The metal salt is formed by a single sulfonate group derived from the diazo component.

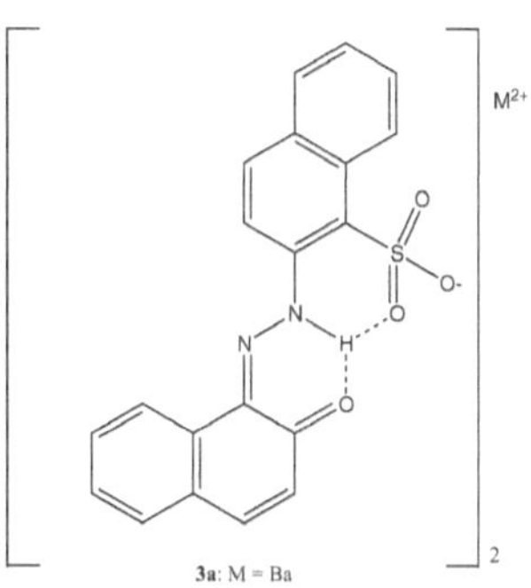

3a: M = Ba
3b: M = Ca

Figure 48.5: Structure of the Lithol Reds (CI Pigment Red 49).

Metal salt azo pigments derived from 3-hydroxy-2-naphthoic acid, (**5a**) – (**5e**), CI Pigment Red 63:1, (**6a**), and CI Pigment Red 63:2 (**6b**) constitute the most important of this group of pigments, as illustrated in Figure 48.7 and listed in Table 48.4. CI Pigment Red 57:1 (**5e**), commonly referred to as the Calcium 4B toner or Lithol Rubine,

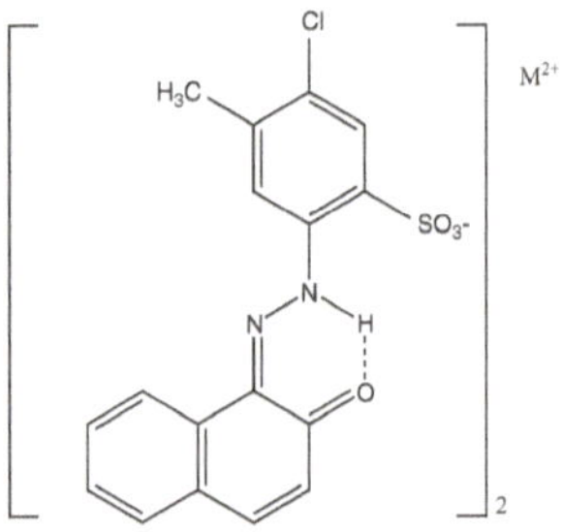

4a: M = Ba
4b: M = Ca

Figure 48.6: Structure of CI Pigment Reds 53:1 (4a) and 53:2).

is of particular importance as the worldwide standard magenta printing ink pigment. These products have a complex crystal structure [13]. The metal atoms bond to oxygen atoms derived from carboxylate, keto and sulfonate groups in the ligands and to water molecules with some bridging between the ligands, all of which leads to a polymeric ladder structure. This is no doubt a feature that promotes insolubility. An interesting observation in this series of pigments is that the manganese salts generally show better lightfastness than the alkaline earth derivatives. It would be of interest to establish the nature of co-ordination of the transition metal Mn^{2+} ion in these products, which appears to enhance the photostability of the chromophore, a feature commonly encountered in transition metal complex pigments.

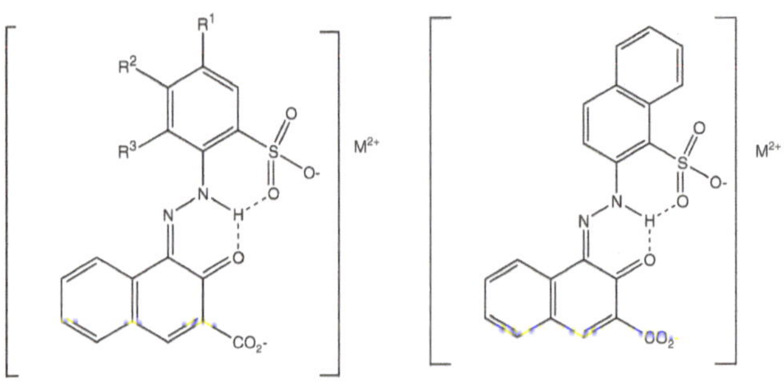

5

6a: M = Ca
6b: M = Mn

Figure 48.7: Metal salt azo pigments from 3-hydroxy-2-naphthoic acid.

A third small group of metal salt pigments is derived from Naphthol AS coupling components. A typical product is CI Pigment Red 151 (**7**) (Figure 48.8), a barium salt that finds some use in plastics.

Table 48.3: Substituent pattern in azonaphtharylamide (Naphthol AS) Pigments (Group 2).

Compound	CI Pigment Red or Orange	R^1	R^2	R^3	R^4	R^5	R^6
2k	R5	OCH_3	H	$SO_2N(C_2H_5)_2$	OCH_3	OCH_3	Cl
2l	R146	OCH_3	H	$CONHC_6H_5$	OCH_3	Cl	OCH_3
2m	R147	OCH_3	H	$CONHC_6H_5$	CH_3	H	Cl
2n	R170	H	$CONH_2$	H	OC_2H_5	H	H
2o	R187	OCH_3	H	$CONHC_6H_5$-*p*-$CONH_2$	OCH_3	OCH_3	Cl
2p	R188	CO_2CH_3	H	$CONHC_6H_5$-Cl_2(2,5)	OCH_3	H	H
2q	038	Cl	H	$CONH_2$	H	$NHCOCH_3$	H

Table 48.4: Substituent pattern in metal salt azo pigments from 3-hydroxy-2-naphthoic acid.

Compound	C.I. Pigment Red	R^1	R^2	R^3	M
5a	48:1	CH_3	Cl	H	Ba
5b	48:2	CH_3	Cl	H	Ca
5c	48:3	CH_3	Cl	H	Sr
5d	48:4	CH_3	Cl	H	Mn
5e	57:1	CH_3	H	H	Ca

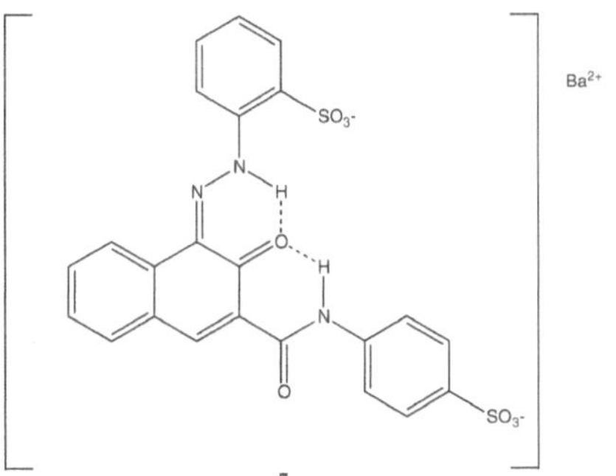

7

Figure 48.8: CI Pigment Red 151 (**7**).

48.4 Synthesis and manufacture

The synthesis of 2-naphthol based pigments involves the reaction of a diazotized aromatic amine with 2-naphthol, a naphtharylamide, or 2-hydroxy-3-naphthoic acid (βONA), as illustrated schematically in Figure 48.9. Generally mild alkaline conditions are used for the azo coupling reaction using the diazotized amine, under which conditions the naphthol-based coupling component is present substantially in the more reactive anionic form. Usually the traditional conditioning process, after the synthesis stage is complete, involves heating the aqueous pigment slurry to the boil for an appropriate time (the "boil-up"). In the case of some of the higher performing products special conditioning procedures are required. In the manufacture of CI Pigment Red 170, for example, in the azo coupling, a weakly crystalline α-form is often formed initially and is transformed during the synthesis into either the β- or γ- forms by appropriate choice of conditions. For example, the β-form results when the azo coupling reaction is carried out in the presence of a nonionic surfactant and the resulting product is subjected to subsequent heat treatment. The γ- form results when the α- form is heated in water under pressure at 130 °C. In the case of the metal salt pigments, the divalent inorganic salt may be added at the end of the coupling process or included simultaneously with the coupling reaction. In the case of CI Pigment 57:1 (**5e**), the pigment is initially formed from aqueous preparation as a trihydrate. It is subsequently converted by drying at around 100 °C to a monohydrate which has considerably higher color strength [14].

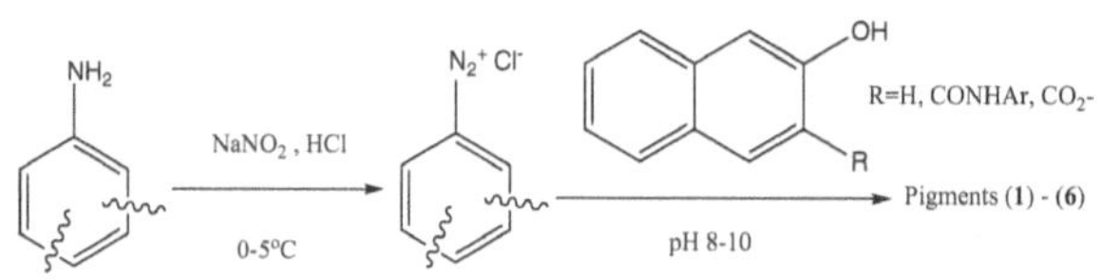

Figure 48.9: Synthesis of Monoazo Pigments based on 2-naphthol and derivatives.

48.5 Applications

48.5.1 Monoazo pigments based on 2-naphthol

Simple monoazonaphthol pigments generally have, at best, moderate levels of fastness properties, but find use in decorative paints, mainly in full shades as the lightfastness decreases in white reductions (with titanium dioxide). Although much diminished in importance, a few of these pigments are still commercially marketed, usually where there are low demands on fastness properties and where economy plays an important role in pigment selection.

48.5.1.1 CI Pigment Red 1 (1a)

There are no manufacturers of this pigment currently listed in the Colour Index, although until recently there may have been some use in the US. The lack of fastness to light, solvents and chemicals means that it of little interest, except in a historical sense.

48.5.1.2 CI Pigment Orange 5 (1b)

Often referred to as dinitroaniline (DNA) orange, this pigment can provide either a pure mid-orange shade or, in larger particle size grades, a much redder shade (often designated as red), although tinctorially weaker in white reductions than the smaller particle size grades. The coarser grade has slightly better lightfastness and its main use is in full or strong shades. The finer grade gives more attractive shades in reduction with titanium dioxide, but its lightfastness deteriorates significantly as the depth of shade reduces. It finds some use in printing inks, mainly water-based flexographic, where its good opacity and bright shade make it useful. It is rarely resinated for economic reasons and the resulting poor gloss limits its value. It is rarely used in paste inks due partly to poor distribution properties on the press. It is not suitable for most plastics.

48.5.1.3 CI Pigment Red 3 (1c)

This pigment was the original red used in the UK for mailboxes, telephone boxes and London buses. It covers a wide range of shades dependent on the particle size. As its particle size is increased, the color becomes bluer and tinctorial strength reduces. In recent years its only significant use has been in oil-based and water-based decorative paints, where it has been used in deep shades. In full shade its color is considered attractive; when reduced with white pigment it becomes bluer, with a less bright pink undertone. It is not suitable for metered tinting systems as its stability to recrystallization is inadequate, leading to unstable color strength and poor color reproduction when blended with other pigments. In inks it is confined to aqueous flexographic and occasionally textile screen printing inks. It is not especially useful for paste inks as it leads to poor gloss and printing properties. Although resination of the pigment can improve the gloss and printing properties, it weakens the color strength of an already weak pigment, thus making it commercially unattractive. It blooms in most plastics.

48.5.1.4 CI Pigment Red 4 (1d)

This pigment, discovered by Hertberg and Spengler in 1907, is known as Chlorinated Para Red or Fire Red on account of its scarlet flame-like color, intermediate between CI Pigment Orange 5 and CI Pigment Red 3. It lacks color strength but in deep shades it provides medium to good lightfastness. Its main use has traditionally been in printing inks, provided that good solvent fastness is not required. It has significantly better fastness to soap than the other pigments in the series. It is occasionally used in paste inks where it offers the highest color strength within this group of pigments. It finds some application in cosmetics (designated as D&C Red 36), meeting the application requirements provided that chemical purity criteria are satisfied, although the products cannot be used around the eyes and it is limited to concentrations below 3% in lipsticks [15].

48.5.1.5 CI Pigment Red 6 (1e)

This pigment, discovered in 1906 by Schrauber and Scheicher, is known as Parachlor Red. It provides a yellowish-red shade with significantly better lightfastness than most pigments in the series and reasonably good soap resistance, but it is currently little used.

48.5.2 Monoazonaphtharylamide (Naphthol AS) Pigments

Naphthol AS pigments provide a high proportion of the classical organic pigments suitable for paint and inks and include some of the highest volume pigments currently in use. They have only limited application in plastics.

48.5.2.1 Group 1 Pigments

48.5.2.1.1 CI Pigment Red 2 (2a)

Included in the original 1911 patent from Winter, Laska and Zitcher, this pigment is still widely used on account of its bright mid-red shade. It is mainly used in printing inks, in which it is yellower and stronger than CI Pigment Red 112 (**2j**), although the latter has superior fastness properties, both to light and solvents. CI Pigment Red 2 is reasonably fast to soap but is not fast to sterilization and suffers from poor stability to recrystallization at elevated temperatures, which can occur during the milling operation, even in water-based systems. Although it finds some use in paints, increasing demands for higher fastness mean that it is less suitable than CI Pigment Red 112. It is used in textile printing if no demands are made for fastness to dry cleaning. It can be used for spin coloration of viscose fibers and film. The paper industry uses large quantities of the pigment where deep red pulp colorations are required. It has traditionally been used for red paper napkins and can also be used as an inexpensive red pigment for decorative paper laminates, applications which do not demand good lightfastness.

48.5.2.1.2 CI Pigment Red 8 (2b)

Another of the original pigments, it is also used primarily in the ink industry. It is much bluer in shade than CI Pigment Red 2, with similar fastness properties but with higher color strength, and often shows good transparency. It provides moderate solvent fastness and can be affected by oils and fats. It has good fastness to soap. It also suffers from poor fastness to recrystallization. Other uses include mass coloration of paper, paper coatings and artists colors.

48.5.2.1.3 CI Pigment Red 9 (2c)

This pigment was discovered in 1922 by Laska and Zitschler at Griesheim Elektron [16]. It provides a very bright, clean, yellowish shade with significantly improved lightfastness compared with CI Pigment Red 2 and CI Pigment Red 8. Its lack of stability to solvents, particularly aromatic hydrocarbons, limits its use to offset and aqueous flexographic inks. It has good fastness to soap, fats, and phthalate plasticizers. It is at best only semi-transparent. It finds application in water-based decorative paints, where its lightfastness is considered good enough for interior applications.

48.5.2.1.4 CI Pigment Red 12 (2e)

This pigment, introduced by IG Farben, has a bordeaux shade and is relatively inexpensive and consequently has limited competition. It can be used in offset, packaging gravure and flexographic inks. It has good lightfastness although this decreases quickly as the depth of shade is reduced. It has only moderate fastness to solvents and undergoes a change of crystal modification in the presence of chlorohydrocarbons and ketones. This change can also occur during dispersion into oils, even under only moderate shear conditions. It is sensitive to fats and alkalis. In paint, it was widely used for deep bordeaux shades, but this use has declined in the light of modern expectations and the availability of alternatives, albeit at higher cost. It is not suitable for most industrial stoving paints due to its limited solvent fastness and tendency to bloom. It has some limited use in textile printing if fastness to dry cleaning is not required and if the print is not coated with PVC.

48.5.2.1.5 CI Pigment Red 14 (2f)

This pigment is less blue than CI Pigment Red 12 but shows superior lightfastness. It can convert to a different crystal modification in the presence of strong solvents such as ketones and aromatic hydrocarbons. Its main market is the paint industry where it can be used in decorative and industrial paints stoved up to 160 °C, above which it starts to bloom, especially at low concentrations. It also bleeds at these temperatures when over-coated. Its superior lightfastness is most evident at full and near full shades but, as with most pigments, it decreases in white reductions. It has been used in combination with CI Pigment Red 112 (**2j**) to match the bluer shades of CI Pigment Red 3, when higher lightfastness is required.

48.5.2.1.6 CI Pigment Red 23 (2h)

This pigment was one of the original series, but never gained much importance although close to the magenta standard for printing. It is bluish red but lacks brightness. Its shade is similar to CI Pigment Red 146, but its lightfastness is lower although it has good fastness to soap and alkalis. It tends to recrystallize, which limits its usefulness for packaging gravure inks. Therefore, it finds most of its use in water based flexographic inks and in textile printing. Its use in textiles appeared to be under threat when a positive test was reported for 4-methoxy-*m*-phenylenediamine, a banned amine listed in the

German 5th amendment to the Consumer Goods Ordinance of April 1997. However, this proved to be a false positive resulting from an artefact of the testing procedure [17].

48.5.2.1.7 CI Pigment Red 112 (2j)

This pigment provides the best performance that can be achieved by Naphthol AS Group 1 pigments. The diazo component supplies three chlorine atoms, which enhance lightfastness and reduce solubility. Although first patented in 1939, it was some years until it became a commercial product. it became established as the standard red pigment for decorative paints, including use in metering systems, almost totally replacing CI Pigment Red 3. It is used to produce the classic signal red shade and the UK Post Office red. Although its use in industrial stoving paints has declined, it can be used successfully up to 160 °C and even above when used at a higher concentration. Its fastness to bleeding is only moderate, but fastness to light is excellent in full shade and remains good in white reduction. It was for some time employed for automotive finishes. In the US, some producers of decorative paints using solvents with high aromatic hydrocarbon content preferred the use of CI Pigment Red 170 (**2n**) as a better option for solvent fastness, while accepting its lower lightfastness. In printing inks, it is mainly used in offset, packaging gravure and flexographic inks, where it offers high color strength with excellent lightfastness. It is fast to soap but is not completely fast to fats or sterilization. It is the best pigment in the group 1 series for use in the coloration of rigid PVC, where it offers excellent lightfastness in full shade, but reducing as the proportion of titanium dioxide white increases. It is also suitable for the spin coloration of viscose, apart from applications with the most demanding requirements. Its moderate fastness to dry cleaning makes it less than ideal for textile printing or for prints that are coated with PVC. However, it is widely used for artists colors, felt tip pens, finger paints and poster paints. Grades of this pigment have been developed with larger particle size, giving high opacity and even better lightfastness, but sacrificing color strength. However, the diketopyrrolopyrrole (DPP) pigment CI Pigment Red 254 is beginning to challenge its dominant position, as its price continues to fall. An unusual use is to color the coatings for seeds to distinguish those that have been given special treatments.

Many other Naphthol AS (Group I) pigments in this series have disappeared over the years or have low current use. These include CI Pigment Orange 24, CI Pigment Red 7, 10, 16, 17, 21, and 22, and CI Pigment Brown 1.

48.5.2.2 Group 2 Naphthol AS Pigments

48.5.2.2.1 CI Pigment Red 5 (2k)

The first in the Group 2 Naphthol AS pigment class, this pigment was discovered in 1931 by Neimeier and Lamberg of IG Farben. The pigment is a bluish red with good fastness properties. It is used both in inks (mainly in Europe) and paints (mainly in the

US and Japan). In inks, it has excellent resistance properties, and good lightfastness. It offers good fastness to soap and chemicals, but its high price has deterred greater use. It finds itself in competition with CI Pigment Red 146 (**2l**) but offers higher lightfastness. It is used mainly in offset inks, but also finds some use in packaging gravure and flexographic inks. In paints, it is used for general industrial finishes, as it does not bloom, but it does bleed. It is an important pigment for textile printing where it shows good lightfastness. It can also be used for the mass coloration of cellulose films and spin coloration of viscose. It is authorized for the coloration of certain cosmetics, as well as being recommended for the coloration of toiletries, and home and fabric care products, in liquid, paste or powder form.

48.5.2.2.2 CI Pigment Red 146 (2l)

This pigment is slightly bluer and about 15% stronger than CI Pigment Red 5 (**2k**), with rather better solvent fastness but somewhat lower lightfastness. Although a little too yellow for the magenta standard, its main use is in printing inks where it is used in offset, packaging gravure and flexographic inks. It appears to be a favorite pigment for security inks. Its use in decorative paints is limited to interior finishes because of modest lightfastness. It is used to a limited extent in industrial paints, even stoving paints, as it is stable to 180 °C, but starts to bleed at 120 °C. In plastics, it can only be used for rigid PVC as it bleeds in plasticizers and is insufficiently heat stable in polyolefins.

48.5.2.2.3 CI Pigment Red 147 (2m)

This pigment is limited to printing inks as fastness properties are lower than CI Pigment Red 146.

48.5.2.2.4 CI Pigment Red 170 (2n)

This pigment was developed in the early 1960s by Hoechst and marketed as Red F5RK. It is a mid-shade red in full shade, moving towards a bluer shade when reduced. When introduced, it set new standards for fastness to solvents in the Naphthol AS pigments series. The pigment is polymorphic. This original product is the β-form. The main use of this form is in industrial finishes, including automotive refinishes, but was rarely considered as lightfast enough for original automotive equipment. A few years after its original introduction, an opaque grade was introduced. This pigment, designated F3RK70, is the γ-crystal modification with a larger particle size which offered even better solvent fastness. This product gained favor as an alternative to molybdate red when a lead-free paint was required. One of its first applications was for agricultural finish paints, offering better solvent fastness than CI Pigment Red 112 (**2j**). The lightfastness of CI Pigment Red 170 is good, with the opaque γ-from being somewhat better than the original β-form product, but it has slightly lower lightfastness than CI Pigment Red 112. It is fast to blooming and stable to 160 °C, but it bleeds slightly above 120 °C, and quite seriously at 160 °C in systems containing esters and especially ketones. It is more economical in applications than perylene pigments but has lost some market share to the high-performance diketopyrrolopyrrole (DPP) CI Pigment Red 254, offering superior fastness but at a higher price, even though the price of DPP pigments dropped dramatically when they no longer had patent protection. While the paint industry predominantly uses the high opacity product, the more transparent grade is preferred for inks on account of its higher tinctorial strength. CI Pigment Red 170 is fast to fats, acids, and alkalis, and reasonably fast to sterilization. It finds use in metal decorative printing, stable for 10 minutes at 200 °C. It finds minor use in decorative paper laminates. The β-form is one of the standard pigments for textile printing, as the most economical, dry-clean fast, mid-red pigment. Its use in plastics has limitations. It is suitable for rigid PVC but tends to bleed in plasticized PVC. Its use in polyolefins is marginal. It can be used in polyethylene, where it shows high tinctorial strength provided that the temperature is kept to the minimum feasible and the dwell time kept as short as possible. It also has a low tendency to cause warping of the final plastic product.

48.5.2.2.5 CI Pigment Red 184

This pigment is derived from a mixed coupling using the same diazo component as is used for CI Pigment Red 146 and CI Pigment Red 147 with the different coupling components used for these two pigments. It was specially developed to meet the standard magenta color requirements for inks while fast to water, acids, and alkali, unlike most metal salts of the CI Pigment Red 57:1 type. It is bluer than both CI Pigment Red 146 and CI Pigment Red 147 with fastness properties similar to those of CI Pigment Red 146. It is heat stable to 170 °C for 10 minutes, or 160 °C for somewhat longer times. Although designed for inks, it is also recommended for textile printing and rubber coloration.

48.5.2.2.6 CI Pigment Red 210

This pigment is formed from a mixed coupling, which employs as one of coupling components the Naphthol AS used for CI Pigment Red 170, and the other with a methoxy group in place of the ethoxy group. The resulting pigment is bluer than CI Pigment Red 170, but shows significantly lower fastness properties, thus limiting its use mainly to inks.

48.5.2.2.7 CI Pigment Red 187 (2o)

The high performance of this pigment, introduced by Hoechst as Red HF4B, owes much to the presence of three amide groups. The pigment exists in two crystal modifications, but only the bluer version is a commercial product. Although the shade is close to the magenta standard, the main application is in plastics. It is completely stable to 260 °C in polyethylene and 250 °C in polypropylene, even at low concentrations, and has a low tendency to cause warping. It can be used for the spin coloration of polypropylene and acrylic fibers. It is suitable for use in polystyrene and polycarbonate up to 300 °C, and in ABS it is stable up to 290 °C. It has very good to excellent lightfastness in all of these polymers. It can be used for rigid and plasticized PVC. It is specifically recommended for artificial leathercloth, where it produces a range of brown shades with a natural appearance when used in combination with iron oxide pigments. It has a few uses in specialist printing inks that require high heat stability, up to 220 °C for 10 minutes. It is suitable for metal decorative and laminate printing. Although eminently suitable for most paint applications, it does not usually fit the economic demands. It is transparent, and so finds applications in decorative pearlescent and metallic finishes, but outside the automotive sector. It is used for coil coatings, where its good heat stability and fastness to chemicals is an asset, although it does not always meet the extremely high lightfastness demands from such applications.

48.5.2.2.8 CI Pigment Red 188 (2p)

The color of this pigment, originally introduced by Hoechst in the 1960s, is intermediate between bright red and orange. It finds use mainly in paints, but with some specialist ink applications. It can be used in high quality industrial finishes where good lightfastness is required, but does not meet the highest demands, such as those of automotive finishes. It can be stoved at up to 200 °C and has very good fastness to most solvents including ketones, making it resistant to bleeding up to 160 °C. A special niche area is in decorative paint tinting systems, mainly in the US, where the pigment is required to have better fastness to solvents, due to the high level of aromatic solvents that have been used in US oil-based paints. The original pigment was joined on the market around 1990 by a variant that offers an even yellower shade with high opacity. This pigment has superior lightfastness, making it more suitable for demanding application areas. These include the lower end of automotive paint applications, where it can be used to produce full shade paints with excellent gloss and good lightfastness, offering an alternative to lead-based pigments. CI Pigment Red 188 is less frequently used for inks, as it lacks the necessary tinctorial strength, even the more transparent version. It has little application in plastics, due to its thermal instability above 220 °C, and its tendency to bloom in both plasticized and rigid PVC.

48.5.2.2.9 CI Pigment Orange 38 (2q)

Although not a high-volume product, this pigment is one of the few monoazo pigments, other than metal salts and benzimidazolones, that are recommended for plastics. Its good heat stability is associated with the presence of three amide groups. The pigment is a bright reddish-orange and finds use in high quality printing inks, such as wallcoverings, laminates and metal decorative prints. It shows good solvent fastness and is fast to fats, acids, alkalis, and sterilization. Its lightfastness is very good, just falling short of excellent. In plastics it is used for all types of PVC, both rigid and plasticized, and can also be used in polystyrene and polyolefins, in which it can be stable up to 280 °C [18] although some literature claims lower heat stability [6]. It also finds use in spin coloration of secondary acetate. It can be used in industrial stoving paints, up to 180 °C, but its lightfastness in pale reductions is inadequate. This limits its use in these applications and precludes its use in the automotive and refinish sectors. It has traditionally offered a compromise between the very expensive perinone pigment, CI Pigment Orange 43 and the disazopyrazolone pigments, CI Pigment Orange 13 and 34, which, although more economic, do not possess sufficient lightfastness or heat stability in polymers. Compared with the main benzimidazolone azo pigment in this area of the spectrum, CI Pigment Orange 36, it has lower fastness to solvents and light, but is much cleaner in shade.

Other Naphthol AS pigments of lesser importance include CI Pigment Red 31, 32, 222, 238, 245, 253, and 256

48.5.3 Metal Salt Azo Pigments

A general feature of metal salt azo pigments is that they are sensitive to even dilute acids and alkalis. Thus, they are not fast to soap. In fact, several are unstable in most aqueous systems. However, they are inexpensive and generally show good fastness to organic solvents, while many have good heat stability. Resination of metal salts is frequently employed to enhance dispersibility, tinctorial strength and transparency.

48.5.3.1 Metal salt azo pigments based on 2-naphthol

48.5.3.1.1 CI Pigment Red 49 (Sodium salt) (3a)

This pigment has high tinctorial strength and a very yellowish- red shade. This is a low-cost pigment that suffers from relatively low lightfastness and solvent fastness and is used in low value solvent-based inks.

48.5.3.1.2 CI Pigment Red 49:1 (Barium salt) (3b)

Often referred to as Lithol Red, this pigment is by far the most important in the series. It is yellower in shade than the magenta standard process color, rather close to the shade of Lake Red C. It has good opacity and can be used to produce inks with good flow. It is reasonably stable towards water and, perhaps more surprisingly, to mild alkali. It is used in water-based inks, offering brightness and economy. In the US, it is also used for publication gravure inks. Its poor heat stability virtually excludes its use in the coloration of plastics.

48.5.3.1.3 CI Pigment Red 49:2 (Calcium salt) (3c)

This pigment is the second most important in the series, experiencing increasing use at the expense of the barium salt on toxicological grounds. It provides a much bluer shade, towards maroon.

48.5.3.1.4 CI Pigment Red 53 (Sodium salt)

This pigment salt has a very yellow shade with the poor fastness to light and solvents expected from a sodium salt. Although many companies have registered the sodium salt in the Color Index, it appears that many of these products are in fact the barium salt.

48.5.3.1.5 CI Pigment Red 53:1 (Barium salt)

This pigment is used on account of its low cost and bright yellowish red shade, often referred to as a “warm red”. It lacks lightfastness and is limited to applications where its toxicological properties are not an impediment to its use. Although its importance is diminishing, it remains significant for printing inks, mainly offset, flexographic and packaging gravure inks. It shows good resistance to solvents and can be used in water-based inks without showing the level of viscosity instability that results in thickening often associated with metal salt pigments, making it a favored choice for water-based flexographic inks. Its fastness to soap is limited. The pigment also has some limited use in plastics, provided that its low lightfastness is not an issue. It bleeds into PVC, but can be used in polystyrene, in which it is stable up to 280 °C and particularly in HDPE, in which at medium depths it is stable to 260 °C for 5 minutes dwell time. It has virtually no warping effect in HDPE injection moldings. An alternative orange crystal modification of the pigment was introduced by Hoechst, but its higher cost prevented it from becoming established. It is no longer available.

48.5.3.1.6 CI Pigment Red 53:2 Calcium salt

This pigment is of little current importance although there are two manufacturers registered in the Colour Index.

There are some other 2-naphthol based metal salt pigments that are of little or no commercial importance, including CI Pigment Red 50:1, 51, 68, and CI Pigment Orange 17 and 46.

48.5.4 Metal salt azo pigments derived from 3-hydroxy-2-naphthoic acid (βONA)

Metal salt azo pigments based on βONA tend to be bluer in shade than the 2-naphthol pigments. They were originally developed for paints as there was little use of color printing when they were first introduced. Their superior lightfastness to the 2-naphthol metal salts increased their appeal.

48.5.4.1 CI Pigment Red 48

These pigments are often referred to as 2B Toners (the diazo component is known as 2B acid). The first commercial product was known as Permanent Red 2B, the sodium salt. Some of the commercial products registered in the Color Index are incorrectly designated as they are not sodium salts. Indeed, many suppliers state the metal ion in their constitution in part of their commercial name!

48.5.4.2 CI Pigment Red 48:1 Barium salt (5a)

This pigment is yellower and more opaque than the corresponding calcium and manganese salts. It has good solvent fastness but poor stability to acids and alkalis, including soap. It finds application mainly in inks and, to some extent, in plastics. In shade and color strength it is intermediate between CI Pigment Red 53:1 (**4a**) and CI Pigment 57:1 (**5e**). It has significantly lower lightfastness than similar metal salts but has good heat stability. Hence, it is also used in some plastics, including plasticized PVC. In LDPE, some product grades are recommended as stable up to 240 °C, while others only to 200 °C. At higher temperatures, the pigment changes to a bluer shade. In paints it finds some application in inexpensive industrial finishes where it shows good fastness to bleeding but is not recommended for external use due to its moderate lightfastness.

48.5.4.3 CI Pigment Red 48:2 Calcium salt (5b)

This pigment is a mid-ruby red shade, with good fastness to light. Although significantly yellower than CI Pigment Red 57:1 (**5e**), it is considered the standard red for packaging inks and is widely used in the US. The pigment has better lightfastness than the barium salt and the CI Pigment Red 49 varieties. In the manufacturing process for paste ink grades, such as offset inks, the pigments are usually resinated to improve transparency, tinctorial strength, print quality and gloss. For liquid packaging inks, where opacity and good flow are often required, the unresinated pigment is used, but resinated versions are preferred if transparency and higher tinctorial strength are required. Most grades are sensitive to water in which they tend to yellow, although more water-resistant grades have been developed, especially for metal decorative inks. If purity requirements are met, this pigment can meet most world-wide food packaging and toy regulations. The pigment is also used in plastics, where it has high tinctorial strength, and heat stability up to 230 °C for a 5-minute dwell time in HDPE. It is more widely used in LDPE, where it shows better lightfastness than the barium salt. Its use in paints is limited because, although fast to bleeding in stoving paints, it often lacks the lightfastness required, especially in reduced shades.

48.5.4.4 CI Pigment Red 48:3 Strontium salt (5c)

This is one of the few strontium salt pigments that has been successful. Its shade is intermediate between the barium and calcium salts in plastics, its main application area. Its fastness to bleeding in plasticized PVC is the highest of all CI Pigment Red 48 products. It is stable to 240 °C, above which it turns bluer. Its main use is in LDPE, but it can also be used in the lower melting temperature grades of polypropylene fibers. It can be used in most applications that require a safer toxicological profile, such as toys and food packaging.

48.5.4.5 CI Pigment Red 48:4 Manganese salt (5d)

This is a bluish red pigment that can be used in many applications, especially general industrial paints because of its good fastness to bleeding. It was traditionally used in combination with lead chrome molybdate reds to provide intermediate shades and can fulfil this role currently with opaque organic red pigments. It is the most lightfast of all the CI Pigment Red 48 products and previously had a limited use in automotive refinish paints, although it tends to darken in light. Compared with other CI Pigment Red 48 products, it is less fast to solvents, acids, and alkalis. Manganese ions often act as drying agents in paints, and the manganese content of this pigment can accelerate the oxidative drying of oil-based paints. In plastics, it finds application in PVC, in which it does not bloom and shows virtually no bleeding. In polyolefins, it provides high color strength especially in polyethylene, where it can tolerate temperatures of 280 °C in certain circumstances. It has a low tendency to cause warping. However, it causes premature ageing in polypropylene, leading to the polymer becoming brittle. It can be used in the mass coloration of fibers such as secondary acetate. Its use in inks has limitations, due to its acceleration of drying, although this effect can be compensated by reducing the concentration of drying agents in the formulation. The pigment can decompose in polyamide resins, leading to an unpleasant odor and printed sheets sticking together.

48.5.4.6 CI Pigment Red 57:1 Calcium salt (5e)

This is the most important monoazo metal salt pigment, which today provides many of the standard magenta colors used in 3- and 4-color printing processes. It is sometimes referred to as Calcium 4B Toner (the diazo component is known as 4B acid), Rubine Toner, or Lithol Rubine (a BASF registered trademark in Europe, although the registration no longer applies in the US). It is sold in massive volumes for process inks. It is also used for publication gravure inks and as a spot color in packaging inks. It provides good printing properties and is highly economical in use. There are many modifications that can be made during manufacture to influence its color, flow, dispersibility, gloss, and transparency. The shade may be modified by narrowing the particle size distribution or by incorporating small amounts of auxiliary diazo component additives, such as Tobias acid (2-amino-1-naphthalenesulfonic acid), which inhibits crystal growth.

The pigment provides high tinctorial strength, and lightfastness that is sufficient for most printing inks. Likewise, its fastness to solvents is sufficient for most applications. It has poor fastness to acids and alkalis, including soap. Resination is generally used to improve its transparency, gloss and dispersibility. The pigment is not usually suitable for water-based inks as it tends to cause thickening on storage, although modifications can improve this property. CI Pigment Red 57:1 can withstand temperatures of up to 100 °C without recrystallizing during dispersion in paste inks, where high shear can develop such temperatures, although under such conditions, the pigment may undergo a shade change. It can also be used in plastics if there is no demand beyond its moderate lightfastness. Its good dielectric properties make it suitable for PVC cables. Its heat stability in polyolefins can reach 250 °C, but it can slightly affect the dimensional stability of the final polymer article although some treatments applied to the pigment can reduce this tendency. Inadequate lightfastness limits its suitability for polypropylene fibers. Its lightfastness is also not usually sufficient for paints, but the pigment occasionally finds use where this property is not considered important. It is important for cosmetics and food packaging if it meets purity requirements. Special grades are made for cheese rind, especially for Dutch cheeses such as Edam.

Other metal salt azo pigments based on βONA are either used in only small amounts or are no longer available, including CI Pigment Red 52:1, 52:2, 52:3, 58:1, 58:2, 58:3, 58:4, 63:1, 63:2, 63:3, 64, and 64:1.

48.5.5 Naphtharylamide (Naphthol AS) metal salt azo pigments

There are a few metal salt azo pigments based on naphtharylamide coupling components. The pigments provide very good fastness properties, especially solvent fastness, compared with conventional metal-free Naphthol AS pigments. This has been an area of significant research especially aimed at heat stable products for the coloration of plastics. It is possible that some recent pigments with undisclosed structures added to the Colour Index are of this type.

48.5.5.1 CI Pigment Red 151 Barium salt (7)

This pigment was developed for the coloration of polymers. It has acidic groups derived from both the diazo and Naphthol AS coupling components. It provides a medium red shade and is characterized by excellent fastness to migration in plasticized PVC that has good lightfastness in full shade but decreasing in white reductions. In rigid PVC it has higher lightfastness. In polyethylene it has moderate tinctorial strength but has high heat stability, stable up to 290 °C. However, it adversely affects the dimensional stability and so is not suitable for injection molded articles such as crates. The pigment produces strong coloration in polystyrene above 260 °C, as it dissolves and behaves like a solvent dye. It is also suitable for ABS.

48.5.5.2 CI Pigment Red 237 Sodium salt

The acid group is on the diazo component. It has a bright yellowish red shade and is recommended for PVC on account of its good fastness to bleeding. It is less heat stable than Pigment Red 151, reaching up to 260 °C. It can be used in paints.

48.5.5.3 CI Pigment Red 243 Barium salt

This is **a** rather dull yellowish-red shade pigment discovered by Toyo, but now other companies also manufacture. It is recommended for the coloration of PVC and polyolefins but has only moderate to good lightfastness. It can be used for the coloration of polyolefins, but it causes shrinkage and thereby dimensional instability.

48.5.5.4 CI Pigment Red 247 Calcium salt

There are two different crystal modifications of this pigment with quite different shades, one a yellowish red (Red HG) and the other a bluish red (Red HB). The two pigments are characterized by excellent heat stability, reaching 300 °C. In polyolefins it does not have a significant effect on dimensional stability, and, because of its excellent heat stability, it is suitable for the spin coloration of polypropylene fibers. Lightfastness is good to very good. It is considered suitable for polystyrene and ABS. The yellower shade modification has slightly lower lightfastness.

References

1. Fox M. Dyemakers of Great Britain 1856-1976. Imperial Chem Ind PLC. 1987;72.
2. Winther A, Arthur AL, Zitschger A. Azo dyes for lakes. US 1034853. 16 Aug 1912.
3. Christie RM. Colour chemistry. 2nd ed. London: RSC, Ch 3 2015.
4. Zollinger H. Color chemistry. 3rd ed. Weinheim: Wiley-VCH Verlag GmbH, Ch 7 2003.
5. Christie RM. The organic and inorganic chemistry of pigments, surface coatings reviews. London: OCCA, 2002.
6. Hunger K, Schmidt MU. Industrial Organic Pigments. 4th ed. Weinheim: Wiley-VCH Verlag GmbH, Ch 2 2019.
7. Whitaker A. The crystal structure of C. I. Pigment Red 3, 4-methyl-2-nitrophenylazo-2-naphthol. Z Kristallogr. 1978;147:99.
8. Whitaker A. The crystal structure of C.I. Pigment Red 6, 4-chloro-2-nitrophenylazo-2-naphthol. Z Kristallogr. 1977;145:271.
9. Whitaker A. Crystal structure analysis of Azo Pigments involving β-Naphthol: a review. J Soc Dyers Colour. 1978;94:431–5.
10. Whitaker A. The crystal structure of C.I. Pigment Red 2,1′-(2,5-dichlorophenyl)azo-2′hydroxy-3′-phenylamidonaphthalene. Z Kristallogr. 1978;146:173.
11. Chang CH. An Investigation into Polymorphism and Structure/Property Relationships in Naphthol Azo Red Pigments. PhD Thesis Heriot-Watt University, 2002.
12. Christie RM, Mackay J. Metal Salt Azo Pigments. Color Technol. 2008;124:133–44.

13. Bekö SL, Hammer SM, Schmidt MU. Crystal structures of the hydration states of Pigment Red 57:1. Angew Chem Int Ed. 2012;51:4735.
14. Christie RM, Moss S. Metal Salt Pigments derived from 3-Hydroxy-2-naphthohydroxamic acid. Dyes Pigm. 1987;8:211.
15. https://www.fda.gov/cosmetics/cosmetic-ingredient-names/color-additives-permitted-use-cosmetics. Accessed: 31 Jan 2021.
16. Laska A, Zitscher A. Process for the preparation of azo dyes. DP 390,627. 14 Feb 1922.
17. Weideli HJ. Pigment Red 23: zu Unrecht gebrandmarkt. Farbe Lack. 1998;104:126.
18. https://www.clariant.com/en/Solutions/Products/2014/03/18/17/09/Graphtol-Red-HFG. Accessed: 31 Jan 2021.

Andrew Towns

49 Naphthopyran dyes

Abstract: This article covers the most commercially important type of photochromic dye. It outlines which of its subclasses are of most use and how they are manufactured. The history of the industrial exploitation of this kind of colorant is briefly given before illustrating its versatility with examples of the most useful ways in which photochromic properties can be manipulated by making adjustments to molecular structure.

Keywords: colorant, photochromism, photochromic, dye

The naphthopyran dye class is the most industrially important type of photochromic colorant [1]. Its members are used extensively in the production of sun-responsive eyewear. This chapter will outline why such dyes became commercially prominent. In doing so, it will give an overview of the chemistry of the class, focusing on those aspects of greatest industrial interest, and highlight the most important modifications to colorant structure employed by manufacturers to manipulate photochromic properties.

49.1 Commercially important naphthopyrans and their general properties

Each of the world's largest producers of polymer-based photochromic ophthalmic lenses employs naphthopyran dyes in their construction. Of the three possible naphthopyran subclasses, only two are commercially useful: 3*H*-naphtho[2,1-*b*]pyrans **1** and 2*H*-naphtho[1,2-*b*]pyrans **2** (see Figure 49.1). Both comprise a naphthalene system to which is fused a pyran ring that bears a spiro-carbon adjacent to its oxygen atom. Such dyes are also less commonly referred to as benzochromenes or simply chromenes. It is rare for the two spiro substituents to be anything other than phenyl or substituted-phenyl groups. (Note that they do not form a spiroheterocyclic system – derivatives comprising this kind of moiety are members of the spiropyran class [2]). Other rings may be fused onto the naphthalene system, especially in the case of [1,2-*b*]pyrans.

The success of these two sub-types lies in their versatility. Naphthopyran chemistry readily enables cost-effective modification of the parent structures **1–2** (X = Y = R = R^1 = R^2 = H) in many ways to manipulate their photochromic properties. Much variation in substitution on the naphthalene and phenyl rings is possible in a manner

This article has previously been published in the journal Physical Sciences Reviews. Please cite as: A. Towns, Naphthopyran Dyes *Physical Sciences Reviews* [Online] 2020, 5. DOI: 10.1515/psr-2019-0085

https://doi.org/10.1515/9783110587128-049

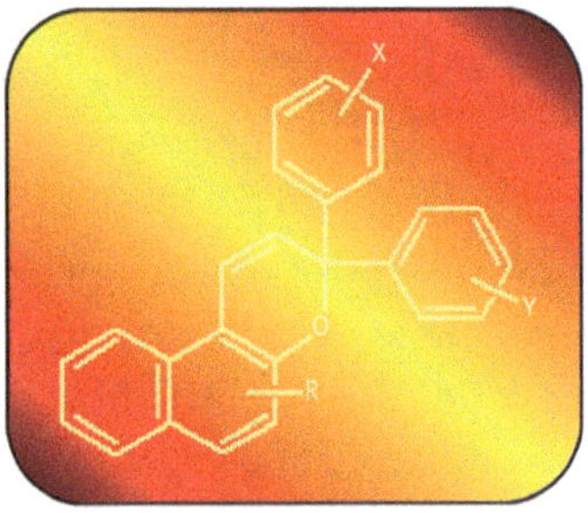

X, Y = H, alkoxy, amino, alkyl, etc.
R = H, amino, alkoxy, etc.

1
industrially important for yellow, orange and red photocoloration

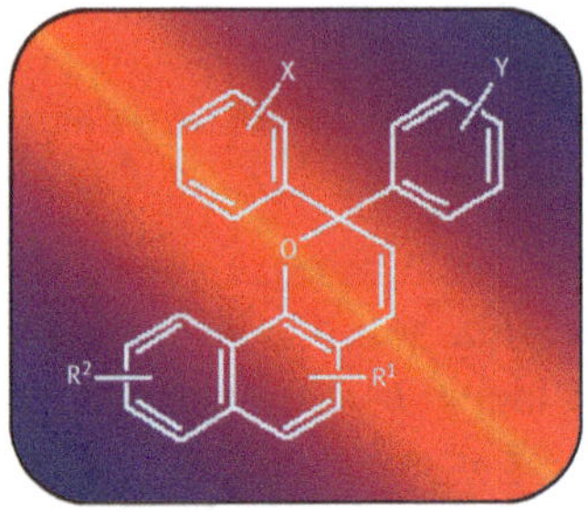

X, Y = H, alkoxy, amino, alkyl, etc.
R^1, R^2 = H, alkyl, aryl, ester, amino, etc.

2
industrially importantly for red, purple, blue and neutral photocoloration

Figure 49.1: General structures of typical industrial naphthopyrans.

that greatly affects not only the hue of photochromism, but also its kinetics, such as half-life ($t_{1/2}$, the time taken for intensity of photocoloration to halve once the stimulating radiation is removed). Numerous different functional groups can be introduced to create ranges of derivatives with similar kinetics whose photocoloration spans the visible spectrum from yellows through to oranges, reds, purples and blues.

A big factor in the current importance of this class is the accessibility of colorants that give neutral shades. As we shall see later in this chapter, appropriately substituted [1,2-*b*]pyrans **2** produce photocoloration with dull and murky shades of a greenish, bluish, purplish, grayish and even brownish nature owing to the presence of more than one broad absorption band in their visible spectra. Colorants furnishing tertiary hues are prized because they serve as the basis for mixtures that produce gray and brown photocoloration from fewer individual dye components – these shades are by far the most important commercially for corrective lenses. Their neutral colors thus simplify the task of formulation. Producing gray mixtures from just three colorants, even when at least one of which has a neutral shade, remains a challenge: the kinetics, temperature-dependence and stability of the components must be matched to maintain hue during the photoactivation, photofading, and lifetime of a lens. However, this task is less formidable than formulation with greater numbers of dyes.

Another contributor to the success of naphthopyran dyes is their response to photostabilizing additives: photochromic systems formulated with combinations of the two are generally as fast to light as those from any other known class. Also of importance to lens performance is a minimized tendency for photocoloration intensity to be affected adversely by high ambient temperatures – members of the naphthopyran class generally exhibit relatively low sensitivity to variations in temperature compared to spirooxazines [3] and other photochromic dye types.

The aforementioned advantages enable formulators to create neutral-colored mixtures solely from naphthopyrans. Dye blends drawn from same chemical type constituted a substantial advance because they minimize differences in the temperature dependency of the photochromism of individual components. Consequently, the colors of such blends show less tendency to vary with changes in ambient temperature than in the case of mixtures formulated with component dyes of more than one class.

49.2 From neglect to dominance

Industry overlooked naphthopyran dyes for many years owing to misinformation and lack of clarity concerning their properties published in the academic and patent literature [4]. The realization that relatively simple naphthopyran derivatives delivered pronounced photochromism at typical ambient temperatures sparked much interest in them as colorants for lenses [5]. Soon after the recognition of this potential, naphthopyran technology made its commercial debut during the early 1990s to provide the hypsochromic component of neutral shade mixtures in conjunction with blue oxazine-based colorants [6]. Knowledge concerning manipulation of naphthopyran color and kinetics grew rapidly during that decade through extensive research activity taking place in the USA, Europe and Japan, which led to the filing of many patent applications covering structural variations. Their dominance was cemented twenty years ago by the development of [1,2-*b*]naphthopyrans whose photocoloration was bathochromically shifted sufficiently for them to replace temperature-sensitive blue spirooxazines and enable the marketing of "all-pyran" neutral-coloring lenses in 1996 within the USA under the Transitions III ® brand [7, 8]. Since then, naphthopyran-derived dyes continued to attract considerable industrial attention in the search for more robust colorants with improved photochromic properties. An example is the attachment of oligomeric chains, e. g. poly (siloxanes) or poly(ethers), to naphthopyran skeletons in order to enhance photocoloration intensity [9, 10] and make usage practical in polymeric media that would otherwise be too hostile [11, 12] – the first examples of dyes functionalized in this way were commercialized over a decade ago.

Over 95% of the volume of naphthopyrans consumed annually is for the manufacture of eyewear. Their usage spills out into novelty and specialized applications, such as toys, textiles, security inks and even fishing gear [12]. Despite being relatively photostable in the context of photochromic dyes, naphthopyrans are orders of magnitude less lightfast than conventional non-photochromic colorants, so potential uses requiring great robustness like architectural coatings and glazing remain out of reach. Their cost, typically quoted in units of "dollars per gram" rather than the "dollars per kilogram" of commodity dyes and pigments, also curtails more widespread uptake.

While many papers on the synthesis and properties of naphthopyran derivatives have been published [13, 14], only a minor proportion of peer-reviewed research literature on photochromism is devoted to such colorants. Reasons for this may be their commercial sensitivity as well as the tendency of academia to focus on P-type photochromic systems rather T-types like naphthopyrans (see Section 49.3 and [1] for definitions of these types). Even amongst academic studies dedicated to T-type dyes, the focus tends to be on spiropyran [2] and to a lesser extent spirooxazine [3] derivatives, perhaps as a consequence of an under-appreciation of the flexibility in synthesis, molecular design, and photochromism that the naphthopyran class offers. In contrast, while they have not totally displaced spiroindolinonaphthoxazine dyes in the industrial arena, naphthopyrans have become dominant in the primary photochromic dye application of ophthalmic lens manufacture.

49.3 Photochromism of naphthopyrans

Industrially important naphthopyrans exhibit heliochromism [15, 16]. They are colorless or weakly colored molecules that readily take on intense photocoloration when solutions of them are irradiated with wavelengths of UV radiation typically present in unfiltered sunshine striking the Earth's surface. The photochromism arises as a consequence of a reversible photoisomerization reaction. Figure 49.2 illustrates the UV-induced photoconversion in the case of a simple naphthopyran **3.** The light excites the dye molecule, leading to a ring-opening reaction and the formation of a strongly colored mixture of photomerocyanines. These polymethine [17] dye structures tend not to be stable and readily isomerize back to the colorless form of the molecule by a thermally-driven ring-closure reaction, hence naphthopyran photochromism being designated as "T-type". (Photochromic substances that only return to their original state by means of irradiation with light are referred to as "P-type" materials.) In certain naphthopyrans, some of the photomerocyanine isomers are driven photochemically back to the cyclic isomer, so they have a minor element of P-type character.

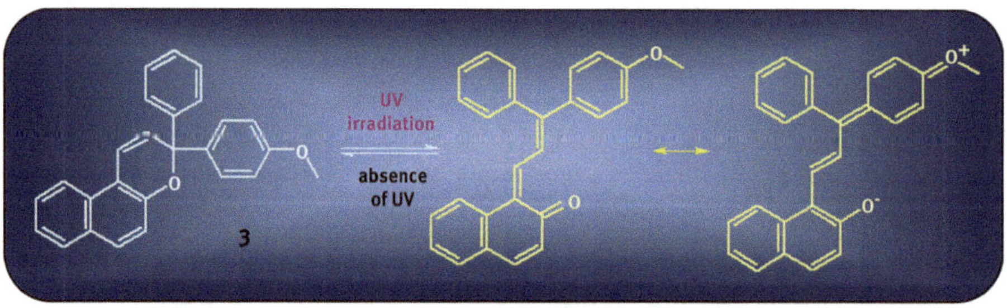

Figure 49.2: Light-induced photoisomerization of a simple [2,1-*b*]naphthopyran **3.**

Commercial naphthopyrans respond well to wavelengths in the UV-A region, especially in the range 360–380 nm. A mixture of several photoisomers results. In polymeric lens matrices, the photocoloration of industrially useful dyes typically reaches 80 % of final intensity within 2 min of constant exposure to a broadband UV source, and decays to half of the intensity attained in around the same duration upon its removal. These timescales are shorter when dyes are dissolved in solvents owing to the environment being less restrictive of the changes in geometry following photoexcitation. In addition, whatever the medium, some of the photoisomeric products are more thermally stable than others [13]. In these instances, half-life figures must be treated with caution, since the decrease in intensity from thermal fading is not a simple first-order decay. Long-lived photoisomers present a problem in ophthalmic lens design because they lead to a residual color that persists for many minutes longer than the time taken for the bulk of photoactivated naphthopyran to ring-close thermally [18]. The ground states of the photomerocyanine forms possess increased charge density on their naphtholic oxygen atoms owing to contributions from zwitterionic canonical forms such as that shown in Figure 49.2. First excited states are associated with migration of electron density towards the naphthalene system. Substitution patterns favoring this transfer, e.g. electron donors on the phenyl rings such as the methoxy group in the [2,1-*b*]naphthopyran shown in Figure 49.2, therefore furnish bathochromic shifts in absorption.

Given the change in molecular geometry that occurs upon ring-opening and ring-closure, commercial naphthopyrans do not exhibit visual photocoloration unless dissolved in a material that provides sufficient free-volume not to restrict these transformations, e.g. a solvent or polymeric matrix such as poly(ethylene). The environment greatly influences not only the hue of photocoloration, since the photomerocyanine forms of naphthopyrans are strongly solvatochromic, but also the kinetics of photochromism. However, dye structure is the main driver of these properties. The rate of thermal fade is readily manipulated by appropriate molecular design. Well-established rules of thumb exist for ways of slowing the rate of the ring-closure reaction, i.e. decreasing the rate of thermal fade (thereby increasing $t_{1/2}$), which shifts the photostationary state equilibrium between ring-opened and -closed isomers towards the former. Industrially important design strategies are covered in Section 49.5, but commercially useful general means of producing naphthopyrans will be described next.

49.4 Synthesis of naphthopyrans

Many naphthopyran dyes are manufactured by combination of naphthols (and their corresponding analogs) with diarylalkynols as illustrated in Figure 49.3.

Owing to the wide range of naphthols and alkynols that are readily synthetically accessible, a diverse array of substituents has been introduced into industrial

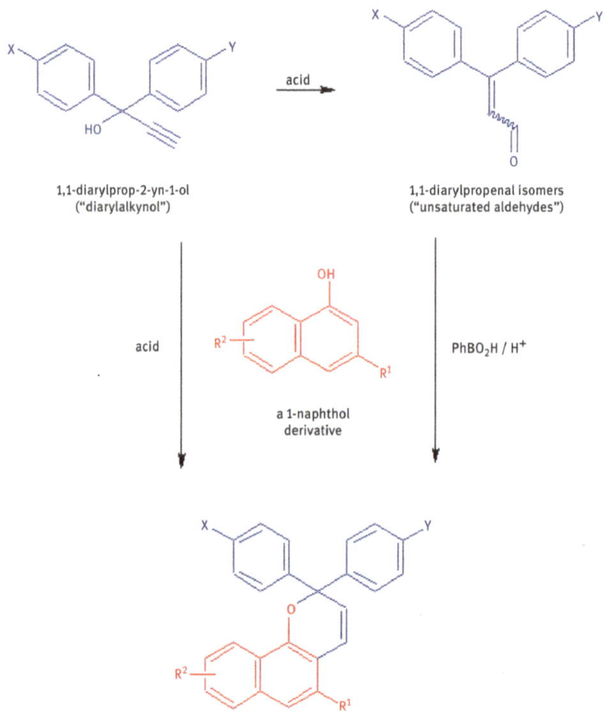

Figure 49.3: Commercial synthetic routes to naphthopyrans using generalized [1,2-*b*] derivatives as examples.

naphthopyrans to fine-tune color, kinetics and other properties. In certain cases, further chemistry is performed on naphthopyran dyes themselves. This may involve functional group transformations to adjust shade and kinetics, or the addition of oligomeric "tails" or "linkers" to improve the photochromism and/or compatibility of dyes with polymeric media that do not otherwise work well together.

The routes shown in Figure 49.3 are two of several pathways to [1,2-*b*]naphthopyrans [13, 14], but they constitute the most important means of assembling the core naphthopyran skeleton of photochromic colorants of this type. This chemistry is also employed to synthesize [2,1-*b*]naphthopyrans for use directly as photochromic dyes or following further chemical transformations. Reaction is typically effected by heating the starting materials with an acidic substance, for example, an arylsulfonic acid or an aluminum oxide-based material, in a non-polar solvent, such as toluene. The best choice of acid tends to be the weakest which still gives complete reaction in order to maximize selectivity and yield. Alkynols bearing either weak donor substituents or none at all tend to require stronger acids than those functionalized with amine groups. However, certain naphthol-derived components do not respond well to use of relatively strong acids, so reaction with alkynols that contain only weak donors necessarily involves compromise in acid choice, leading to lower purities and yields. In these cases, it is better to convert the alkynol to the corresponding propenal by reaction with an acid, then combine the resultant mixture of unsaturated aldehydes with the naphthol in the presence of phenylboronic acid – the extra step can in some instances more than double yield and/or furnish product of greater purity.

Aside from the simplest cases, preparation of the alkynol and naphthol precursors generally involves multiple-step syntheses from commercially available raw materials (see [13] for examples). Formation of the naphthopyran system from alkynol and naphthol constitutes the last synthetic step in many instances. An exception is the manufacture of oligomer-functionalized dyes of the sort mentioned in Section 49.2 whose pendant chains are attached in the last stage of the manufacturing sequence. In the case of the industrially very important "indeno-fused" [1,2-*b*]naphthopyran types (see Section 49.5.2), both pathways have been employed. Their synthesis either involves creation of naphthol-derived intermediates comprising the final targeted functionality (benzofluorenols), which allows direct production of the indeno-fused dye in the last step, or performing chemical transformations on naphthopyran-based intermediates to create the indene motif.

From an economic perspective, industrial manufacturing processes to naphthopyrans preferably afford the colorant in suitable purity and acceptable yield directly by filtration from cooled reaction mass. Very often, recrystallization of isolated dye cannot be avoided owing to the need to remove colored and/or non-photochromic side products. Column chromatography is very much a last resort owing to the considerable cost it entails when operated at a multi-kilo scale. However, the technique may be unavoidable if a problematic impurity cannot be removed by means of recrystallization or if the naphthopyran product is not a solid, as is often the case for dyes that have been functionalized with oligomeric chains.

Most industrial naphthopyran dyes are supplied as powders, although they are also provided to lens fabricators in the form of dye blends dissolved in monomer mixtures for subsequent polymerization. As neat materials, they are generally classified as non-hazardous for transport and tend to be non-toxic. Naphthopyran dyes usually have shelf-lives of years when kept neat in cool, dark, dry conditions. The physical appearances of some dyes are highly dependent upon their impurity profile and isolation method even when of good purity (> 99.0 % area as determined by high performance liquid chromatography), although such variations do not tend to affect their photochromism. All commercial naphthopyran dyes are essentially water-insoluble with most having poor solubility in alcohols (≤ 1 %w/v at 20 °C). They tend to dissolve far more readily in solvents of low to medium polarities. For example, both [1,2-*b*] and [2,1-*b*] derivatives generally have good solubilities in ethers (e.g. typically 2–40 %w/v in THF at 20 °C) and aromatic hydrocarbons (e.g. usually 2–20 %w/v in toluene at 20 °C) as well as moderate solubilities in common aliphatic esters and ketones (e.g. around 1–3 %w/v at 20 °C in ethyl acetate or acetone).

49.5 Structure–photochromism relationships in naphthopyrans

This Section describes some of the most important design strategies utilized by industry when creating naphthopyran colorants with commercially useful properties. While manufacturers heavily exploit approaches that apply to both [2,1-*b*] and [1,2-*b*] classes, others are only appropriate for one of the two types. We will therefore look at the relationship between structure and photochromism of both kinds separately.

49.5.1 Design strategies for [2,1-*b*]naphthopyrans

The parent [2,1-*b*]naphthopyran **4** exhibits yellow photocoloration with sufficiently rapid thermal fade to be of commercial interest (see 4′ = 4″ = H in Table 49.1). However, this relatively fast rate of fade necessarily equates to relatively weak photocoloration. In order to boost its intensity and adjust its hue, several strategies involving the location of substituents onto the phenyl and/or naphthalene rings of **4**, especially at the 4′-, 4″- and 6-positions, have been developed.

Table 49.1: Impact of varying 4′,4″-groups on absorption maximum and half-life of parent [2,1-*b*] naphthopyran **4** dissolved in toluene (20 °C) [19].

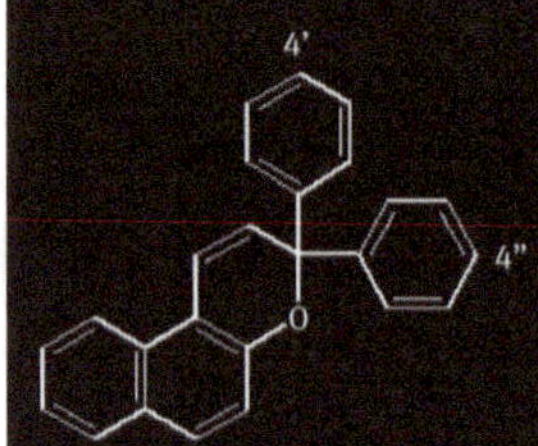

4

4'-group	4''-group	λ_{max}(nm)	$t_{1/2}$ (s)
H	H	429	14
OMe	H	460	5
OMe	OMe	472	2
NEt_2	H	544	2
NEt_2	NEt_2	576	1

Introduction of electron donors into the 4′- and 4″-positions is a very important means of creating commercially valuable dyes that produce orange and red photocoloration. They produce substantial bathochromic shifts in the absorption of the yellow photocoloration of the parent naphthopyran. There is however a catch. As donor strength at these two positions increases, $t_{1/2}$ drops, so photocoloration intensity also tends to decrease. Placing a methoxy or tertiary amino function into the 4′-position leads to a bathochromic shift and shortening in half-life as does a second such group, albeit to a lesser degree (see Table 49.1).

Because of their small $t_{1/2}$ values, all of the substituted dyes listed in Table 49.1 (i.e. 4′ ≠ H) are too weakly photochromic to be of much industrial interest in lens manufacture. A second strategy is therefore often employed to counteract the decrease in half-life without sacrificing much bathochromism. Placing an amino or alkoxy function into the 6-position of the naphthalene moiety lengthens half-life considerably which is accompanied by an increase in photocoloration intensity. While the change also causes a hypsochromic shift, it allows the introduction of an electron donor into the 4′-position in order to push photocoloration out to longer wavelengths without producing a dye that has too short a half-life and whose photocoloration is too weak. The approach is illustrated with the data in Table 49.2 for naphthopyrans in which a morpholino substituent is located at the 6-position. Industry has been utilizing this means of creating valued yellow, orange and red colorants for more than twenty years [20] – all of the example derivatives **5** listed in Table 49.2 have been commercialized.

Table 49.2: Absorption maxima and half-lives of 6-morpholino derivatives of [2,1-*b*]naphthopyran **5** in toluene solution at 20 °C [19].

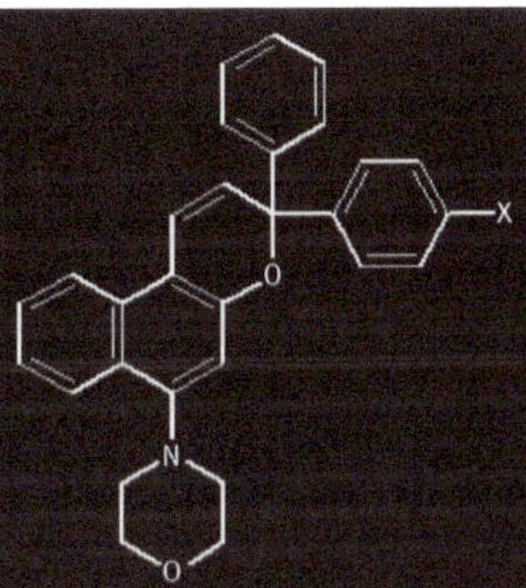

5

X	λ_{max} (nm)	$t_{1/2}$ (s)
H	418	103
OMe	433	54
	470	42
	488	33

The intensity of photochromism can be further boosted by placing a substituent in the 2′-position as shown in Table 49.3, i.e. *ortho* to the spiro carbon. This type of pattern renders the thermal ring closure less favorable through steric hindrance. The slower rate of back reaction translates to a photostationary state in which the photomerocyanine species is present at higher concentration and stronger coloration is perceived. Methyl and methoxy groups located at the 2′-position of **6** (i.e. X = Me or OMe) produce only small bathochromic shifts, owing to their moderate electron-donor character, but markedly lengthen half-lives. Table 49.3 shows examples where $t_{1/2}$ is increased by two orders of magnitude through use of this ploy, with only a minor impact on shade – the position of absorption maxima are within a few nm (see Table 49.3). The approach has been used industrially for many years in conjunction with both [2,1-*b*]- and [1,2-*b*]-naphthopyrans in order to increase depth of photocoloration for a given flux of UV radiation, whilst not largely disturbing their hue.

Even when taking advantage of the types of strategy outlined above, the general relationship between color and rate of thermal fade for [2,1-*b*]naphthopyrans

Table 49.3: Effect on absorption maximum and half-life of 2′-substitution on photochromic properties of [2,1-*b*]naphthopyrans **6** in toluene solution at 20 °C [19].

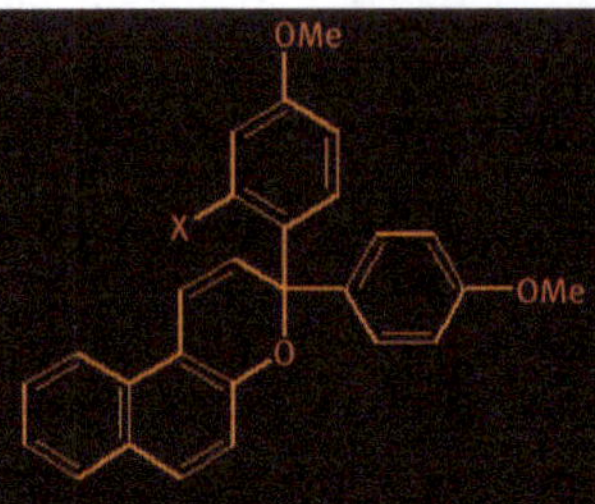

6

X	λ_{max} (nm)	$t_{1/2}$ (s)
H	472	2
Me	470	400
OMe	476	320

restricts this class to being a source of commercially important yellow, orange and red dyes only. For more detailed information, see [4, 13], and [19].

49.5.2 Design strategies for [1,2-*b*]naphthopyrans

The parent [1,2-*b*]naphthopyran **7** (see Figure 49.4) is of no commercial significance because of its very slow fading red-orange photocoloration (λ_{max} 508 nm and $t_{1/2}$ > 400 s in toluene at 20 °C). The introduction of appropriate functions onto its phenyl rings (positions 2′-4′) and naphthalene system (positions 5–10) produces a plethora of industrially useful photochromic colorants by speeding up rate of thermal fade. These design strategies constituted a major advance in photochromic colorant technology and went on to be extensively patented. This section gives a flavor of those exploited in industrial [1,2-*b*] colorants.

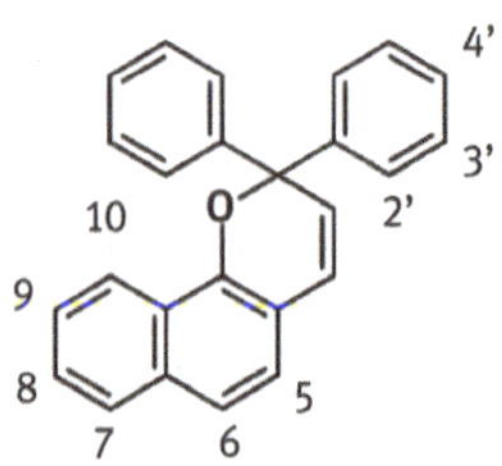

7

Figure 49.4: Molecular structure and position numbering for parent [1,2-*b*]naphthopyran.

Some of the approaches utilized with this class are shared with [2,1-*b*]naphthopyrans. For example, the introduction of strong electron donors in the *para* positions of the phenyl rings attached to the spiro carbon remains an indispensable tool for manipulating color and kinetics. These substitution patterns shift photocoloration into the bluish-red, purple and blue regions, whilst simultaneously shortening half-life to commercially useful durations.

The 5- and 6-positions of [1,2-*b*]naphthopyrans are of key importance in the design of industrially useful photochromic dyes. The utility of many commercial colorants of this class depends greatly on the placement of functional groups or the fusion of ring systems at these locations. In contrast to the location of a methyl group at a 2′-position (which as we have seen dramatically retards the thermal back reactions of [2,1-*b*]-naphthopyrans – and does so too for [1,2-*b*] derivatives), introduction of 5-methyl functionality increases the rate of ring closure. By significantly reducing half-life, this strategy transforms the slow-fading parent compound **7** into derivatives (such as **8** where R = Me and R′ = H) whose rates of thermal reversion back to colorless ring-closed isomers approach industrial utility. The effect is caused by steric crowding within the photomerocyanines reducing their stability without greatly affecting color – the 5′-substituent clashes with the proton on the γ-carbon as shown in Table 49.4. Addition of a second methyl function at the adjacent 6-position further shortens half-life to a duration that is industrially useful, whilst having little impact on shade: the 5,6-dimethyl-substituted dye **8** (R = R′ = Me) in Table 49.4, a colorant which photoactivates to give red hues, is one of several with this pattern that have been commercialized.

Table 49.4: Effect of substitution at 5- and 6-positions on photocoloration of [1,2-*b*]naphthopyrans **8** (data tabulated relate to toluene solutions at 20 °C) [21].

8

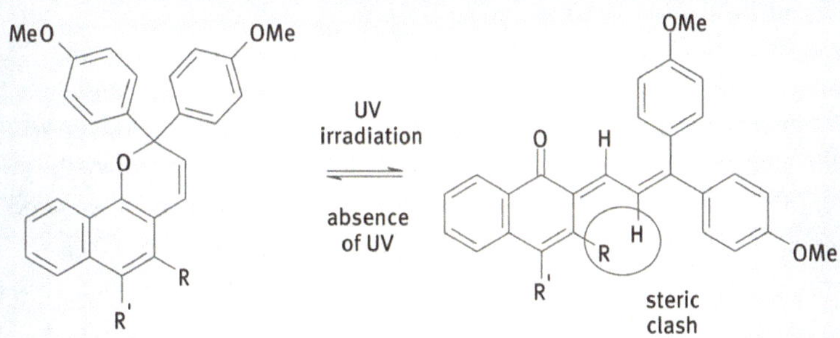

R	R'	λ_{max} (nm)	$t_{1/2}$ (s)
H	H	508, 412	>1800
Me	H	496, 416	178
Me	Me	492, 418	65
CO_2Me	H	492	2

The pattern of 5-methyl-6-aryl substitution has also proved commercially useful, especially in conjunction with aryl-substitution at the 9-position, i. e. dyes **9**. The introduction of donor groups onto 6- and 9-phenyl rings not only slows thermal ring closure, increasing the intensity of photocoloration, but also furnishes dyes with dull purplish or bluish hues that serve as a good basis for neutral colored mixtures. The donor functions increase the intensity of a second absorption band in the visible region (see Table 49.5), leading to dull colors being produced on irradiation that require fewer shading components to achieve gray photocoloration (e. g. dyes **9B-9D** have all been commercialized with **9B** being particularly successful in the 2000s).

Table 49.5: Effect of 6,9-diaryl substitution on photocoloration of [1,2-*b*]naphthopyrans (data tabulated relate to toluene solutions at 20 °C) [19].

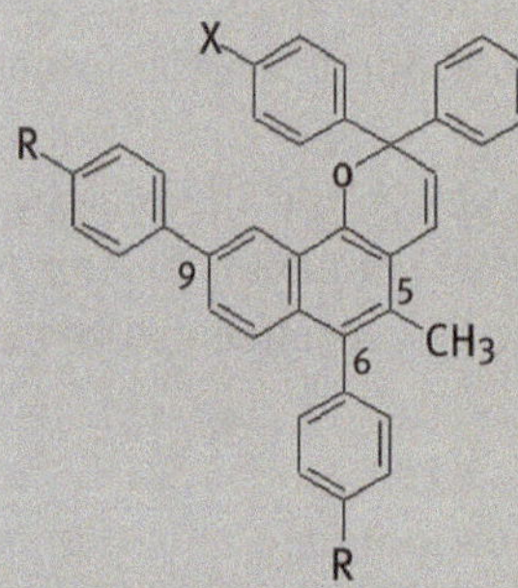

9

	X	Y	R	λ_{max} (nm)	$t_{1/2}$ (s)
A	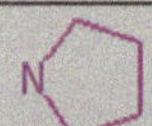	H	H	550	42
B	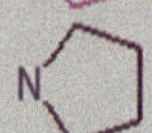	H	OMe	560, 485	52
C	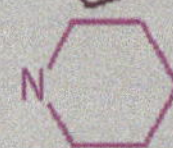	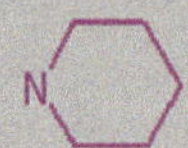	H	551	14
D	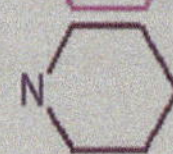	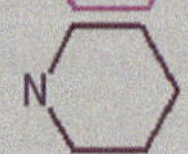	OMe	560, 465	22

Location of stronger donors at the 9-position alone also furnishes dyes whose photocoloration features more than one broad absorption band in the visible region to give colorants that can act as the base for formulating neutral-colored lenses. Derivatives of this kind were first marketed two decades ago, e.g. **10** (X = OMe) and **10** (X = piperidino) which furnish brownish [22] and grayish [23] photocoloration, respectively (see Figure 49.5). These examples both possess 5-ester substitution as a means of speeding up the thermally-driven cyclisation to a commercially useful rate. As illustrated in Table 49.4, the increased bulk of the carboxymethyl function creates greater steric hindrance in photomerocyanine species than a 5-methyl group, which promotes reversion to the ring-closed form of the dye.

A commercially important means of tuning both the shade and fade rate of a dye's photocoloration is annelation at the 5- and 6-positions. Indeno-fused dyes **11** have been utilized in lens manufacturing since the 1990s (see Figure 49.5). Attachment of an aryl function at the equivalent of the 6-position (held in plane with the rest of the π-system by the indeno methylene bridge), coupled with appropriate 4′- and 4″-substitution, furnishes bathochromic and intense photomerocyanine

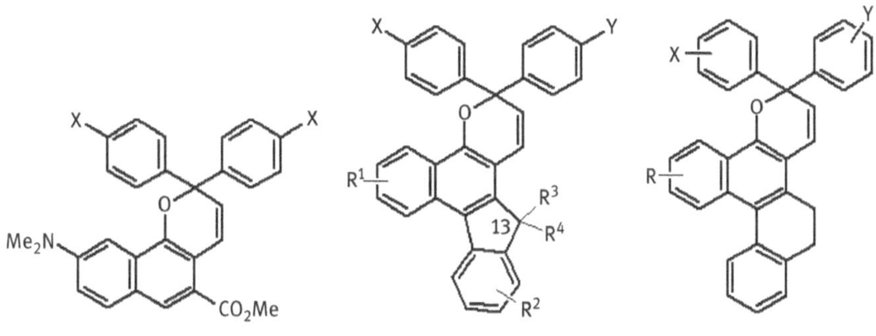

10 11 12

Figure 49.5: Generalised structures of commercially-successful neutral-coloring types of [1,2-*b*] naphthopyran dye.

species. Donor groups located on the naphthalene system contribute to dull but strong bluish and greenish photocoloration by enhancing the intensity of absorption at shorter wavelengths relative to that of the longest wavelength absorption band. Kinetics, rather than color, are tuned by choice of group(s) at the 13-position [19]. Dyes with 13,13-dimethyl and 13-alkoxy-13-alkyl patterns have been industrially important, while commercial success has also been achieved with dyes bearing cycloaliphatic rings at this position [24, 25]. Another commercially useful motif employed in the design of dyes producing photocoloration of neutral shades is shown in structure **12** [26] (see Figure 49.5).

49.6 Future developments

Work aimed at developing more responsive, intense and robust colorants for existing commercial applications continues in industrial settings. Given its sensitive nature, reporting of the fruits of this work has become increasingly rare outside of the patent literature. While naphthopyrans featured prominently in the first few triennial International Symposium on Photochromism (ISOP) events held during the 1990s and around the turn of the Millennium, much of the content arising from industrial research, they hardly appeared in ISOP proceedings over the past decade. Nevertheless, work is ongoing that seeks to capitalize on the advantages of naphthopyrans containing oligomeric tails or linkers, for example, by enabling use of dyes in polymeric lens matrices that are not otherwise conducive to good photochromic performance but which offer desirable properties such as high refractive indices and mechanical robustness [27, 28]. Other directions that aim to limit disadvantageous matrix effects on naphthopyran photochromism may join this approach as a commercial success. These strategies include use of polyrotaxanes to increase free-volume available for photoisomerization [29]. Another involves dispersing into a substrate nanocapsules that enclose solutions of naphthopyran dye: the colorant's environment favors fast switching, while encapsulate dimensions mean that incident light is not scattered to a noticeable extent so optical clarity is not impaired [30]. Achieving better photochromic response by adapting existing techniques and lens formulations remains of interest to industry too, for example by modification of the monomer composition of photochromic coatings applied to lenses [31]. Academia is also exploring means of improving colorant performance in lenses as well as other applications: for example, one strand of research aims to prevent the formation of photoisomers of relatively high thermal stability, thereby eliminating the problem that they present of causing residual color which persists long after the disappearance of stimulating radiation. Such work has shown that, through modification of naphthopyran structure, creation of slow-fading photoisomers can be suppressed [32] and even rendered impossible [18] – however, further development is needed to take proof of concept through to commercial reality. The main thrust of academic research into naphthopyrans is the exploration of their use as functional dyes [33] in application areas such as information technology, electronics, and nanomachinery. A summary can be found in the review [34], although the authors' claim that naphthopyrans have inferior photostability as well as resistance to fatigue compared to P-type colorants, and diarylethenes in particular, is false. Currently, these potential applications remain distant from becoming a commercial reality. They would also likely exploit another class of photochromic colorant, most probably P-type. Nonetheless, it is conceivable, albeit still a remote possibility, that industrialization of just one of these applications could transform the role played by naphthopyrans of dominating a sector of the optics market to becoming a key part of a technology that pervades many parts of everyday life.

49.7 Summary

The photochromism of naphthopyran colorants is strongly dependent on their molecular structure so modifications to them afford ready manipulation of hue and kinetics. They are delicate and expensive in comparison to non-photochromic commercial dyes and pigments, so their usage will stay narrow. Nevertheless, their robustness, flexibility and temperature-insensitivity relative to other photochromic colorant classes led to them rising swiftly from obscurity to industrial prominence. Of particular importance has been the development of hypsochromic [2,1-*b*]naphthopyran dyes, followed by that of bathochromic colorants based on [1,2-*b*]naphthopyrans giving photocoloration of dull shades. The naphthopyran class will remain the industrially dominant type of photochromic colorant unless one of two things happens: (i) another application for photochromism usurps ophthalmic lenses as the principal commercial use of such colorants, or (ii) the development of another class of dye with better kinetics, robustness and/or economics. Both eventualities are remote in the short term.

References

1. A Towns. Photochromic dyes, to be submitted.
2. A Towns. Spiropyran dyes, to be submitted.
3. A Towns. Spirooxazine dyes, to be submitted.
4. Van Gemert B. Benzo and naphthopyrans (chromenes). In: Crano JC, Guglielmetti RJ, editors. Organic photochromic and thermochromic compounds, vol. 1. Chapter 3. New York: Plenum, 1999:111–40.
5. Van Gemert B. The commercialization of plastic photochromic lenses: a tribute to John Crano. Mol Cryst Liq Cryst Sci Technol Sect A Mol Cryst Liq Cryst. 2000;344:57–62.
6. Crano JC, Flood T, Knowles D, Kumar A, Van Gemert B. Photochromic compounds: chemistry and application in ophthalmic lenses. Pure Appl Chem. 1996;68:1395–8.
7. Van Gemert B, Kish DG. The intricacies of color matching organic photochromic dyes. ChemiChromics '99 conference, New Orleans, 1999.
8. Van Gemert B. (PPG Industries), Novel photochromic indeno-fused naphthopyrans. World Patent Application WO 96/14596 A1, 1996.
9. Ercole F, Malic N, Davis TP, Evans RA. Optimizing the photochromic performance of naphthopyrans in a rigid host matrix using poly(dimethylsiloxane) conjugation. J Mater Chem. 2009;19:5612–23.
10. Malic N, Evans RA. The use of poly(alkylene oxide)s to achieve fast and controlled photochromic switching in rigid matrices. J Polym Sci A Polym Chem. 2012;50:1434–44.
11. Malic N, Campbell JA, Evans RA. Superior photochromic performance of naphthopyrans in a rigid host matrix using polymer conjugation: fast, dark, and tunable. Macromol. 2008;41:1206–14.
12. Corns SN, Partington SM, Towns AD. Industrial organic photochromic dyes. Color Technol. 2009;125:249–61.
13. Hepworth JD, Heron BM. Synthesis and photochromic properties of naphthopyrans. In: Gribble GW, Joule JA, editors. Progress in heterocyclic chemistry, vol. 17, Chapter 2. Oxford: Elsevier Science, 2005:33–62.
14. Hepworth JD, Heron BM. Photochromic naphthopyrans. In: Kim S-H, editor. Functional dyes, Chapter 3. Amsterdam: Elsevier, 2006:85–135.
15. Heller HG, Oliver SN, Whittal J, Johncock W, Darcy PJ, Trundle C. (The Plessey Company plc), Photochromic fused-ring organic compounds and their use in photoreactive lenses. UK Patent GB2146327B, 1986.
16. Braslavsky SE. Glossary of terms used in photochemistry 3rd edition (IUPAC recommendations 2006). Pure Appl Chem. 2007;79:293–465.
17. Mustroph H. Polymethine dyes 2020. Phys. Sci. Rev. DOI:10.1515/psr-2019-0084.
18. Sousa CM, Berthet J, Delbaere S, Coelho PJ. Enhancement of the color intensity of photochromic fused-naphthopyrans. Dyes Pigm. 2019;169:118–24.
19. Towns AD. Industrial Photochromism. In: Bergamini G, Silvi S, editors. Applied photochemistry, Chapter 5. Switzerland: Springer, 2016:227–79.
20. Rickwood M, Smith KE, Gabbutt CD, Hepworth J. (Pilkington PLC), Photochromic naphthopyran compounds. World Patent Application WO 94/22850 A1, 1994.
21. Anon. Research disclosure, no. 361. May 1994:267–8.
22. Clarke DA, Heron BM, Gabbutt CD, Hepworth JD, Partington SM, Corns SN. (James Robinson Ltd), Neutral colouring photochromic 2H-naphtho[1,2-b]pyrans and heterocyclic pyrans. World Patent Application WO 98/42693 A2, 1998.

23. Clarke DA, Heron BM, Gabbutt CD, Hepworth JD, Partington SM, Corns SN. (James Robinson Ltd), Grey colouring photochromic fused pyrans. World Patent Application WO 00/18755 A1, 2000.
24. Teranishi K, Takahashi T, Miyazaki M, Takenaka J. (Tokuyama Corp), Chromene compound and curable composition containing the chromene compound. Japanese Patent Application JP 2018-62496 A, 2018.
25. Mann C, Weigand U, Melzig M. (Optische Werke G Rodenstock), Photochromic spirofluorenopyran compounds. US Patent US6331625 B1, 2001.
26. Chan YP, Jean PC, Breyne O. (Corning SA), Naphthopyrans annelated in C5-C6, their preparation and compositions and (co) polymer matrices containing them. World Patent Application WO 00/15268 A1, 2000.
27. Kasori Y, Kadowaki S, Kawato N, . (Mitsui Chemicals Inc.), Polymerizable composition for optical materials, material obtained from said composition, and plastic lens. World Patent Application WO 2017/047745 A1.
28. Kadowaki S, Kawaguchi M, Kawato N. (Mitsui Chemicals Inc.), Polymerizable composition for optical materials, and molded body. World Patent Application WO 2018/062384 A1.
29. Izumi S, Ohara A, Shimizu Y, Momoda J. (Tokuyama Corp.), Photochromic curable composition, use thereof, and polyrotaxane monomers. World Patent Application WO 2018/030257 A1, 2018.
30. Vázquez-Mera N, Roscini C, Hernando J, Ruiz-Molina D. Liquid-filled capsules as fast responsive photochromic materials. Adv Optical Mater. 2013;1:631–6.
31. Niemi KM, Ambler DM. (Younger Optics), Optical articles comprising photochromic poly(urea-urethane). World Patent Application WO 2018/160885 A1.
32. Inagaki Y, Kobayashi Y, Mutoh K, Abe J. A simple and versatile strategy for rapid color fading and intense coloration of photochromic naphthopyran families. J Am Chem Soc. 2017;139:13429–41.
33. Towns A. Colorants: General survey. Phys Sci Rev. 2019;4. DOI:10.1515/psr-2019-0008.
34. Mukhopadhyay A, Moorthy JN. Phenomenon to functions: photochromism of diarylpyrans, spectrokinetic properties and functional materials. J Photochem Photobiol C: Photochem Rev. 2016;29:73–106.

Robert Christie and Adrian Abel

50 Organic pigments: general principles

Abstract: This introductory chapter presents an overview of the general principles underlying the structural chemistry, manufacturing processes, and application technology of organic pigments. The coverage provides a fundamental theoretical and practical basis for the chapters that follow in this series that are devoted to specific chemical classes of industrially significant organic pigments of the azo, phthalocyanine, carbonyl, dioxazine, and metal complex classes. The initial sections cover the fundamental differences which mean that dyes and pigments are considered universally as two separate types of colorant, based on their solubility characteristics. They also provide discussions of the contrasting chemical, technological, and performance features of organic and inorganic pigments. An outline of the most important historical features in the development of the synthetic organic pigment industry is then presented, from its origins in the 19th century that followed soon after the development of the industrial synthetic dye industry, through its expansion in the 20th century, to its current position as a mature global industry. A section then follows that describes the functions that organic pigments are required to perform in their application, mainly their optical functions that include not only color properties, including hue, strength, brightness, but also the contrasting requirements for transparency or opacity as demanded by specific applications. The pigments are also required to resist the conditions and agencies that they might encounter in applications, assessed as fastness properties, such as fastness to light, heat, solvents and chemicals, amongst many others, to an extent that specific applications demand. The principles, in broad terms, of the ways in which chemical structures determine colour and performance of organic pigments are discussed, with focus not only on the influence of molecular structure, but also on the effect of the crystal structural arrangement and the particulate structure, including particle size and shape and its distribution, on application performance. This is important as these pigments are applied as a dispersion of finely divided crystalline solid particles that are insoluble and are ultimately trapped mechanically in their application medium, often a polymer. The manufacture of organic pigments is discussed in broad terms. The overall process may be considered in stages, initiated by the chemical synthetic sequence in which the pigment is formed, followed by a conditioning stage where the crude product thus obtained is modified to optimise its performance properties, and finally finishing where the product is processed into a form, or preparation, that is suitable for its intended applications. Finally, the technological principles underlying a broad range of the most important

This article has previously been published in the journal Physical Sciences Reviews. Please cite as: R. Christie, A. Abel, Organic Pigments: General Principles *Physical Sciences Reviews* [Online] 2021, 6. DOI: 10.1515/psr-2020-0187

https://doi.org/10.1515/9783110587128-050

application areas for organic pigments, which are mainly in paints, inks, and plastics, are discussed.

Keywords: pigments, dyes, organic pigments, inorganic pigments, colorant, lake, azo pigments, phthalocyanine pigments, carbonyl pigments, dioxazine pigments, metal salt pigments, hue, brightness, color strength, transparency, opacity, fastness, polymorphism, dispersibility, amide group, printing inks, surface coatings, plastics, rubber, paper

50.1 Pigments *vs* dyes

The term colorant is used to describe the two broad product types that are used to add color to our lives, namely dyes and pigments [1, 2]. Both product types may be supplied as colored powders, and in this form, products of each type with the same color might well appear indistinguishable to the eye. However, they are distinctly different materials in their physical and technical properties and in the ways that they are incorporated into their applications. Dyes and pigments are distinguished based on their solubility characteristics. Essentially, dyes are required to dissolve, for example, in water to assist their main application, which is to textiles, since they are usually applied from aqueous solution. However, certain dye classes are insoluble, for example, vat dyes in the case of the cellulosic fibers to which they are applied, but are capable of solubilization by a chemical conversion process. In contrast, pigments are broadly required to be essentially insoluble in the medium into which they are incorporated. To be more inclusive, a more rigorous distinction may be proposed; dyes are intentionally in solution at some point in their application, while pigments are not intentionally in solution at any point during their application. The principal traditional applications of organic pigments are in paints, printing inks and plastics, although they are used more widely, for example, in the coloration of building materials, such as concrete and cement, and in ceramics, paper, textiles, rubber, cosmetics, crayons, and glass. In most cases, the application of pigments involves their incorporation by dispersion into a liquid medium, for example, a wet paint or ink or a molten thermoplastic material. The medium is then allowed to solidify, for example, by solvent evaporation, physical solidification or polymerization, and the pigment particles become trapped in the solid polymeric matrix. In contrast to most textile dyes where individual dye molecules, or their solubilized forms, are strongly attracted to individual polymer molecules of the fibers to which they are applied, pigments are considered to have only a weak affinity for their application medium, and only at the surface where the pigment particle is in contact with the medium.

50.2 Organic *vs* inorganic pigments

In chemical terms, pigments are classified as either inorganic or organic depending on the branch of chemistry from which they are derived [2–5]. Certain generalizations may be made in comparing the technical performance of inorganic and organic pigments in their applications (although exceptions are found). Inorganic pigments are generally characterized by excellent durability, for example, resistance to heat, light, weathering, solvents, and chemicals and, in those respects, they can offer technical advantage over most organic pigments. Also, inorganic pigments are often significantly less expensive than organics. In contrast, organic pigments show superior intensity and brightness of color compared with inorganic pigments. Although characterized by these good color properties, organic pigments are variable in the range of fastness properties that they offer. The ability either to provide opacity or to ensure transparency in their application provides a further contrast between inorganic and organic pigments. Inorganic pigments are, in general, high refractive index materials which leads to high opacity in application, whereas organic pigments are low refractive index materials and consequently provide transparency. However, higher opacity grades of certain organic pigments are also available by modifying their particle size and shape. Conversely, certain inorganic pigments (e.g., iron oxides) can be manufactured to provide transparency, also by modifying their particle size and shape to optimize this property.

50.3 A brief history of organic pigments

People have made use of pigments since prehistoric times, for example, in body decoration and in paintings that decorate cave dwellings [6, 7]. In those days, colors were obtained from natural sources, mostly inorganic pigments from minerals, dug from the earth, ground to a powder and combined with a crude binder. Organic colorants from natural sources such as plants were also used in antiquity, mainly for their bright colors, and the term "plant pigments" is still in use today. However, these colors were fugitive and would not be categorized as "pigments" in the current commercial sense. Synthetic inorganic pigments have been in use for over four centuries since Egyptian blue, a glass-like substance based on a mixed calcium/copper silicate, $CaCuSi_4O_{10}$, was manufactured in Egypt [8]. The oldest synthetic pigment still manufactured and used today is Prussian blue, dating originally from the early eighteenth century. Synthetic organic pigments have a more recent history. They emerged towards the end of the nineteenth century, developed largely from the synthetic textile dyes that had become well-established. The origin of the synthetic dye industry is attributed to Sir William Henry Perkin on account of his discovery in 1856 of a purple dye, originally named aniline purple, but later to become known as mauveine [9]. Perkin, a 19-year-old British organic chemist, was carrying out research aiming to develop a synthetic route to the antimalarial drug, quinine. A particular objective in his investigation was to prepare quinine by oxidation of allyltoluidine. His attempts proved unsuccessful, not surprisingly in hindsight based on current knowledge of the complex heteroalicyclic structure of quinine. As an extension of the research, his attention turned to oxidation of aniline with potassium dichromate. The reaction gave a black product from which he extracted a small amount of a purple dye that could dye silk. Within a year, Perkin had opened a factory to produce the dye in Greenford, London. Following the rapid success of Mauveine, many more synthetic organic textile dyes of various types were introduced.

Among the earliest organic pigments were products with bright, intense colors commonly referred to as lakes, which were products prepared from known water-soluble textile dyes rendered insoluble by precipitation on to an inert, colorless, and insoluble inorganic substrate. The name derives from the lac insect that contained a red colorant from which a pigment lake was prepared by precipitation on to such a substrate in a process that had been used for centuries with natural dyes. This process could be applied to a wide range of synthetic textile dyes, especially anionic dyes precipitated on to substrates such as alumina and barium sulfate. It was also found that soluble cationic (basic) dyes treated with tannin or antimony potassium tartrate provided insoluble pigments. Substrates such as alumina hydrate provided a porous structure into which the dye could be physically absorbed and could also react to form insoluble aluminum salts. Although they were often derived from brightly colored dyes, the resulting lake was much duller in color, as anyone with a postage stamp collection from the nineteenth century can testify! Another early development in industrial organic pigments was the introduction of pigments of the azo chemical class. Many of these, referred to as classical azo pigments, remain the most important commodity yellow, orange, and red organic pigments in use today, providing bright intense colors at relatively low cost, although exhibiting variable technical performance. The re-discovery of copper phthalocyanine blue in 1928 and its subsequent development into a commercial pigment was decisive in the development of the organic pigment industry. This pigment, introduced commercially in the mid-1930s, was the first product that offered outstanding intensity and brightness of color, typical of organic pigments, combined with an excellent range of fastness properties, comparable with many inorganic pigments. The discovery stimulated the quest in subsequent decades for other chemical types of organic pigment that could emulate, in the yellow, orange, red and violet shade areas, the high performance provided by copper phthalocyanine blue and green pigments. The research intensity acquired further impetus from the growth throughout the twentieth century of applications that demanded high levels of technical performance, especially lightfastness and thermal stability such as automotive paints, plastics, and synthetic fibers. From the mid-twentieth century onwards, a range of high-performance organic pigments emerged, encompassing a variety of chemical types including carbonyl and dioxazine pigments (alternatively referred to as polycyclic pigments) together with some high-performance azo pigments [6]. In the latter part of the twentieth century, the introduction of new organic pigments into the marketplace declined, the last significant group being the diketopyrrolopyrrole (DPP) pigments developed in the 1980s. This trend reflected the maturing of the pigment industry, which has seen more emphasis on process and product improvement, and the need to address a range of environmental issues.

50.4 Classification of organic pigments

Commonly, organic pigments are referred to broadly as either classical or high-performance. The classical organic pigments are long-established, relatively low cost, industrial commodity products, which benefit from excellent coloristics, but which do not uniformly offer the excellent durability and resistance properties provided by high-performance pigments. Alternatively, organic pigments can be classified according to common chemical structural features, the most important chemical classes being azo, phthalocyanine, carbonyl and dioxazine pigments. Classical organic pigments include the monoazo and disazo pigments that have been in production for around a century. Copper phthalocyanines are also generally considered as classical organic pigments. High-performance organic pigments combine strong bright colors with excellent technical performance that suits them to the most demanding applications, but generally at higher cost. Examples of products developed specifically to satisfy the demands include carbonyl and dioxazine types. However, two groups of azo pigments have been developed to provide high performance, namely the benzimidazolone and disazo condensation pigments. In the range of organic pigments, copper phthalocyanines stand out. They offer high intensity and brilliance of color, together with an excellent set of fastness properties and are reasonably low cost. Thus, copper phthalocyanines could equally be considered as high-performance products. Arguably, they come closest to being the "perfect" pigments. The most important chemical classes encountered in industrial organic pigments are shown in Figure 50.1, which also illustrates their relationships with the classical and high-performance categories. Organic pigments are also commonly grouped together according to color. In broad terms, azo pigments mostly fill the yellow, orange, and red shade areas, while copper phthalocyanines dominate the blues and greens. The high-performance carbonyl pigments provide colors covering most of the color range, while dioxazines provide violet pigments.

50.5 Properties and functions of organic pigments

Pigments are solid colorants that are essentially insoluble in their application media. Mostly, pigments are incorporated into a liquid medium e.g., a wet paint or

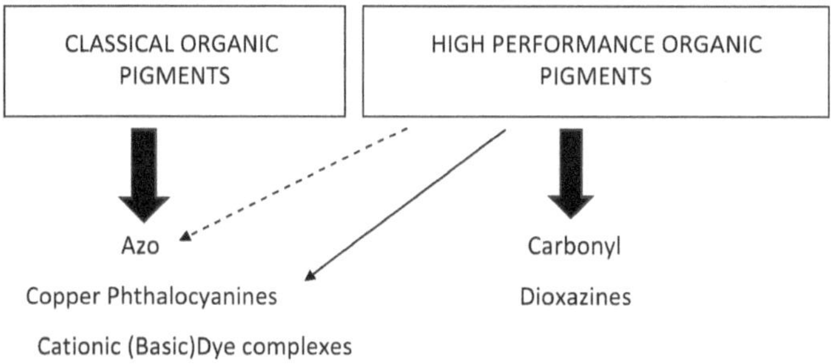

Figure 50.1: Organic pigment classification.

ink or a molten thermoplastic material, by a dispersion process. In this process, clusters or agglomerates of pigment particles are broken down into smaller entities, primary particles and small aggregates. The pigmented media then solidify by solvent evaporation, physical solidification or polymerization, and the dispersed particles become mechanically fixed in the solid matrix. The principal role of pigments is to modify the optical properties of the paint, printing ink, plastic, or other material into which they are dispersed. The most obvious is the ability to impart the desired color. However, a pigment may also be required to influence other optical properties, notably opacity, arguably most importantly in paints that are required to obscure the surface to which they are applied. Alternatively, and in complete contrast, high transparency may be essential. Multicolor printing most commonly uses inks of the subtractive primary colors, yellow, magenta, and cyan, together with black as a fourth color. In this case, the optical effect of the first color printed must not be obscured by subsequently printed colors, and so this application requires the use of transparent pigments. A pigment owes its optical properties to a combination of two effects as pigment particles interact with visible light: light absorption and light scattering. Color is determined mainly by light absorption, while opacity depends principally on light scattering. An object appears colored if certain wavelengths of incident light are selectively absorbed. The color of an organic pigment is dependent on its molecular constitution although two further factors influence their color, to varying extents, because they are used as solid, crystalline particles. These factors are the crystal structure, i.e., the way in which the molecules pack in their crystal lattice, and the particle structure, predominantly size and shape distribution. The degree of light-scattering by particles is dependent on the refractive index of the pigment, high refractive index materials leading to high opacity. This property is also dependent to an extent on the crystal structure of the pigment and its particle size and shape.

Organic pigments must also withstand the effects of the environment that they encounter during processing and in their useful lifetimes. These technical performance features are referred to and assessed as fastness or resistance towards specific agencies. A pigment will be selected for an application based on its optical and fastness properties, on toxicological and environmental considerations and, inevitably, on cost. The fastness properties of most relevance to pigments are resistance to light, weathering, heat, and to solvents and other chemicals to which they may be exposed. The properties of lightfastness and weatherfastness are related although, on occasions, organic pigments which show good resistance to fading when exposed to light can perform less well under the combined attack of sunlight and moisture. These performance properties are determined by the molecular structure of the pigment, the crystal lattice structure, and to an extent by its particle size, concentration in the application medium and on the nature of the medium itself. Lightfastness requirements vary with the demands of the application, e.g., the levels of external exposure that the product will be required to resist. For example, outstanding fastness to light is required for pigments used in automotive paints, such as those applied to cars and other vehicles, and in plastics used in the construction of the vehicle. Lighter shades, especially in combination with white pigments, commonly referred to as reductions, tend to show poorer lightfastness than deeper shades. In contrast, lightfastness is much less important for low cost, colored printed materials. The heat stability of a pigment refers to its resistance towards color change when exposed to high temperatures. In the coloration of thermoplastics, heat stability is a vital feature. The degree of heat stability required for such applications depends on the processing temperature for the polymer in question, which can range from 150–350 °C, and on the time of exposure. Since pigments in their application are required to remain as discrete particles held within a polymer, they must resist dissolving in solvents with which they may come into contact. The extent to which solvent resistance is required depends on the nature of the solvent to which the pigment is exposed, and to the temperature of exposure, since solubility increases as the temperature is raised. A tendency for the pigment to dissolve can lead to problems, for example, in paints, of bleeding of color into overpainted layers [10]. In plastics, a tendency to dissolve in the polymer may lead to problems of contact bleed or migration, whereby a colored plastic material causes staining of a dissimilarly colored material that it contacts, plate-out, referring to an accumulation of additive on the surfaces of plastics processing equipment [11], or solvent bleed occurring when color leaches from a plastic article immersed in a solvent. Bloom can also be observed whereby a powdery deposit of colorant appears on the surface of a paint, ink film or plastic product, leading to poor fastness to rubbing. This results when a pigment with limited solubility dissolves in a hot application medium forming a supersaturated solution when cooled. In time, pigment molecules diffuse to the surface and crystallize. Resistance towards changing color on exposure to acids, alkalis or other chemicals occasionally features in pigment selection for applications where this may be encountered. Toxicological and environmental issues have assumed immense importance, especially in recent years. Toxicity is of critical importance for applications where ingestion is a possibility, such as in paints, printing inks, and plastics which may contact food and also in toys and graphic instruments. The presence and concentrations of heavy metals, notably lead and cadmium, are strictly controlled, although this applies more often to inorganic pigments. In general, it is reassuring that organic pigments may be considered as relatively non-toxic materials. It may be argued that the insolubility of pigments means that they can pass through the digestive system without absorption into the bloodstream and thus present little hazard. However, the presence in certain pigments of toxic impurities such as residual reactants, by-products in the manufacturing processes, and degradation products, can be of concern. Levels of such compounds identified are controlled by national legislation and regulations.

A feature commonly encountered in the crystalline solid-state structures of pigments is polymorphism. A material exhibits polymorphism if it exists in forms with identical molecular composition but different molecular arrangements in the crystal structures. Different polymorphic forms may show significantly different color and technical properties, features that determine their industrial applications. Numerous pigments exhibit this phenomenon, including copper phthalocyanine, quinacridones and many azo pigments. Particle size, and its distribution, are also factors in determining the properties of a pigment. Organic pigments are often produced in as fine a particle size as is technically feasible, commonly below 0.05 µm, to maximize color (tinctorial) strength and transparency. Printing inks provide a prime example of an application for which such properties are essential. In contrast, for industrial surface coatings, pigments with somewhat larger particle size are desirable for their improved opacity and enhanced fastness properties, although these properties are achieved at the expense of color strength.

It is essential that an organic pigment is well dispersed into its application medium. Pigment powders consist of clusters of agglomerated particles that are subjected to conditions that break down large aggregates into small aggregates and primary particles to the degree required to give the optimum properties. The term dispersibility refers to the ease with which the required degree of dispersion is achieved. The dispersibility of commercial pigments is optimized so as to minimize the energy requirements of the dispersion process. Enhanced dispersibility is often achieved by surface treatment with additives. The chemical nature of the pigment particle surface is of importance in their performance because these surfaces are in contact with the application medium. The surfaces of organic pigment particles which form directly in their synthesis are often highly polar (hydrophilic) so that they may show some incompatibility with non-polar (hydrophobic) media with which they are required to interact. The pigments are therefore often treated with materials such as surfactants or resins, either during or after the synthesis phase of their manufacture to change the nature of the surfaces. These surface-active agents may be considered as consisting of a balance of polar groups, such as sulfonic or carboxylic acid groups or amino groups, and non-polar groups such alkyl chains. When the pigment is treated with the agent, the polar group attaches to the pigment surface, due to dipolar or ion-dipole forces of attraction. Thus, the pigment acquires a new surface which is more hydrophobic and hence more compatible with an organic application medium. In addition, because of the steric barrier presented by the attached agents, the pigment particles are kept apart and less susceptible to aggregation, while those aggregates that are formed are more easily broken down by the dispersion process. The use of resin surface treatments is particularly important with certain pigment classes. Frequently, the resin of choice is rosin or colophonium (derived from pine trees), or a chemically modified derivative. Resination softens the texture of the pigment so that it is easier to disperse. When used early in the manufacturing process, the resin coats the pigment particles as they are formed, thus limiting crystal growth and leading to a minimized particle size, enhancing tinctorial strength and transparency. The increased color strength often more than compensates for the mass of the colorless resin on the pigment, thus improving economics. Resinated pigments often provide improved gloss in applications where this feature is important. However, they are limited to applications where the resin will not harm the film forming properties of the medium in which the pigment is dispersed.

50.6 Synthesis and manufacture of organic pigments

The process by which an organic pigment is manufactured may be considered as involving three stages, as illustrated in Figure 50.2. The first is the synthesis in which the sequence of chemical reactions that form the pigment is conducted. However, simply carrying out this chemistry does not usually provide the product in a form optimized for its intended application. In a second phase of the process, referred to as conditioning, the product is developed into an appropriate crystalline form, with a controlled particle size distribution, and the surfaces of the particles may be altered by suitable treatments. Depending on the pigment, these two phases may be either distinctly separate or combined into a single process. Finally, the pigment is subjected to finishing in which it is converted into a form that is acceptable to the user. The methods of synthesis of organic pigments are quite diverse and dependent on the chemical class of pigment. The conditions of synthesis of those organic pigments that are prepared by precipitation from reaction in water may often be controlled to ensure that the pigments are formed directly at an appropriate particle size. Commonly they will precipitate from this aqueous medium in a poorly defined crystal form, and so they are subsequently treated at elevated temperatures, either in the presence or absence of solvents (water or organic solvents), to develop high crystallinity and ensure that the products show optimum optical properties. Some high-performance organic pigments are synthesized at high temperatures often by reaction in organic solvents and are produced initially in a large particle size form, which must be reduced to a fine particle size to make them useful pigmentary products. One way in which this may be achieved is by mechanical grinding, sometimes in the presence of abrasive materials such as inorganic salts. Alternatively, the pigment may be precipitated in a fine particle size by drowning out solutions of the crude pigments, for example, in concentrated sulfuric acid, into water.

Figure 50.2: Organic pigment manufacturing sequence.

Pigment manufacturers commonly produce a range of grades of the same pigment for use in different applications. Products with the same CI (Colour Index) designation but with slightly different shades, either from the same or different manufacturers are regularly observed. A common way to modify the shade of pigments is to co-synthesize using small amounts of alternative reactants with similar chemical structures, for example, auxiliary diazo or coupling components in the case of azo pigments. Pigments are supplied to the customer in a variety of forms, determined by the finishing process in their manufacture. For convenience, pigments are often supplied as dry powders. In the case of pigments prepared in water, they are isolated by filtration and dried at elevated temperatures. A problem with this stage of the manufacturing process is that drying can cause hydrophilic aggregation and "sintering" of the particles. The resulting clusters of particles may be hard in texture and difficult to disperse. This problem is minimized by the surface treatments in which a surfactant or resin layer at the pigment surface presents a steric barrier that keeps the pigment particles apart as the water evaporates, and also by careful control of drying conditions.

There has been concern in some situations with the level of dust generated in the handling of dry pigment powders, to which pigment manufacturers have responded by developing low-dusting, non-dusting and various types of granular forms of the dry pigments. Alternatively, pigments may be supplied in the form of pastes in which the problems associated with drying may have been avoided. These products include aqueous pastes, flushed colors (mainly for inks), and masterbatches (for plastics). Further details are discussed in the next section. Such concentrates are then easily incorporated by mixing into the final polymer composition at an appropriate stage of the overall process.

Other chapters in this series discuss the structural chemistry of the individual organic pigment chemical classes, their methods of synthesis and their applications. In dye chemistry, the relationship between molecular structure and technical performance, including color and fastness properties is reasonably well-established [2, 12, 13]. Computer-aided molecular modelling techniques have advanced significantly in sophistication so that the properties of a particular dye, especially its color, may be predicted with some confidence and without the need to resort to synthesis. With organic pigments, such structure-property relationships are much more complex because of complications based on the dependence of the coloristic and technical performance of pigments not only on molecular structure but also on crystal structure arrangement and the nature of the pigment particles, particularly their size and shape distribution. The application of increasingly sophisticated solid-state analytical techniques, notably single crystal X-ray diffraction studies and advanced methodology in structure determination using X-ray powder diffraction methods, have been effective in providing information on the crystalline structures of organic pigments. In addition, the science of crystal engineering, which involves the application of computer-based modelling techniques to crystal packing arrangements and morphology, has advanced rapidly leading to enhanced understanding of the properties of organic pigments and offers potential for application in the design of new products for improved performance [14].

Structure-property relationships are discussed in the chapters on individual pigment classes, although a few general features are of some relevance at this point. Organic pigments generally provide higher intensity and brightness of color than inorganic pigments. The colors are primarily attributed to the π–π * electronic transitions associated with the extensively conjugated aromatic systems of the colorants [12, 13]. Organic pigments are generally unable to provide the level of opacity offered by most inorganic pigments because molecular organic crystals have lower refractive index. However, the combination of high color strength, brightness, and transparency means that they are well suited to printing ink applications. The fastness characteristics of industrial organic pigments are variable and dependent on molecular structures and crystal structural arrangements. organic pigments frequently contain structural features that enhance solvent-fastness. These features are designed to increase the strength of intermolecular interactions within the crystal structures of the pigments, leading consequently to reduced solubility. Increasing molecular size is one way in which this may be achieved and invariably leads to improved solvent-resistance, but this feature alone need not necessarily enhance other fastness properties, especially lightfastness. Hydrogen bonding, both int*ra*molecular and int*er*molecular, is commonly encountered in many organic pigment classes, and play a prominent role in determining lightfastness. It is suggested that this is due to electronic and steric protection of the chromophore towards photochemical degradation. The hydrogen bonding reduces electron density at the chromophore, thus reducing susceptibility towards photochemical oxidation. Also, many high-performance pigments are large coplanar molecules, facilitating close packing of the molecules in stacks and thus strengthening intermolecular interactions.

Organic pigment structures are invariably conjugated, aromatic structures, a feature that leads to high stability. A range of substituents are found to modify color (in which case they are acting as auxochromes) and enhance fastness properties. The amide (-NHCO-), or carbonamide, group is encountered in many organic pigments. Its presence in an organic molecule generally enhances fastness to solvents, light and heat, a feature attributed to the ability of the amide group to participate in strong dipolar interactions and hydrogen bonding, as illustrated in Figure 50.3.

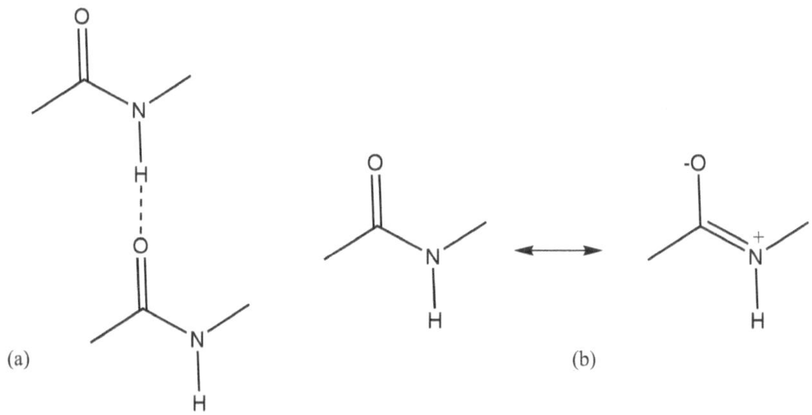

Figure 50.3: The origin of (a) intermolecular hydrogen bonding and (b) dipolar interactions due to the amide group.

In addition, hydrogen bonding between the amide group and other molecular substituents, e.g., carbonyl and nitro groups, commonly occurs. A related structural feature is found in many heterocyclic pigments (e.g., quinacridones and DPPs), which contain both N–H and C = O groups in ring systems. This arrangement provides a similar improvement in fastness properties as a result of strengthening intermolecular association throughout the crystal structure. The sulfonamide ($-SO_2NH-$) group is also occasionally used to enhance the performance by providing intramolecular and intermolecular hydrogen bonding, in a similar manner to the amide group. These groups also provide opportunities, by bridging, to incorporate additional aromatic rings, thus increasing molecular size. Other substituents encountered include halogens, nitro, alkyl and alkoxy groups. The incorporation of halogen substituents (F, Cl, Br) is often found to enhance technical performance, as well as to provide desirable color modifications. The improvement in durability in this case may be explained by the increase in molecular size and consequent reinforcing of intermolecular association, and to the strength of the carbon–halogen bond. The incorporation of certain metal ions, notably of alkaline earth elements and transition metals, introducing inorganic character into the organic pigments, can have a beneficial effect on insolubility and other fastness properties. Relatively low molecular weight molecules containing "salt-forming" groups, notably sulfonate ($-SO_3^-$) and carboxylate ($-CO_2^-$) groups, can be rendered insoluble by metal salt formation. In relevant chapters, the products are referred to as metal salt pigments. It can be confusing that the terms "toners" or "lakes", are often used to describe such pigments, mostly for historical reasons. In Europe, and especially in the UK, they are often referred to as toners, although this term means something quite different in other parts of the world, especially the US where it is often used as a synonym for organic pigments generally. In the US, metal salt pigments are generally known as lakes, a term that, historically, refers to the early pigmentary products in which soluble dyes were attached to an absorbent base. Examples of each of these structural effects will be regularly encountered and discussed further in the chapters on the individual organic pigment classes.

50.7 Application of organic pigments

Pigments are retained as particles in their applications in a polymer or resin. The main applications of organic pigments are printing inks, paints, plastics, rubber, paper, artists colors, textile printing, and other miscellaneous applications. Applications may use specific suitable techniques for dispersing powder pigments, while other applications use pigment preparations (elaborations), where the pigment has been pre-dispersed in a medium or system compatible with the binder or polymer. Examples include dispersions in aqueous surfactant solutions for water-based systems and in plasticizer for plasticized PVC. These preparations usually have a high pigment content. For polymers, concentrates, known as masterbatches, are often supplied using pigment mixtures blended to a standard color, so that a single addition can be incorporated into a polymer to provide consistent colors. An alternative, frequently used in inks, are flush pastes whose preparation rely on the fact that most organic pigments are lipophilic (hydrophobic), so that they have higher affinity for oil-based media than water [15]. After the synthesis phase of the manufacturing process, the excess liquid is squeezed out using a filter press and a press-cake is formed that may contain 20–50% pigment. The press-cake is then intimately mixed with a resin in a kneader in the presence of a cationic surface-active agent. After some time, the pigment "breaks" whereby it transfers from the aqueous phase in which it was produced, into the oil phase of the resin. This process typically takes thirty minutes to an hour. The process may have to be repeated two or three times to achieve optimum results. The comparatively clear aqueous phase can then be drained off. At this stage, the absence of hard agglomerates makes the pigment much easier to disperse, resulting in savings in energy and time. This process is popular in the paste ink sector, but when used in offset inks, the ink manufacturer usually requires a system free from water, and thus drying may be required, usually conducted under vacuum.

In terms of volume, almost 50% of organic pigments are used in printing inks, 25% in paints and coatings, 15% in plastics, with other applications making up the balance. In terms of value, the share in printing inks is less, while conversely the share in plastics is higher, based on the application demands. Specific applications place different demands on the pigment, in terms of color (hue, strength, and brightness), opacity or transparency, fastness properties, and the increasing demands of environmental regulations, during production, processing, and in the final product including its eventual disposal, all balanced against cost.

50.7.1 Printing inks

The volume of material produced by printing, especially color printing, has increased dramatically in recent years. The wide range of substrates on to which printing is now carried out includes paper (books, newspapers, magazines, banknotes, stamps, posters, packaging, brochures, advertising material, wallcoverings, etc.), board, plastic film and containers, metal foils and textiles. The printing ink is the fluid suspension printed as a surface coating on to the substrate, which then dries as a thin solid film. Like paints, a printing ink must show good adhesion to the substrate to which it is applied and retain its mechanical and color properties for the expected lifetime of the printed product. However, unlike paints, which have both a decorative and protective function, printing inks generally serve only the purpose of reproducing text and images while providing a decorative effect. Commonly this means that the durability requirements of inks are less than is the case with paints. It is important to understand the nature of the various printing methods used, since how the ink is transferred to the substrate is of prime importance in ink formulation.

It may initially seem surprising that printing inks are the largest user of organic pigments. This is explained by the volume of printed products for which the high color intensity requires vast amounts of pigment. Although a wide variety of pigments are employed, the pigments used in three- and four-color printing predominate, printed one on top of another while using a white substrate as a base. The three-color process uses inks of the three subtractive primary colors, yellow, magenta, and cyan. A fourth color, black, is often employed to enhance the sharpness of lines and is a less expensive way to provide black than by mixing the three primaries. In general, as much color as possible is required from as thin a print film as possible, and therefore a major requirement is as high color strength as possible, and a pure shade to provide bright colors. Duller shades may be achieved by adding black or by printing a complementary color. If a pigment is printed on top of another color, it is essential that it is transparent, although the first color laid down is often opaque. In the US, yellow is generally printed first, whereas in Europe blue is printed first, so that transparent magentas and yellows are required. While high lightfastness is often not necessary (color fading on our newspapers is not a concern), there are exceptions. Pigments for packaging, which may be part of a shop window display, and advertising hoardings must have reasonable lightfastness, while wallcoverings need excellent fastness. The requirement for fastness to solvents depends largely on the formulation since the binder is often dissolved in an organic solvent or thinned with solvents. For example, linseed oil presents little demand on the solvent resistance of a pigment, whereas nitrocellulose uses powerful alcohols and sometimes esters/ketones, requiring pigments with higher fastness to solvents. Heat stability is not usually an issue as ink drying processes are rapid, although the use of higher temperatures increases the tendency of the pigment to dissolve. A particular application where good heat stability is required is metal decorating (metal dec) inks which are dried at elevated temperatures, and the cans to which they are applied may have to undergo heat sterilization. Alternative newer processes are being used more and more. UV curing is used increasingly, with high temperatures rarely a problem, although pigments may on occasions interfere with the curing process. Fastness to chemicals is often essential for packaging. Soap is highly alkaline and can degrade some pigments.

The yellow primary color generally uses diarylide yellows (disazoacetoacetanilides) and the magenta primary by using a calcium salt azo pigment (Rubine toner). The cyan primary color is almost exclusively obtained using β-copper phthalocyanine blue, which also provides excellent lightfastness. Alternative pigments are available where the technical demands are higher, but they are almost always more expensive. Quinacridones may be used for magenta and selected high-performance azo pigments for yellow. The black ink uses carbon black. Secondary primary colors, orange, red, violet, and green may be achieved by mixing primary colors, but the result is often less bright than desired. Secondary primary colors can be obtained from alternative pigments, orange using disazopyrazolone pigments, red from azonaphthols or cationic dye complexes, violet from dioxazine pigments and green from halogenated copper phthalocyanines. Consideration of toxicological properties is vital for printing inks. In food and drink packaging, the colored side is usually not in direct contact with the contents. However, during the printing of a sheet there is often undesirable transfer of wet ink on to the reverse side of a sheet that is stacked on top. Likewise, in the case of print collected on a reel (web) there can often be some transfer to the back of the substrate, an effect known as set-off. Thus, even when the printed side is not in direct contact with the food, the pigment used must not present a hazard to the consumer. Pigments for coloration of toys, graphic instruments and comics are also covered by toxicology regulations which vary between nations and which are constantly updated. The toxicity of impurities in the pigment must also be considered, for example, the presence of polychlorinated biphenyls (PCBs) and aromatic amines. Therefore, it is clearly necessary for the user to follow the advice given on up-to-date safety information provided by the pigment manufacturer.

There are five well-established traditional printing processes currently in use, namely lithography, flexography, gravure, letterpress, and screen printing. To carry the image to the substrate, the first four use printing plates which may be either flat or, for continuous printing, cylindrical. On these plates there are two distinct physical areas: the image (printing) areas, and the non-image (non-printing) areas. In relief printing (letterpress and flexography), the image area of the printing plate is raised above the non-image area, while in gravure printing the image area is recessed. In the lithographic process, the image and non-image areas are in the same plane, and the process relies on these areas being chemically different. Screen printing involves a stenciling process. With most gravure, flexography and letterpress, and all screen printing, ink transfer takes place directly from the image carrier to the substrate. In offset printing, the image is transferred from the image carrier to an intermediate cylinder, from which it is subsequently transferred to the substrate. Most lithography is carried out in this way.

Letterpress is an old printing method which no longer holds the importance it once did. It is a form of relief printing where a raised metal surface transfers the ink onto the substrate. The type is set by a compositor, who locks moveable type into a bed. This is a manually demanding process and is the main reason why the technique has lost popularity, especially for newspaper printing. The process generally makes no special demands on the pigments. Flexographic (flexo) printing is also a form of relief printing. It uses a flexible rubber plate and employs a rotary printing method. Such processes can run at remarkably high speeds and print on almost any surface, especially non-porous surfaces. Advances in the production of photopolymer plates brought about in the late twentieth century opened new markets to flexography. Also, anilox rollers were developed made of either steel or aluminum. They are coated with a special grade of ceramic, laser-etched to produce thousands of tiny cells, thus precisely controlling how much ink is presented to the raised image areas of the printing plate. The finer the detail required, the higher number of screen dots need to be adopted on the roller. Flexographic inks have a low viscosity and so, together with gravure inks, they are referred to as liquid inks. Four ink binder systems may be employed. Water-based inks, although sometimes used in combination with alcohols, present little demand on the solvent resistance of pigments. Solvent-based inks require pigments that are resistant to the solvents used, mainly alcohols, sometimes with esters or ketones. Printing systems based on UV or electron beam cured resins have a high viscosity and pigments with high oil absorption, i.e., with large surface areas, may lead to flow (rheology) problems. Oil-based inks where the oil is dissolved in hydrocarbon solvents require pigments that resist dissolving in those solvents. These inks are the least desirable from an environmental protection viewpoint. Over half of all flexographic printing is used for packaging, although there has been a tendency for small runs to move to digital printing. Offset lithography is an indirect printing process, whereby the ink is transferred onto the paper by means of an intermediate rubber cylinder, reducing the wear on the printing plate. Lithography works on the principle that "oil and water do not mix", making use of hydrophobic and hydrophilic areas on the printing plate. In the original process, an image would be drawn with oil or wax on to a smooth limestone plate. The stone was then treated with acid which etched the untreated parts of the stone. The stone was then dampened with water, which is only retained in the etched areas. An oil-based ink was then applied to the stone and was attracted only to the original image areas, that were not damp, as the areas that retained water repelled the oil. This technique is still used for some fine art reprints. However, in the modern process, polymer coatings are used as the image areas, applied to a plastic or metal plate. This plate is wetted with a so-called fountain solution, and then inked. The ink is transferred to an offset cylinder, and then further transferred by the rubber offset roller onto the substrate. Earlier image creation processes depended on photography. A photographic negative of the desired print would be placed in contact with the plate treated with a photosensitive emulsion. Exposure of the plate to UV light sets the polymer where the UV has passed through the negative. The unset polymer is then washed off leaving the reverse of the negative image, i.e., duplicating the positive image. Currently, the image is produced by direct laser imaging. Offset lithography requires inks with high color strength as it generally prints very thin films. Therefore, pigments with a high tinctorial strength are required. They are also required to produce inks with good flow properties, thus presenting the pigment manufacturer with a dilemma. Pigments with small particle size give high color strength but have high surface area and hence high oil absorption, thus increasing the viscosity of the ink. Many pigments used in offset printing are resinated during their manufactured. The use of surface-active agents during pigment manufacture can upset the ink/water balance on the printing plate, thus affecting the ability of the plates to pick up the ink correctly. The substrate can be sheet-fed, which lowers costs associated with setting up the process, minimizes paper waste, and gives good quality. Web offset operates with huge reels of paper and occurs at incredibly fast speeds, yet it is set up relatively quickly, ideal for processes that involve long runs. Most newspapers now use web offset. Currently, offset lithography is the most common printing process, but some packaging printing is moving to flexography and shorter run promotional material to digital printing. Gravure printing involves engraving an image on a printing cylinder [16]. The printing plate in rotary gravure is a cylinder made of copper-plated steel, or occasionally ceramic. The engraving creates cells which determine how much ink is transferred to the substrate. The deeper and larger the cell, the more ink will be transferred, hence deepening the color. Three methods of engraving may be used to make the gravure cylinders – diamond drill, chemical etching, or laser engraving. The ink is of low viscosity, i.e., a liquid ink, contrasting with offset and letterpress inks, which are as paste inks. The ink is applied to the cylinder and enters the recesses. Excess is removed from the non-image areas by a doctor blade, so that ink is only in the etchings. The substrate is squeezed between the printing roller and an impression roller and the ink transfers to the substrate. In multicolor processes, each ink must dry before the next colored ink is applied. The gravure process allows more ink transfer to the paper than offset or flexography, so that deeper colors can be obtained. Gravure printing can operate at high speed, up to 14 ms^{-1}, and provides unsurpassed print quality on many different surface types. As well as paper, it can print polyolefin and polyester films. Due to the high costs in preparing the plates, it is used only for high volume work, such as newspapers, magazines, and packaging. It is currently losing some market share in publications to offset and in packaging to flexography.

The resins used for packaging inks are selected based on the substrate and the properties required from the print. Resins used include nitrocellulose, polyamide, polyurethane, acrylic, and others (including polyester, maleic, ketonic, and polyvinyl). There has been some success in introducing water-based inks, although the inability to dry at the speed of the presses has held back developments. For publication gravure, the resins include maleic acid-modified rosin and phenolic resins, calcium, and zinc resinates, hydrocarbon resins and others. Polyethylene-based waxes are often added to improve abrasion resistance. The volatility of the hydrocarbon solvents used to dissolve the resins is critical to the process. Although aromatic hydrocarbons are generally more volatile, they also introduce environmental and toxicological hazards, and the trend has been towards aliphatic solvents [17]. There have also been determined efforts to develop methods to recover solvents, both to reduce hazards and cost. The solvents in turn limit which pigments can be used, further exacerbated by the heat used to dry the ink rapidly. Many pigment manufacturers offer special grades of pigments for publication gravure, providing enhanced color strength and ease of dispersibility, as coarse particles can damage the printing plates.

Screen printing is one of the oldest printing processes, thought to go back in a crude form to prehistoric cave painting, with developments brought about through history, especially in China and Japan. It was frequently referred to as silk screen printing, although in screen construction silk has been largely replaced by polyester. The major advantage of screen printing is that it can be used on many surface types, for example, curved or irregular surfaces for textiles, balloons, decals, etc., that do not lend themselves to other forms of printing. It is also used for printed circuit boards. The process can use many different types of ink, usually more viscous than gravure or flexographic inks, but less viscous than letterpress or offset inks.

Printing methods are in continuous evolution. Recent decades have seen the successful introduction of processes encompassed by the term digital printing, whereby a computer-generated image is directly delivered to create the image on the substrate. Digital printing produces short print runs quickly and cost-effectively on demand, lending itself to personalized items. Inkjet printing is the best known of these processes, familiar to us as inkjet printers at home or in the office. Inkjet printing is a non-impact process which involves directing small droplets of ink in rapid succession, under computer control, on to the substrate. There are two main types: continuous (CIJ) and drop-on-demand (DOD). In the first method, a continuous stream of ink droplets is forced out of a nozzle under pressure. Some of the droplets, electrostatically charged, pass through a charged electrode and are deflected at the substrate to form the image, while the remainder are collected and either recycled or discarded. In DOD inkjet, pressure on the ink is applied when the droplet is needed to form part of the image. An array of nozzles is used to eject the ink either mechanically or by an electric charge to generate the image. Digital printing allows instant modification of images as required and the quality and speed are constantly improving. Desk top machines generally use dyes rather than pigments, which present some issues with fastness to water, but in general little attention is paid to the performance of the prints other than reasonably good color reproduction. As we know, ink cartridges are expensive, although as the process has become established, the costs have reduced a little. Commercial inkjet printers often use pigments, which provide much superior lightfastness and more water-fast prints. The pigment must be free from large particles that can block the jet nozzles, and so needs to be ultra-fine. Laser printing, a form of toner printing, bears similarities to the photocopying process, not surprisingly as the process was first developed by Xerox Corporation. Both systems use optical or electrical methods to form an electrostatic latent image to which the toner is attracted and subsequently transferred to the substrate. In a laser printer, the first step involves providing a photo-conducting drum with a uniform electric charge. In a second step, a laser writes the information stored in the memory of the printer on to the charged drum. The toner carries a charge that is the same as on the drum and so repelled from the non-image areas on to uncharged image areas. This is followed by transfer of the image from the drum to the substrate, usually paper, by applying to the back of the paper an electrical potential of opposite charge to the toner particles. The next step fixes the image to the paper, achieved by a heat treatment that melts the toner resin and fuses it to the paper. The toner comprises a resin (usually a polyester), pigment and a charge control agent. Traditionally, pigment is dispersed into the molten resin and the solid material micronized using air jet mills, using a high-speed jet of compressed air or inert gas to provide the required particle size. The pigments used normally correspond to the four process colors, requiring high tinctorial strength and good lightfastness, so that many of the recommended products are high-performance pigments.

50.7.2 Coatings (paints)

Paints are probably the application most associated with pigments. Interestingly, the term "pigment" is derived from the Latin "pigmentum", itself derived from "pingere", to color or to paint. Historically, paints were the first medium to use pigments, including the cave paintings in Cueva de El Castillo, Cantabria, Spain which date back over 40,000 years. Indeed, the use of coatings in the form of body painting may go back as far as 100,000 years [18].

There are many types of paint, broadly classified as decorative (architectural) or industrial. However, within these there are many sub-groups, each with its own characteristics and demands, with implications for the pigments used in their coloration. Major considerations are fastness to light and weather, solvent fastness, heat stability and transparency/opacity. For many applications, toxicological issues and environmental considerations are increasingly important. Water-based decorative architectural paints are taking an ever-increasing share of this market. Improvements in the technical properties have combined with addressing environmental concerns to extend the applications that can be satisfied with water-based paints. In countries where there is a UK influence, most water-based paints are described as *emulsion* paints. Technically this is incorrect, as the resins are not emulsions (colloids of two or more immiscible liquids) but more correctly dispersions (fine solid polymer particles in a continuous liquid phase). In the US, such products are called latex paints and in the German language "Dispersionsfarbe". The resins are usually of vinyl (mainly polyvinylacetate) or acrylic types. Traditionally, they provide either a matt or semi-gloss finish. Some modern emulsion paints provide good gloss, but not at the level obtained from solvent-borne systems. Although traditionally water-based paints were limited to interior coatings, external paints are increasingly water-based, especially for wood. The absence of organic solvent allows the use of pigments that have low resistance to solvents, including several monoazo pigments. The demand for fastness to light varies both geographically and in line with economic demands. This demand has increased over the years and many pigments that were traditionally used are no longer considered good enough. Opacity is not usually required as this is normally achieved using titanium dioxide white pigments, but occasionally deep colors, especially yellows, may be based on pigments that improve covering power. Pigments for these paints should resist conditions that they are likely to encounter in application, such as detergent-based cleaning agents, and condensation. Since dispersion of pigments into the latex particles is not feasible, most water-based decorative paints are colored using pigment dispersions. In these products, the pigment is dispersed in multicompatible solutions of non-ionic or anionic surface-active agent, together with humectants that give protection against drying out and freezing. Biocides are used to provide protection against growth of molds and bacteria. An alternative way to color such paints is to use tinting systems, whereby colored pigment pastes are metered into base paints. This allows retailers to offer a wide range of colors without having to keep excessive stocks. Often these tinting systems may to be used for both water-based and oil-based paints. For oil-based paints, only pigments that are stable to the solvents used are suitable. The solvent is usually based on aliphatic hydrocarbons but may have some aromatic hydrocarbon content. Efforts to reduce the volatile organic content (VOC) of paints is leading to reduction in the use of organic solvents, thus increasing the market share of water-based paints. While the gap between the gloss levels of water-based and traditional oil-based gloss paints has narrowed, the higher gloss of the latter is still generally preferred. Water-based gloss paints are based on acrylic resin technology. As the gloss increases, so also is the tendency for "blocking". This phenomenon occurs when a second coat is applied over a freshly painted area and dries, and the two coats adhere so strongly to one another that the first coat may peel away from the surface. In solvent-based decorative architectural paints, alkyd resins have replaced the traditional linseed oil-based paints. Alkyds are polyesters chemically modified with fatty acid groups which lead to a more flexible film. Air-drying alkyds used in decorative paints dry by a cross linking reaction between the unsaturated groups in branches of the molecules, effected using atmospheric oxygen. Other developments have led to faster drying, e.g., by reacting residual OH groups on the alkyd with isocyanate (-NCO) groups. A variety of oils can be used, often selected according to their tendency to avoid yellowing, a defect of linseed oil, and to economic considerations. Dehydrated castor oil is often used with sunflower oil an alternative. Alkyd resins for decorative paints are dissolved in mineral spirits (white spirit in the UK). Apart from the higher demands for solvent fastness, the requirements for pigments used in solvent-based decorative paints are not dissimilar to those for water-based paints, but the demand for lightfastness may be higher as they are more likely to be used externally.

Industrial paints is a general term used to encompass products that are used to coat articles as part of the manufacturing process. The classic application is automotive paints, which require pigments with the highest levels of fastness properties and heat stability. In contrast, many engineering items, such as gardening tools, are coated mainly for sales appeal and can utilize pigments with lower levels of fastness properties. There may also be requirements for the pigment to avoid migration, especially into food, and to satisfy relevant toxicological regulations. The requirement for opacity to obscure the surface is met as described for decorative paints. However, some paint systems, such as those widely used in automotive finishes to provide metallic and pearl effects, require transparent colored pigments, as it is essential that light penetrates the film and is reflected back from the metallic or pearlescent pigment particles. This places even greater demand on the lightfastness and weatherfastness of the colored pigment.

Most industrial paints must dry rapidly, which is achieved in one of three main ways. One is air-dried (or force-dried) finishes, which use highly volatile solvents. If there is no chemical reaction involved, they are known as non-convertible coatings. The media for these paints include nitrocellulose or vinyl co-polymers, which are soluble in ketones and esters, thus requiring the most solvent fast pigments to avoid migration and bleeding problems. Another way of bringing about rapid drying is to stove the paint, which not only drives off the solvent but also brings about cross linking in the resin, forming so called convertible coatings. In general, the stoving schedule involves temperatures of around 120–180 °C, typically for 30 minutes. Some pigments are unstable at such temperatures, and many more may dissolve in the solvents used to dissolve the resin. Typical resins used are alkyds drying by a cross-linking reaction with urea-formaldehyde or melamine-formaldehyde resins. The main solvents for such paints are aromatic hydrocarbons, typically xylene, with an additional solvent for the cross-linking resin, typically butanol. Finally, some finishes are so-called two pack systems, whereby the paint is supplied in two containers and the contents are mixed before application. Typical of these paints are polyurethanes, consisting of an isocyanate (-NCO) monomer with two or three isocyanate groups, which reacts with a polyol that has two or more OH groups. They provide coatings with good finishes, resistant to acids, alkalis, and many solvents, and are often used on account of their chemical resistance. However, high cost and to some extent safety concerns limit their wider use. Polyurethane coatings are widely used for wood finishes and vehicle refinishing. While the pace of replacing solvents in industrial paints with water-based systems has been slower than in architectural paints, progress is now rapid. Acrylic resins are also commonly used often combined with vinyl neodecanoate, which gives good adhesion to metals and is highly hydrophobic, providing a coating that is highly resistant to alkali degradation and to UV light [19].

Several automotive paints employ water-based technology for the primer and base coat, often using acrylic/urethane resins. A final clear coat (topcoat) is often used to give high gloss, which contains UV absorbers in order to protect the colored paint layers. The paints must be cured at elevated temperatures. However, whereas traditional paint systems were stoved at around 135 °C, many vehicles now have plastic parts, so that lower temperatures must be used, typically around 80 °C.

An alternative way of reducing emissions of VOCs is to use powder coatings. The paint in powder form is applied electrostatically using a spray or fluidized bed. The coated item is then subjected to heat, melting the powder, and fixing to the substrate. Powder coatings were originally employed mainly on metals, and one of their first successes was in the coating of pipes using epoxy resins. Advances have meant that powder coatings are used on non-metallic surfaces, including plastics, medium density fiberboard, and even wood. Many resins used are thermosetting and require a cross-linking agent incorporated into the formulation. This is often an epoxy resin. However, since the process can have a destructive effect on some organic pigments, care must be taken in pigment selection. Dispersion of the pigments occurs in similar equipment to that used in the plastics industry. The polymer is thoroughly mixed with the pigment and other ingredients. This blend is passed through an extruder, in which the polymer is melted, and shear applied via a rotating screw. The extruded material is then cooled and broken down to a fine powder and sieved, so that when the powder is applied it forms a flat surface.

50.7.3 Plastics (polymers)

In relative historical terms, plastics represent a newer outlet for organic pigments than printing inks and paints, and they are usually highly demanding. Several new pigments were developed during the second half of the twentieth century to satisfy the requirements for heat stability and insolubility throughout the coloration process. Most polymers are thermoplastic that melt when heated. Examples are polyethylene, polypropylene, polystyrene and styrene co-polymers, polyvinyl chloride, nylon, acrylic, etc. As with printing inks and paints, the pigments must be dispersed into the application media. At room temperature the polymers are solid and are melted to their liquid phase before dispersion is carried out. The temperature at which polymers melt varies among the different chemical entities, and even between different grades of the same polymer, depending on the degree of polymerization and in some cases the presence of additives. At too high a temperature, the pigment can be degraded, and the viscosity of the polymer can be reduced thus lowering the shear conditions responsible for achieving optimum dispersion. Originally, dry pigments would be intimately mixed with the polymer beads or powder and then dispersed in an extruder. The extruder heats the polymer so that it melts, while a screw slowly carries the molten polymer through the barrel, in which pigments experience high shear forces, and from there the polymer exits as a tape or strand. The time in the barrel, known as the dwell time, e.g., 10 minutes or 30 minutes, can also be critical. The colored polymer can then be chipped. Often, the material is reheated in an injection molder and injected into a die (mold) to form the required shape of a plastic article. Alternatively, it may be placed into a blow molder, whereby the polymer is ejected in a ring shape, with air blown into the center of the ring, so that it forms a type of balloon, eventually forming a plastic film that can be made into plastic bags or used for packaging. In contrast with paints and inks, only small percentages of pigment are used to color plastics. Thus, accuracy is important when incorporating pigment, particularly pigment mixtures. Since inconsistent color was often experienced, most companies now use pre-dispersed pigments, often blended to a specific color, in the form of a masterbatch. An alternative is to use metered paste dispersions, either in a multi-compatible resin or, especially for polyvinyl chloride, in a plasticizer.

Polyvinyl chloride (PVC) is one of the most versatile polymers, albeit with some misgivings over long-term environmental effects. Plasticized PVC is used in artificial leathercloth, dip molded articles such as gloves and dolls, cable insulation, chair coverings, and floor coverings. Rigid PVC is used for films, pipes, shutters, and bottles. PVC degrades under the influence of light unless it is stabilized, very often with metal soaps, but can be made harder with much better weathering properties, by combination with chlorinated polyethylene, a mix that is now widely used in window profiles. PVC itself is hard and rigid but is often made more flexible using plasticizers. Traditionally, low molecular weight phthalates were used but these were gradually replaced in favor of higher molecular weight grades. However, phthalates are being phased out because of concerns that they may be endocrine disrupters, i.e., that they interfere with hormonal mechanisms, especially during fetal development and in infants [20]. Alternative non-phthalate plasticizers are now being promoted, such as 1,2-cyclohexanedicarboxylic acid diisononyl ester, dioctyl terephthalate and renewable citrate esters. There are claims that they have inferior plasticizing effects and are more expensive, but they are being enthusiastically promoted. PVC is used in three main ways. It can be extruded at temperatures in the range 160–200 °C. The screw ensures thorough wetting and dispersion, breaking down the pigment agglomerates. The extruded tape or strip is then granulated and can be injection molded into required shapes. Many pigment manufacturers offer their pigments in a pre-dispersed state, which require shorter processing and improved dispersion. The use of mixing rollers (kneaders) is the oldest method, pre-dating extruders, and uses a two-roll mill consisting of two heated rollers that rotate in opposite directions at slightly different speeds creating shear. It is this shear at the nip that disperses the pigment into the molten polymer. Some additional shear is generated where the rollers separate. The colored plastic material can then be calendered into film or sheets. In spread coating, the pigment is dispersed into a plasticizer on a three roll mill or bead mill and is then stirred into a plastisol, PVC mixed into a plasticizer in appropriate proportions, typically 50:50 to 80:20, PVC : plasticizer. The colored plastisol can be coated on fabric to produce artificial leather cloth, and to paper or metal. It is then heated in ovens to above 140 °C, or with infrared heating. After cooling, the solid plastisol remains bonded to the substrate. Many companies specialize in dispersing pigments either as a single pigment, but more often as a specific color, into plasticizers and offer these to the end-product manufacturer. PVC is not particularly demanding of pigments, as it is processed at relatively low temperatures. A problem is that plasticized PVC and many plasticizers will act as solvents, especially at elevated temperatures. This can lead to bleeding, blooming, or plate out, as discussed in a previous section. Although low concentrations of pigment are used, lightfastness in PVC is generally superior to that in paints and printing inks. However, as with other applications, lightfastness becomes inferior as the concentration of the colored pigment is lowered, and as the proportion of titanium dioxide present increases.

Polyethylene can be either low density (LDPE) which can be extruded in the range 180–260 °C, or high density (HDPE) which is more crystalline and is processed at a temperature in the range 200–300 °C, although a typical temperature would be nearer the lower end of this range. Polyethylene is typically used for pipes, cable insulation, bottles, and film. Polypropylene (PP) is a semi-crystalline polymer processed at similar temperature to HDPE but more frequently towards the upper end of the temperature range. Its applications are also similar, but in addition it is used for fibers, especially for carpet backing. It is slightly harder than polyethylene and has high resistance to chemicals. Polystyrene (PS) is clear and when unmodified is often known as crystal polystyrene. It is processed in the range 160–260 °C. It can be colored with soluble dyes, which have some affinity for the polymer structure, as well as with pigments. Some pigments actually dissolve at the processing temperatures, but do not migrate as they are held within the PS crystals like dyes. Applications include formers for electric coils, disposable cutlery, and household articles. Unmodified as a homopolymer, it is somewhat brittle. However, it is often co-polymerized with other monomers such as butadiene, to form impact-resistant grades. For even better impact resistance, other copolymers have been developed, such as styrene acrylonitrile (SAN), which can be glass clear, and acrylonitrile butadiene styrene terpolymers (ABS), which are now used extensively for lightweight components, such automotive bonnets and boot (trunk) lids, where their high impact resistance and low weight make them preferable to steel. ABS can be aggressive when molten, to an extent dependent on the ratios of the three components, thus limiting the number of pigments which can be used for its coloration. Foamed polystyrene is mostly uncolored but can be colored with dyes.

Engineering polymers is a generic term for polymers which have better mechanical properties and thermal stability than classical polymers, although they are significantly more expensive. The boundaries between engineering polymers and commodity polymers on one side and on the other side the so-called advanced engineering polymers are not set in stone. The best-known polymers of this type are polyamides, commonly known as Nylon. They can be manufactured with different compositions, such as Nylon 6, Nylon 66, and Nylon 10.10. All produce hard components and hence are used for gear wheels, filaments, and engineering parts. Nylon is a difficult polymer to process and highly hygroscopic; even a small amount of water must be rigorously avoided. Nylon is often processed at a temperature of 300 °C, but above 260 °C it may chemically attack many organic pigments, severely limiting the colorants that can be used. Nylon can be colored with acid dyes, but these products have poorer lightfastness, compared with pigments. Advanced engineering polymers include polyethylene terephthalate (PET), polycarbonate (PC), polyether ketone (PEK), and acetal. They are processed at temperatures above 300 °C, and so only a few organic pigments are suitable. Some organic pigments encourage nucleation in semi-crystalline engineering polymers, reducing technical performance and processability. Like Nylon, many of these polymers are hygroscopic, leading to a reduction in the mechanical properties of polymers during processing. Some typical processing temperatures are PET 270–310 °C, PC 280–340 °C, Nylon 66 280–340 °C [21].

Thermoset polymers are not colored in the same way as thermoplastics. A frequent method used for coloring thermoset resins (e.g., urea-formaldehyde, melamine-formaldehyde, phenolics) is to blend the powdered polymer with pigments without generating excessive heat. Phenolic resin molding compounds can be colored with both dyes and pigments. Epoxy resins are generally low molecular weight liquids, which require a second component (known as hardener) that reacts with the epoxy resin and forms a hard, cross-linked structure. Their main use is as adhesives that have excellent resistance to chemicals. They have an inherent brownish color which makes obtaining pale bright shades difficult. They can be colored using powder pigments which may be pre-dispersed in the epoxy resin or premixed with fillers. The system makes no special demands on pigments. Polyurethanes (PUR) result from reaction of a polyol with at least three hydroxyl groups, with a diisocyanate. They can be formed into either rigid or flexible foams. Colorants are introduced into the polyol using a pigment paste or now more often pastes containing specific dyes. Organic pigments are preferred for the coloration of PUR due to their high color strength and lack of abrasiveness, but as heat is generated during the reaction, care must be taken to use only heat stable pigments.

While heat stability and associated stability to solvents, including molten polymer and plasticizers, is the major challenge, lightfastness may also be important. The highest demand is for plastics that are exposed, such as window profiles. An application that falls somewhere between paint and plastics, is coil coating. In this application a coating is applied directly to steel as the steel is fabricated, and the coated steel is then used for cladding on the outside of industrial buildings, an application demanding the highest lightfastness. Manufacturers claim that the coating will require to be resistant to fading for up to 15 years, which eliminates the use of many organic pigments. However, some inorganic pigments accelerate the degradation of the PVC. Packaging is often required to show good lightfastness as an affected product may lose sales appeal when displayed in shop windows. Many household articles are made from colored plastics. While they must not fade too quickly, economic considerations may reduce the demands for lightfastness. Toxicological properties are a vital consideration. There are some distinct differences between regulations in various countries. Some use a positive list, including pigments that have been approved for use in food packaging, e.g., the French positive list. Such lists are usually accompanied by limits on levels of impurities, such as prohibited heavy metals and aromatic amines. It is common to test colored polymers used in food applications for migration. For this purpose, food simulants such as acids, fatty substances, solvents (alcohol) are held in contact with the plastic. Staining of either the foodstuff or the plastic prohibits its use. Toy regulations demand that the colorant contains less than prescribed limits of heavy metals. Sometimes a limit is set on soluble heavy metals, but in other cases it may be the total level of each heavy metal, which is easier to measure.

50.7.4 Rubber

Rubber can be either natural or synthetic, both having similar properties. The former is obtained from the sap of the rubber tree, which grows in tropical climates. The sap is a latex based on isoprene. It is coagulated and then warmed, resulting in natural rubber. Synthetic rubber now predominates. There are several different types, including styrene butadiene (SBR), nitrile (NBR), and synthetic polyisoprene (IR). Rubber is usually processed in a similar way to flexible PVC, using either a 2-roll mill (calendering) or a kneader. However, temperatures used are generally lower than PVC ranging from 60–150 °C, but some synthetic rubbers may require temperatures up to 220 °C, which requires an extruder. The final article can be made into sheets or injection molded. Rubber is less demanding of pigments than other polymers, largely because of the lower processing temperatures. Bleed resistance is necessary. However, certain grades of rubbers, especially nitrile rubber, are used where stability to certain chemicals is required, and therefore the pigment must also be stable under such conditions. High lightfastness is rarely required. While powder pigments may be used, pre-dispersed pigments are frequently preferred.

50.7.5 Paper

Paper is usually colored with dyes, generally direct dyes. However, if higher levels of fastness properties are required, pigments are the favored option. Pigments do not have direct affinity for paper and so pigment dispersions incorporating anionic surface-active agents are commonly used. Coloration generally requires interaction of the surfactant with aluminum sulfate or some other retention agent. The pigment particles agglomerate and are held mechanically by the paper fibers, before the paper is dried. Pigmented paper can achieve deeper colors than dyed paper but are much less economical. Even pigments with lower levels of fastness properties are generally superior in lightfastness to the dyes used. A special application is decorative laminate papers. After coloration, the paper is resinated with a thermoset resin (e.g., melamine-formaldehyde) and cured at an elevated temperature. The paper is then sandwiched between brown kraft paper backing and a clear overlay paper, made from pure cellulose from well delignified pulp. This layer becomes transparent after impregnation allowing the appearance of the decor paper to become visible, and also improves scratch and abrasion resistance. Such laminates pass through a heating process when the resin may act as a solvent, so that pigments must show moderate resistance to heat and solvents. They are frequently used for furniture, surface tops or flooring where they are required to have good fastness properties. However, in such applications the lightfastness of certain pigments can be inferior compared with their use in other applications. It appears that some pigments may be reduced by the resins used in the decorative laminate, especially carbonyl vat pigments. Another method of coloring paper is to coat the surface with a colored resin. The resins are normally water-based and, apart from during drying, little heat is applied. Therefore, there are low demands for solvent fastness or heat resistance. The lightfastness requirement depends on the conditions that the paper encounters in application but is usually not high. Pigment preparations similar to those used in the mass coloration of paper or emulsion (latex) paints generally suffice.

50.7.6 Artists colors

There are many types of artists colors available on the market. Oil paints use drying oils and are often colored with inorganic pigments. This has often been explained by a prejudice against organic pigments developed due to some bad experiences using unsuitable pigments with insufficient lightfastness. Gradually, high-performance organic pigments, such as phthalocyanines (blue and green) and quinacridones, are gaining some limited acceptance. On the other hand, acrylic paints offer brighter colors, low toxicity, and a wider range of shades. These can be based on either organic or inorganic pigments. Finger paints and poster paints are popular with children because of their bright colors. If organic pigments are used, they generally employ their dispersed forms. Such pigments must meet national requirements for safety, often as part of the toy regulations.

50.7.7 Textile printing

Dyes used in textile printing are selected for the specific textiles for which they are designed, while pigments are incorporated in binders that can be applied to the surface of any fiber, natural or synthetic. The binders used are similar to those used for emulsion (latex) paints, although specific resins have been developed that are flexible and give a softer handle. Pigment printing is generally economical as the washing off process, necessary when using dyes, is not required. Application is relatively simple and higher fastness demands can be met. If the printed article is to be dry cleaned, only pigments with good solvent fastness can be used.

References

1. Christie RM. Colour chemistry. 2nd ed. London: RSC, Ch 9 2015.
2. Zollinger H. Color chemistry. 3rd ed. Weinheim: Wiley-VCH Verlag GmbH, 2003.
3. Buxbaum G, Pfaff G. Industrial inorganic pigments. 3rd ed. Weinheim: Wiley-VCH Verlag GmbH, 2005.
4. Herbst W, Hunger K. Industrial organic pigments. 3rd ed. Weinheim: Wiley-VCH Verlag GmbH, 2004.
5. Faulkner EB, Schwartz RJ. High performance pigments. 2nd ed. Weinheim: Wiley-VCH Verlag GmbH, 2009.
6. Orna MV. The chemical history of color. Heidelberg: Springer, 2013.
7. Christie RM. Colour chemistry. 2nd ed. London: RSC, Ch 1 2015.
8. Brack P. Egyptian blue; more than just a colour. Chemistry world. Oct 2015.
9. Garfield S. Mauve: how one man invented a colour that changed the world. New York: WW Norton, 2001.
10. Abel A. Pigments for paints. In: Lambourne R, Strivens T,editor. Paint and surface coatings. 2nd ed. Cambridge: Woodhead Publishing, 1999:97.
11. Murray B. Pigmentation of colored plastics. In: Patton T, editor. Pigment handbook. vol. II. New York: John Wiley & Son, 1973:281.
12. Griffiths J. The historical development of modern colour and constitution theory. Rev Prog Color. 1984;14:21–32.
13. Gordon P, Gregory P. Organic chemistry in colour. New York: Springer Verlag, 1993.
14. Erk P. Crystal design of high performance pigments. In: Faulkner EB, Schwartz RJ, editors. High performance pigments. 2nd ed. Weinheim: Wiley-VCH Verlag GmbH, 2009. Ch 8.
15. Sanders JD. Pigments for inkmakers. vol. X. London: SITA Technology, 1989.
16. https://www.referenceforbusiness.com/industries/Printing-Publishing-Allied/Commercial-Printing-Gravure.html16.).
17. IARC monograph. Evaluation of Carcinogenic risks to humans: printing processes and printing inks, carbon black and some nitro compounds. vol. 65. Lyon: IARC, 1996:46.
18. https://www.theguardian.com/science/2011/oct/13/stone-age-painting-kits.

19. Arriaga V, Vanaken D. Branched vinyl ester monomers for hydrophobic emulsion polymers. Paint and Coatings Industry Magazine, 2013.
20. Diamanti-Kandarakis E, Bourguignon J-P, Giudice LC, Hauser R, Prins GS, Soto A, et al. Endocrine disrupting chemicals; an Endocrine society scientific statement. Endocr Rev. 2009;30:293–342.
21. The coloration of plastics and rubber. Clariant brochure. 2019:PI 3002 E.

Gerhard Pfaff

51 Oxonitride pigments

Abstract: Oxonitride pigments belong to the inorganic yellow, orange and red colorants. They were developed to complement the yellow, orange and red color range. Oxonitride compositions that are interesting for pigment uses crystallize in perovskite, chalcolamprite and spinel structures. The most suitable compositions for pigment purposes are most probably $LaTaON_2$ and $CaTaO_2N$. The synthesis of oxonitride pigments is technically difficult and takes place using a solid state-gas reaction. The pigments are characterized by excellent optical and application characteristics in particular regarding brilliance and color strength, strong hiding power, good dispersibility, light fastness and temperature stability.

Keywords: oxonitride pigments, solid state-gas reaction

51.1 Fundamentals and properties

Oxonitride pigments were developed to complement the yellow, orange and red color range. Compositions like $LaTaON_2$, $CaTaO_2N$ and $SrTaO_2N$ were investigated for their use as pigments. The development of oxonitride pigments occurred mainly in the 1990s. The intention was the substitution of cadmium- and lead-containing pigments by inorganic high chromatic yellow and red pigments. $CaTaO_2N$ and $LaTaON_2$ seemed to have the greatest potential for practical use amongst the oxonitrides [1, 2].

The color of oxonitride powders can be adjusted from yellow via orange to red in dependence on the composition. Oxonitrides are in chemical terms solid solutions and exhibit semiconducting properties. The band gap between their valence and conductivity band is decisive for the color. The variation of the amount and nature of the cations in the oxonitrides determines the specific color. The ratio of nitrogen to oxygen in the composition can be increased by successive substitution of a cation by another cation with higher valency. Such an ion exchange can be used to shift the color towards the region of longer wavelengths. The substitution of a cation by another cation with lower valency leads to a shortwave shift of the color. Oxonitride compositions that are interesting for pigments crystallize in perovskite, chalcolamprite and spinel structures [3, 4].

This article has previously been published in the journal Physical Sciences Reviews. Please cite as: G. Pfaff Oxonitride Pigments *Physical Sciences Reviews* [Online] 2021, 6. DOI: 10.1515/psr-2020-0188

https://doi.org/10.1515/9783110587128-051

51.2 Production of oxonitride pigments

The synthesis of oxonitride pigments requires a solid state-gas reaction at high temperatures [1, 5, 6]. For the synthesis of $LaTaON_2$, an equimolar blend of Ta_2O_5 and La_2O_3 is ground with the threefold quantity of a NaCl/KCl mixture (mineralizer). The ground powder is thermally treated for several hours at 900 °C in a stream of ammonia.

Equation (50.1) describes the reaction and shows that water is formed besides the target product $LaTaON_2$.

$$Ta_2O_5 + La_2O_3 + 4\,NH_3 \rightarrow 2\,LaTaON_2 + 6H_2O \tag{51.1}$$

Oxonitrides containing alkaline earth metals in their composition are produced with the addition of $CaCO_3$ or $SrCO_3$ to the starting mixture. The synthesis of $CaTaO_2N$, for example, takes place using a mixture of Ta_2O_5 and $CaCO_3$, as shown in eq. (51.2).

$$Ta_2O_5 + 2\,CaCO_3 + 2\,NH_3 \rightarrow 2\,CaTaO_2N + 3\,H_2O + 2CO_2 \tag{51.2}$$

The resulting oxonitrides are washed to remove soluble compounds, filtrated and dried.

The described production process for oxonitride pigments is technically difficult to conduct, in particular on a larger scale. High costs for this production route are almost unavoidable. Another problem are comparatively high costs for the raw materials. These facts explain, why oxonitride pigments have not been manufactured in production scale so far. A commercialization of these pigments has so far not taken place.

51.3 Pigment properties and uses

Outstanding characteristics of the yellow, orange and red oxonitride pigments are their good brilliance and color strength, strong hiding power, good dispersibility, light fastness and temperature stability [1]. The average particle size of products developed in the laboratory are in the range of 1 µm. Oxonitride pigments are stable against alkaline solutions and acids.

The pigments have application potential for easy-to burn in glazes and vitrifiable colors (firing range 500 to 680 °C). The coloration of plastics, coatings and printing inks with oxonitride pigments is basically possible. The high thermal stability of the pigments would even allow their use for the mass pigmentation of plastics with following extrusion and for annealing lacquers [5, 6]. The high manufacturing costs and thus the high price have so far prevented the use of the oxonitride pigments.

Usable toxicological studies with oxonitride pigments have not yet been documented. The pigments basically seem to be promising substitutes for cadmium- and lead-containing compositions, since the constituent elements are regarded as harmless.

The production process for oxonitride pigments is challenging and must consider that a release of gaseous ammonia during the solid state-gas reaction is prevented. A properly-working exhaust air purification system is absolutely necessary for the manufacture of this type of pigments.

References

1. Jansen M, Letschert H-P. Inorganic yellow-red pigments without toxic metals. Nature. 2000;404:980.
2. Pfaff G. Inorganic pigments. Berlin/Boston: Walter de Gruyter GmbH, 2017:159.
3. Günther E, Hagenmayer R, Jansen M. Strukturuntersuchungen an den Oxidnitriden $SrTaO_2N$, $CaTaO_2N$ und $LaTaON_2$ mittels Neutronen- und Röntgenbeugung. Z Anorg Allg Chem. 2000;626:1519.
4. Pfaff G. In Winnacker-Küchler: Chemische Technik, Vol. 7 Industrieprodukte. Dittmeyer R, Keim W, Kreysa G, Oberholz A, editors. 5th ed. Weinheim: Wiley-VCH, 2004:357.
5. Patent EP. 0 697 373 (dmc^2 Degussa Metals Catalysts Cerdec AG). 1996.
6. Patent EP. 0 627 382 (Cerdec AG). 1994.

Heinz Mustroph

52 Oxonol dyes

Abstract: Oxonol dyes are classified as anionic polymethine dyes, which cover a wide variety of structural types. The name of the class originates from the oxygen atoms which terminate each end of the polymethine chains that form the backbone of their structure. In technically useful dyes, these oxygen atoms tend to be substituents of heterocycles. The main technical application of water soluble oxonol dyes was in silver halide photography as filter dyes and antihalation dyes. Lipophilic oxonol dyes are used in bio-analysis and medical diagnostics to stain cells, bacteria or liposomes for example. Their main bioanalytical usage is in the determination of membrane potentials in eukaryotic cells and prokaryotic bacteria.

Keywords: anionic polymethine dyes, antihalation dyes, cell membrane electrostatic potential, cytometry, filter dyes, oxonol dyes, streptooxonol dyes

52.1 Fundamentals

Conjugated dialdehydes like e. g. malondialdehyde and glutacondialdehyde exist in a tautomeric equilibrium between keto and enol form. The enol form **1b** generally predominates over the corresponding keto form **1a** as illustrated with malondialdehyde. By elimination of the proton of the enol form the simplest model compounds of the anionic polymethine dyes are generated with the anions of malondialdehyde **1c** and glutacondialdehyde.

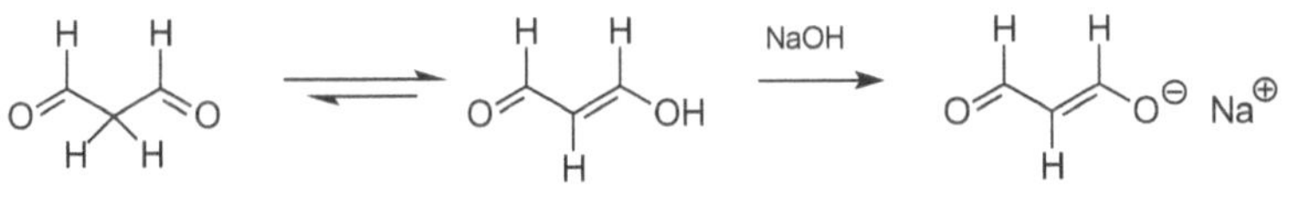

1a **1b** **1c**

As dyes, these open chain compounds have no significance. The terminal oxygen atoms form must form part of unsaturated heterocycles (as in the case of **2**) in order for such compounds to be technically useful anionic dyes. Due to a suggestion of Leslie G. S. Brooker and Grafton H. Keyes these compounds are named as *oxonol dyes* because oxygen is the charge-carrying terminal atom [1].

The simplest possible type of cyanine dye family is the *streptocyanine dye* class, where two charge-carrying terminal nitrogen atoms do not form part of an unsaturated

This article has previously been published in the journal Physical Sciences Reviews. Please cite as: H. Mustroph Oxonol Dyes *Physical Sciences Reviews* [Online] 2021, 6. DOI: 10.1515/psr-2020-0189

https://doi.org/10.1515/9783110587128-052

heterocyclic ring, but they are part of an open chain $N(R^1)R^2$ group or of saturated heterocyclic rings [2–4]. Following this classification in terms of systematic nomenclature, the proper labelling of the compounds, where the terminal oxygen atoms are not constituent of unsaturated heterocycles is as *streptooxonol dyes* [3, 4].

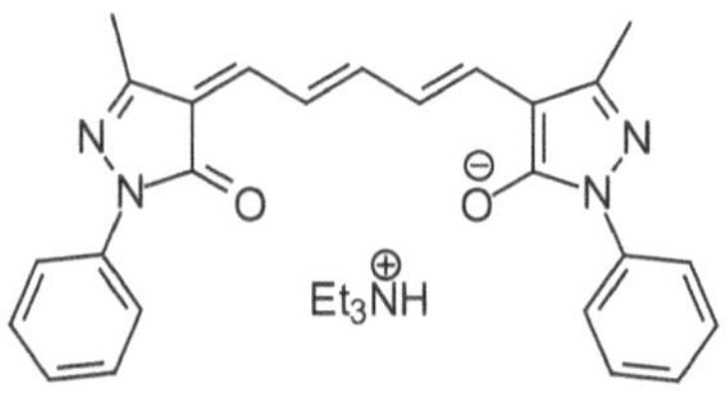

2

Oxonol dyes have not found many technical applications and, therefore, have been little studied. Due to their electronic structure, oxonol dyes are classified as anionic polymethine dyes, which cover a wide variety of structural types [2–8]. As *cyanine dyes* they are characterised by an odd number $2n+3$ of π-centres and $2n+4$ π-electrons (where n is the number of vinylene groups –CH = CH–). This special feature has a marked impact on their electronic structure and thus their equilibrium structure in the electronic ground state [3–8]. The π-charge density distribution along the carbon atoms of the polymethine chain alternates, which can be measured by ^{13}C NMR chemical shifts [9]. In symmetrical oxonol dyes the differences in the carbon–carbon equilibrium bond lengths are small [10, 11]. As is the case with other subclasses of polymethine dyes this feature results in narrow absorption bands in the electronic spectrum.

Also the anionic polymethine dyes exhibits the very typical vinylene shift of extended conjugation as illustrated by the streptooxonol dyes **3** [12] and the commercially available oxonol dyes **4**.

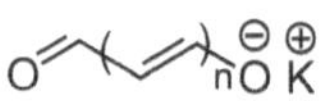

3

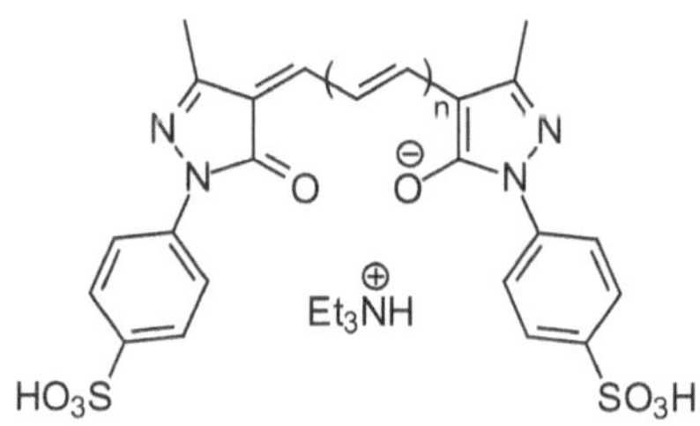

4

$(n = 1)\ \lambda_{max} = 267$ nm	
$(n = 2)\ \lambda_{max} = 362$ nm	
$(n = 3)\ \lambda_{max} = 455$ nm	$(n = 0)\ \lambda_{max} = 434$ nm
$(n = 4)\ \lambda_{max} = 547$ nm	$(n = 1)\ \lambda_{max} = 525$ nm
$(n = 5)\ \lambda_{max} = (644)$ nm	$(n = 2)\ \lambda_{max} = 628$ nm

52.2 History of the oxonol dyes

The first oxonol dye **5** (Pyrazolone Yellow), was synthesized by E. Besthorn at the company Hoechst by the reaction of 1-phenyl-3-methylpyrazol-5-one with chloroform and alkali, which was published by Ludwig Knorr in 1887 [13].

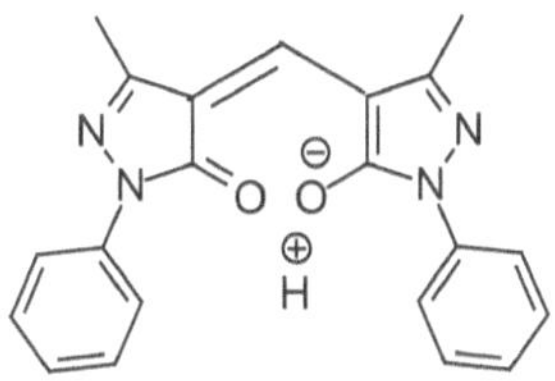

5

Here, silver halide photography provided the motivation for dye chemists to make further developments in the field of oxonol dyes. Monomethine-, trimethine- and pentamethine-oxonol dyes, which are yellow, red and blue in color were prepared. In particular, water solubility was increased by substitution with carboxylic and sulfonic acid groups. The best known representatives are Oxonol Red **6** and Oxonol Blue **7**.

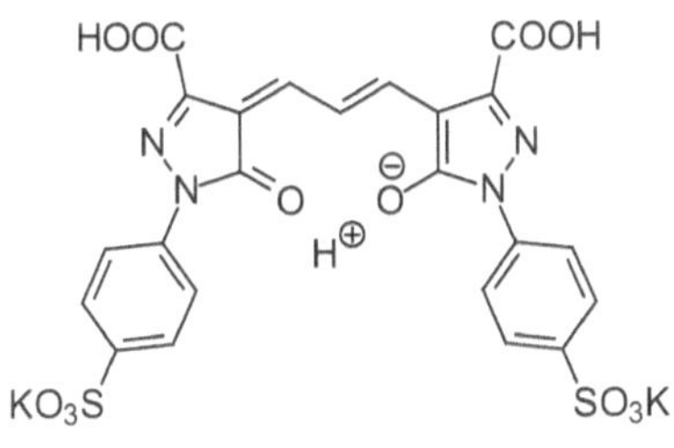

6

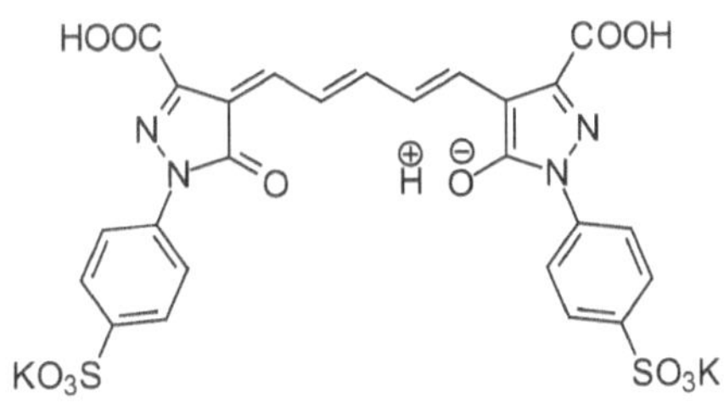

7

In addition to the pyrazolones others cyclic ketomethylene compounds like e. g. barbituric acids, 3-cyano-1-alkyl-4-methyl-6-hydroxypyrid-2-ones, dimedone, hydantoin, isoxazolones, Meldrum´s acid or thiazolinones were used to synthesize oxonol dyes [5, 8]. However, besides the oxonol dyes based on pyrazolones only oxonols from barbituric acids found some special applications.

52.3 General synthetic routes to oxonol dyes

The first oxonol dye **5** was prepared from the reaction between chloroform and 1-phenyl-3-methyl-pyrazol-5-one [13]. Later formaldehyde, dimethylformamide or orthoesters were used to synthesize monomethine oxonols. Nowadays oxonol dyes **10** are prepared from cyclic ketomethylene compounds **8** by reaction of diphenylformamidine hydrochloride (**9**; n = 0), malondialdehyde dianil hydrochloride (**9**; *n* = 1) and glutacondialdehyde hydrochloride (**9**; *n* = 2), which are also used in the synthesis of trimethine, pentamethine and heptamethine cyanine dyes [5, 8, 14].

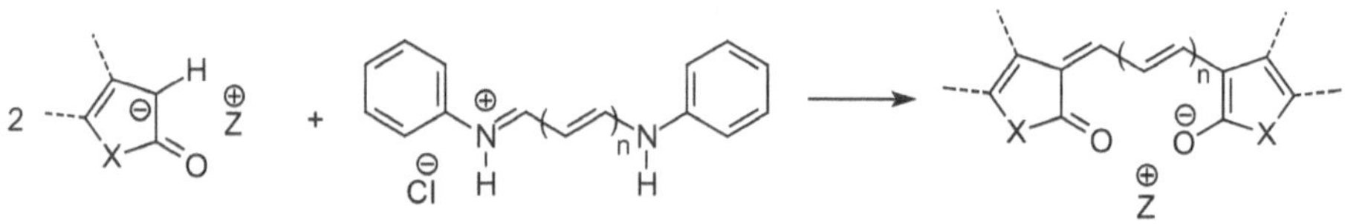

8 **9** **10**

Unsymmetrical monomethine oxonols are synthesized by Vilsmeier formylation of pyrazolone derivatives and reaction with a further molecule of pyrazolone. However, unsymmetrical trimethine and pentamethine oxonols are unusual.

52.4 Commercial uses of oxonol dyes

Oxonols are in general relatively unstable owing to their negative charge. In addition, as the number of vinyl groups *n* increases their stability decreases rapidly. Therefore, in comparison with other dye classes their technical applications are limited.

One of the few stable oxonol dyes is the first synthesized dye **5** in its protonated form (CI Solvent Yellow 93). It is mainly used in textile printing or mass coloration of transparent or opaque thermoplastics such as polystyrene, polyester, poly(methyl methacrylate), polyvinylchloride and styrene-acrylonitrile copolymers .

The main technical application of oxonol dyes was in silver halide photography [15–17]. As *cyanine* and *merocyanine dyes* are widely used as spectral sensitizers many trials were undertaken to do the same with oxonol dyes, but without success. Due to the excess of halide anions at the surface of the silver halide crystals the surface is negatively charged. Therefore, the cationic cyanine and the neutral merocyanine dyes are well adsorbed at the surface of silver halide crystals. However, the anionic oxonol dyes are poorly adsorbed at the surface and, therefore, are not suitable as spectral sensitizers.

Scattering of light at the silver halide crystals in photographic materials decreases the sharpness of the photographic picture by *diffuse reflection*. Sharpness can be improved by added filter dyes to the photographic layer. These dyes absorb in the same range of the spectrum where the silver halide material is spectrally sensitized. But there is a cost: reducing diffuse reflection of light lowers sensitivity simultaneously. So, in photographic materials a compromise between a desired increase in sensitivity and an undesirable decrease in sharpness has to be made.

In addition to diffuse reflection at the silver halide crystals a *reflection halation* at the interface between silver halide layer and the film base and mainly at the back of the film base occurs. It results from total reflection caused by the different refractive indexes. Therefore, *antihalation dyes*, located in various layers, are used to absorb this light and to prevent reflection halation.

Filter and antihalation dyes have to be decolorized without residual staining after processing. For these applications the disadvantage of the oxonol dyes, their low stability, becomes an advantage here. In the processing solutions they are decolorized and dissolved. For the dissolution they have to be readily water soluble and so the oxonol dyes most used were e. g. Oxonol Red **6** and Oxonol Blue **7** or **11**.

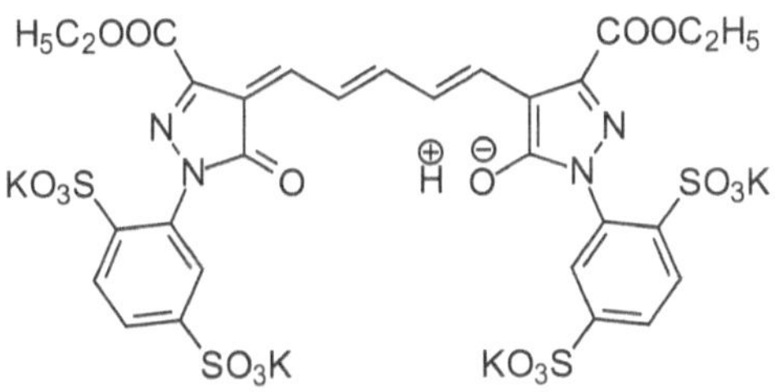

11

Lipophilic oxonol dyes are used in bio-analysis and medical diagnostics to stain cells, bacteria or liposomes for example [18, 19], and references cited therein. However, the main application was cytometry (image cytometry, flow cytometry) for a longer time. This method involves a wide range of techniques to measure the molecular properties of cells. Mostly the signal of fluorescent molecules (lipophilic cationic dyes such as cyanine and rhodamine and lipophilic anionic dyes) is detected.

Across a cell membrane there is an electrostatic potential. This membrane potential is an important physiological parameter of the living cell. Two components of the whole membrane potential are the two surface potentials. That is the potential drop between the extra-cellular space and the extra-cellular membrane surface and the drop between the intra-cellular space and the intra-cellular membrane surface. The third component is the dipole potential generated by the different distribution of ions at the two sides of a cell membrane.

For determination of membrane potentials in eukaryotic cells and prokaryotic bacteria some oxonol dyes as lipophilic anionic dyes were developed such as e. g. **12** [DiBAC$_4$(3)] and **13** [Oxonol VI].

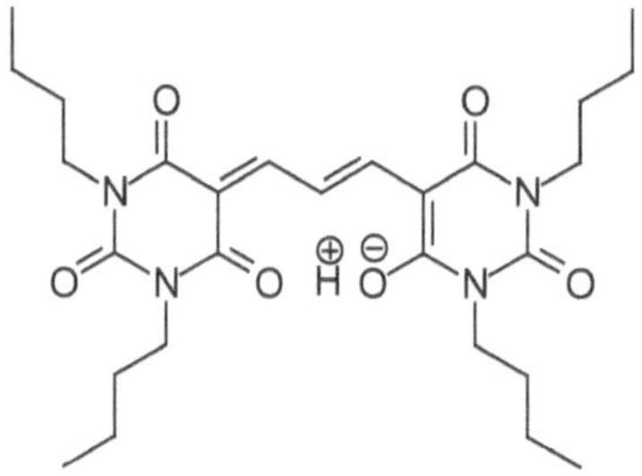

12

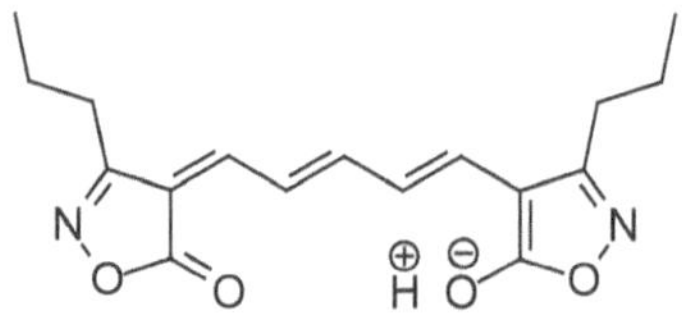

13

With this technique relatively large changes in the membrane potential can be detected. However, changes in membrane structure and changes in protein or lipid concentration in the medium in which cells are suspended can produce changes in cellular fluorescence which may be misinterpreted as changes in membrane potential. Newer radiometric techniques allow precise measurement of membrane potential.

Digital optical data storage with dyes started with the so called WORM (write once read many) using laser diodes (LD) with $\lambda = 830$ nm emission wavelength, followed by the recordable compact disc (CD), the CD-R (recordable) with laser emission wavelength $\lambda = 780$ nm and later for video recording the DVD (digital versatile disc) DVD/R which uses LDs with emission wavelengths 650 nm [20, 21]. For every system special dyes had to be developed. Besides cyanine dyes phthalocyanines in CD-R and azo metal complexes in DVD/R were used commercially [20, 21]. For these applications the light and thermal stability of conventional oxonol dyes was insufficient.

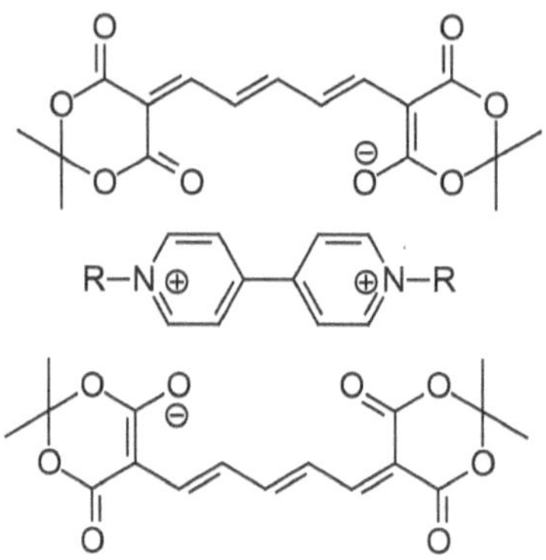

14

After the standards were set with cyanine dyes and azo metal complexes in DVD/R, a new approach with oxonol dyes was published. The new 2: 1 complex salts composed of two oxonol dyes based on Meldrum's acid and one 4,4′-bipyridinium-cation e. g. **14** exhibit an improved light stability [22] as well thermal stability [23].

References

1. Brooker LG, Kreyes GH. US 2 241 238. 29 Apr 1939.
2. König W. Über den Begriff der „Polymethinfarbstoffe" und eine davon ableitbare allgemeine Farbstoff-Formel als Grundlage einer neuen Systematik der Farbenchemie. J Prakt Chem. [2] 1926;112:1.
3. Mustroph H. Polymethine dyes. Phys Sci Rev. 2020;5. DOI:10.1515/psr-2019-0084.
4. Tyutyulkov N, Fabian J, Mehlhorn A, Dietz F, Tadjer A. Polymethine dyes – structure and properties. Sofia: St Kliment Ohridski University Press, 1991.
5. Sturmer DM. Syntheses and properties of cyanine and related dyes. In: Weissberger A, Taylor EC, editors. The chemistry of heterocyclic compounds. vol. 30. New York: Wiley, 1977:441–587.
6. Griffiths J. Colour and constitution of organic molecules. London: Academic Press, 1976.
7. Dähne S. Systematik und Begriffserweiterung der Polymethinfarbstoffe. Z Chem. 1965;5:441.
8. Hamer FM. The cyanine dyes and related compounds. In: Weissberger A, editor. The chemistry of heterocyclic compounds. vol. 18. New York: Interscience, 1964.
9. Radeglia R. ^{13}C-NMR-Spektroskopische Untersuchungen der Elektronenstruktur von einfachen Polymethinen. ^{13}C-Chemische Verschiebungen, ^{13}C-H- Kopplungskonstanten und Eu(DPM)$_3$-Verschiebungseffekte. J Prakt Chem. 1973;315:1121.
10. Etter MC, Kress RB, Bernstein J, Cash DJ. Solid-state chemistry and structures of a new class of mixed dyes. Cyanine-oxonol. J Am Chem Soc. 1984;106:6921.
11. Grossel MC, Edwards DJ, Cheetham AK, Eddy MM, Johnson O, Postle SR. Oxonol dyes: X-ray crystallographic and solid-state 13C nuclear magnetic resonance studies of some organic semiconductors. J Mater Chem. 1991;1:223.
12. Malhotra SS, Whiting MC. Researches on polyenes. Part VII. The preparation and electronic absorption spectra of homologous series of simple cyanines, merocyanines, and oxonols. J Chem Soc. 1960;3812.
13. Knorr L. Synthetische Versuche mit dem Acetessigester. Liebigs Ann Chem. 1887;238:137.
14. Mustroph H. Cyanine dyes. Phys Sci Rev. 2020;5. DOI:10.1515/psr-2020-0145.
15. James TH, editor. The theory of the photographic process. 4th ed. New York: Macmillan, 1977.
16. Keller K, Kampfer H, Matejec R, Lapp O, Krafft W, Frenken H, et al. Photography. In: Elvers B, editor. Ullmann´s encyclopedia of industrial chemistry. Weinheim: Wiley-VCH, 2000.
17. Fujita S. Organic chemistry in photography. Berlin: Springer, 2004.
18. Sabnis RW. handbook of biological dyes and stains. Hoboken: Wiley; 2010.
19. Haugland RP. The handbook: a guide to fluorescent probes and labeling technologies. Eugene: Molecular Probes Inc, 2005.
20. Mustroph H, Stollenwerk M, Bressau V. Current developments in optical data storage with organic dyes. Angew Chem Int Ed. 2006;45:2016.
21. Wochele RE, van Houten H, Duchateau JP, Kloosterboer HJ, Verhoeven JA, van Vlimmeren R, et al. Information storage materials, 2. Optical recording. In: Elvers B, editor. Ullmann's encyclopedia of industrial chemistry. Weinheim: Wiley-VCH, 2011.
22. Inagaki Y, Morishima S, Wariishi K, Saito N, Akiba M. A new class of light-fast oxonol dyes : organic-glass forming salts of oxonol anions and 4,4´-bipyridinium cations. J Mater Chem. 2006;16:345.
23. Morishima S, Wariishi K, Mikoshiba H, Inagaki Y, Shibata M, Hashimoto H, et al. Tuning the thermochemical properties of oxonol dyes for digital versatile disc recordable: reduction of thermal interference in high-speed recording. J Soc Photogr Sci Technol Jpn. 2010;73:252.

Robert Christie and Adrian Abel

53 Perylene and perinone pigments

Abstract: Perylenes and perinones are separate groups of pigments categorized within the carbonyl chemical class. The two pigment groups show similarities, for example, in their chemical structural features and, to an extent, in their technical and application properties as high-performance organic pigments. Perylenes constitute a series of firmly established high-performance pigments, offering red and violet colors, and also extending to black. Synthetically, they are derived from perylene-1,4,5,8-tetracarboxylic acid. The perylenes tend to be quite expensive pigments, but their high levels of fastness properties mean that they are suitable for highly demanding applications. In particular, they offer very high heat stability. Two perinone pigments are used commercially. In their synthesis from naphthalene-1,4,5,8-tetracarboxylic acid, they are formed as mixtures of the two isomers, which can be separated. The trans isomer, CI Pigment Orange 43, is a highly important commercial pigment, especially for plastics, while the cis isomer, CI Pigment Red 194, is bordeaux in color and is of much lesser importance. The perinone, CI Pigment Orange 43, provides a brilliant orange color and has very good fastness properties. Its commercial manufacture involves a challenging multistage procedure and consequently it is one of the most expensive organic pigments on the market.

Keywords: perylene pigments, perinone pigments, perylenes, perinones, high-performance pigments, carbonyl pigments, polycyclic pigments, crystallochromy, cis isomer, trans isomer, CI Pigment Red 149, CI Pigment Red 178, CI Pigment Red 179, CI Pigment Red 190, CI Pigment Red 194, CI Pigment Red 224, CI Pigment Violet 29, CI Pigment Black 31, CI Pigment Black 32, CI Pigment Orange 43

53.1 Fundamentals

Perylenes and perinones are separate groups of carbonyl pigment, although they are related to an extent in terms of their chemical structures, their methods of synthesis, their historical origin, the solid-state color generation mechanisms and, to an extent, their technical and application properties as high-performance organic pigments. They are both categorized here as subclasses of the carbonyl pigment chemical class, although they can also be classed as polycyclic pigments. Perylene pigments are derivatives of perylene-1,4,5,8-tetracarboxylic acid (**1**), while perinone

This article has previously been published in the journal Physical Sciences Reviews. Please cite as: R. Christie, A. Abel, Perylene and Perinone Pigments *Physical Sciences Reviews* [Online] 2021, 6. DOI: 10.1515/psr-2020-0190

https://doi.org/10.1515/9783110587128-053

pigments are derived from naphthalene-1,4,5,8-tetracarboxylic acid (**2**), as illustrated in Figure 53.1 [1–3].

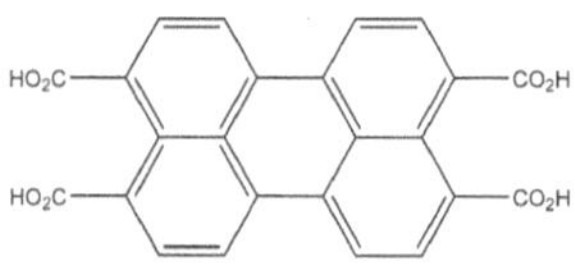

1

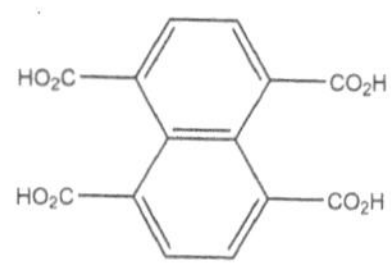

2

Figure 53.1: Tetracarboxylic acids from which perylene and perinone pigments are derived.

53.2 History

In common with many of the anthraquinonoid pigments, a group of carbonyl pigments that are described elsewhere, perylene and perinone pigments have experienced two distinct phases in their histories. The first involves the original discovery of the compounds. Some of these products were originally developed as vat dyes. These textile dyes are insoluble colorants that are applied to cellulosic fibers, such as cotton, via a water-soluble (leuco) form obtained by reduction with sodium dithionite in alkaline solution. The leuco species are then oxidized to regenerate the pigment as insoluble particles trapped within the fibers. This application as dyes for textiles preceded their development as pigments by around half a century. Perylene-1,4,5,8-tetracarboxylic acid (**1**) was discovered in 1912. The diimide (**3f**) derived from compound (**1**) was first synthesized in 1913 by Kardos, of BASF and was found to offer a very dull shade of maroon. It was originally designated as CI Pigment Brown 26, but now appears as CI Pigment Violet 29, as illustrated in Figure 53.2 with the substituents listed in Table 53.1. In the 1950s several of these compounds were converted into commercial pigments. The initial developments were carried out by Harmon Colors in the US, although the first commercial pigment, CI Pigment Red 149 (**3a**), was introduced by Hoechst in 1957. Harmon Colors introduced CI Pigment Red 190 (**3d**) in 1959, which was followed by CI Pigment Red 179 (**3c**) (1959) and CI Pigment Red 178 (**3b**) (1966) by BASF [4]. The dominant perinone colorant was first synthesized by Eckart and Greune in 1924 at Farbwerke Hoechst. The product, which proved to be a mixture of two isomers (**4**) and (**5**) as illustrated in Figure 53.3, was originally developed as a vat dye which offered a cleanliness of shade that was not available from other products. It was introduced as a pigment in 1950 although sold under its vat dye designation until around 1970.

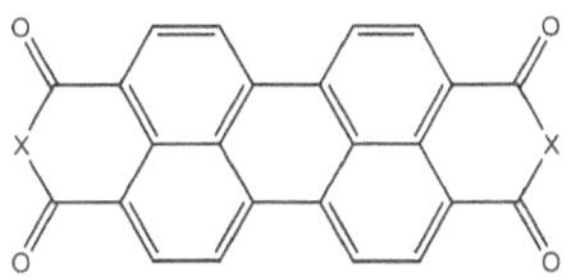

3

Figure 53.2: General structure of perylene pigments.

Table 53.1: Substituents in perylene pigments (**3**).

Compound	CI Pigment	X
3a	Red 149	N-C_6H_3*m*,*m*'-$(CH_3)_2$
3b	Red 178	N-*p*-C_6H_4-N = N-C_6H_5
3c	Red 179	N-CH_3
3d	Red 190	N-C_6H_4*p*-OCH_3
3e	Red 224	O
3f	Violet 29	N-H
3g	Black 31	N-$CH_2CH_2C_6H_4$
3h	Black 32	N-$CH_2C_6H_4$*p*-OCH_3
3i	–	N-C_2H_5

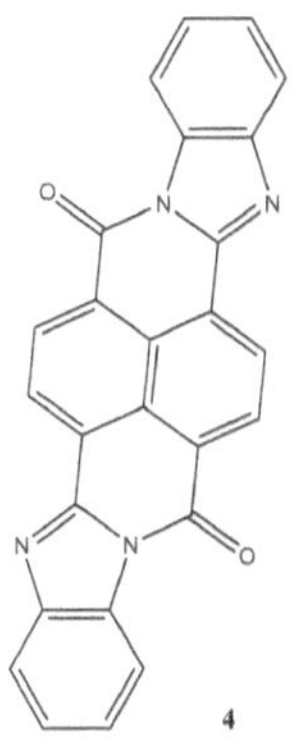

4

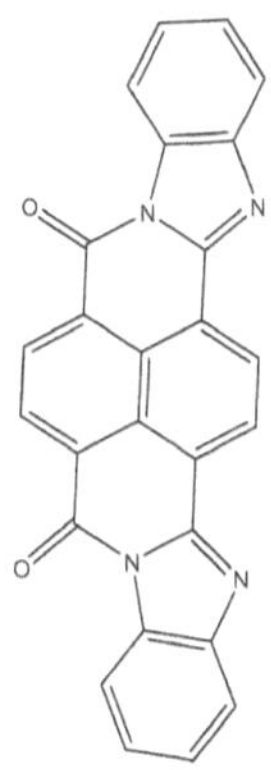

5

Figure 53.3: Structures of perinone pigments.

53.3 Structures and properties

Perylene pigments are mostly diimides of perylene-3,4,9,10-tetracarboxylic acid (**1**), represented by the general structure (**3**)**,** together with the bis-anhydride (**3e**), as illustrated in Figure 53.2, with reference to the substituent patterns given in Table 53.1 [5]. They are industrially important high-performance pigments offering a range of colors mainly covering the red and violet part of the spectrum, but also extending to black. An interesting observation in the perylene series is that small structural changes in the peripheral substituent attached to the imide nitrogen can lead to quite profound color differences. As an example, the dimethyl compound (**3c**) is red while the diethyl derivative (**3i**) is black. X-ray diffraction studies have been applied to a range of perylenes to characterize the effect of differences in the crystal structure on the color of the pigments, a phenomenon referred to as crystallochromy [6, 7]. For example, compound (**3c**) consists of a parallel arrangement of molecules in stacks, whereas in the case of compound (**3i**), the molecules are in stacks twisted with respect to one another, and with considerably more overlapping of the perylene ring systems in neighboring molecules. The color generation mechanism has been investigated for three representative pigments, CI Pigment Red 149 (**3a**), CI Pigment Red 179 (**3c**), and CI Pigment Black 31(**3 g**) with special attention on the influence of intermolecular interactions within the crystal structures [7]. The color in the solid state is determined mostly by two absorption bands in the visible region. One absorption band involves the molecular character, and it is of interest that the absorption spectra of these pigments in solution, where they exist as individual molecules, are similar. The other band is due to interactions between transition dipoles within the crystal structures. The bright red color of CI Pigment Red 149 (**3a**) is reported to be due to the dominance of the molecular absorption and insignificant intermolecular interactions, whereas the maroon color of pigment (**3c**) and the black color of pigment (**3g**) are characterized by medium and strong absorptions, respectively, due to the intermolecular interactions. Certain perylenes are also useful as highly efficient, lightfast fluorescent dyes [8, 9]. In the most important of these perylene dyes, the imide nitrogen atoms are substituted with aryl or bulky alkyl groups such as *t*-butyl, providing intense orange to red fluorescent colors in polymers. They are of special interest as functional dyes because of their intense visible light absorption, high stability, electron accepting ability, and quantum yields close to unity. Because of these properties, they have been extensively investigated for optoelectronic and photovoltaic devices, thermographic processes, energy-transfer cascades, light-emitting diodes, and near-infrared absorbing systems.

There are two isomeric perinone pigments that have been used commercially, the *trans* isomer CI Pigment Orange 43 (**4**), and the *cis* isomer CI Pigment Red 194 (**5**), as illustrated in Figure 53.3. The *trans* pigment is a highly important commercial pigment, especially for plastics, while the *cis* isomer is of much lesser importance. Like the perylenes, they provide high levels of technical performance.

The color generation mechanism is also similar to that proposed for the perylene pigments. The color in the crystalline solid state is determined by molecular absorption bands and additional bands at longer wavelengths due to excitonic interactions between transition dipoles within the crystal structures. In particular, the nature of the molecule pairing along the stacking axis is found to play an important role, giving rise to the orange color of *trans* isomer (**4**) and the bluish red of the *cis* isomer (**5**) [10].

53.4 Synthesis and manufacture

The synthetic procedures leading to perylene and perinone pigments, as described in this section, are long-established. The crude products lack tinctorial strength and brightness of shade and require conditioning procedures to provide the pigments in a form that is useful for application. Procedures have been developed over the years to produce the appropriate particle size distribution and crystal form, enhance wettability by surface treatments that lower interfacial tension, and ensure ease of dispersion [11]. The diimide perylene pigments (**3a-3d**) and (**3f-3h**) may be prepared by reaction of the bis-anhydride (**3e**) (1 mol) with the appropriate primary aliphatic or aromatic amine (2 mol) at elevated temperatures in a high-boiling solvent. There are several ways whereby the crude products thus obtained may be converted into a useful pigmentary form. These include reprecipitation from concentrated sulfuric acid or acid swelling using 80% sulfuric acid, milling in the presence or absence of auxiliaries, and by solvent treatments, with combinations of the techniques often used to provide optimized products. It is also possible to adapt the sequence used in vat dyeing, by reducing the crude product with sodium dithionite to the water-soluble leuco compound, removing impurities, and then oxidizing to provide the pigment in an appropriate pigmentary form.

The synthesis of perinone pigments (Figure 53.4) involves condensation of naphthalene-1,4,5,8-tetracarboxylic acid (**2**), or more commonly the monoanhydride (**6**) derived from compound (**2**), with *o*-phenylenediamine (**7**) in refluxing acetic acid. This process affords a mixture of the two isomers (**4**) and (**5**). The isomers may be separated by treatment with a solution of potassium hydroxide in ethanol, which leads to a pale-yellow precipitate and a yellow solution. The *trans* isomer (**4**) is obtained by hydrolysis of the precipitate. This intermediate has been described as a "potassium hydroxide addition compound", produced on a multi-ton scale for more than 80 years. Very recently, this assumption has been proved incorrect [12]. The intermediate has been characterized by X-ray crystal structure determination, confirmed by NMR studies, as the potassium salt of a double ring-opened species which recyclizes when treated with acid. The *cis* isomer (**5**) is obtained by re-acidification of the yellow solution. Traditional organic pigment conditioning techniques, including milling, acid treatment, and solvent treatment at elevated temperatures, are used to convert the crude products into commercially useful pigments.

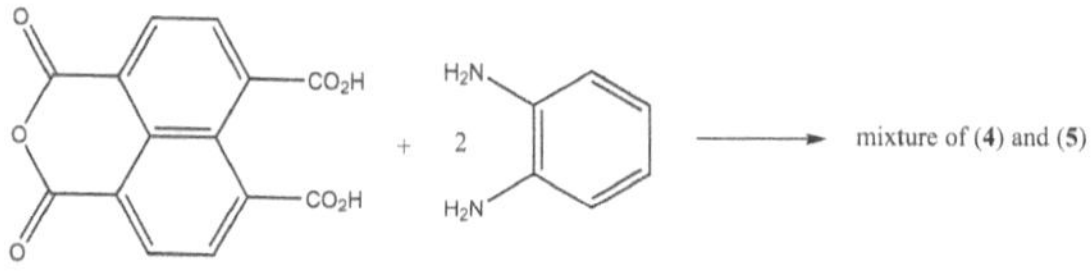

Figure 53.4: Synthesis of perinone pigments.

53.5 Applications

53.5.1 Perylene pigments

The perylene pigments that have been brought onto the market have fared better than many of the anthraquinononoid pigments in that several of them are firmly established as commercial products and have stood the test of time, while the number of commercial anthraquinonoid pigments has declined substantially over the years. The perylenes tend to be quite expensive but their high fastness properties take them into application areas that many other pigments fail to satisfy. In particular, they exhibit very high heat stability. However, the development of diketopyrrolopyrrole (DPP) pigments has impacted their use to an extent in more recent times.

53.5.1.1 CI Pigment Red 149 (3a)

This is a bright mid-red shade pigment, used mainly for the coloration of plastics, where its main feature is exceptional heat stability. This makes it suitable for use in PVC, polyolefins, polystyrene and its copolymers, and even polycarbonate, in which it can be processed at up to 310 °C. It is one of the few organic pigments that can be considered, with certain limitations, as having some suitability for the pigmentation of polyamides. In plasticized PVC, it is fast to migration and has roughly double the tinctorial strength of quinacridone red (CI Pigment Violet 19, γ-modification). It has very good lightfastness, but this is not usually considered sufficient for applications that involve exterior exposure, such as coil coatings. Its dielectric properties are appropriate for use in cable sheathing, although its cost would normally exclude such use. In polyolefins, it is stable to 300 °C and offers excellent lightfastness, even in pale reductions with titanium dioxide. It is not recommended for injection moldings of polyolefins as it causes warpage, although this adverse effect decreases as the temperature of molding increases. Along with other perylene pigments, it adversely affects the protective effect of antioxidants based on hindered amine light stabilizers (HALS), which are polymer stabilizing species, especially important for polyolefins. It is recommended for polystyrene and can be used to color ABS and other styrenic copolymers. It is recommended for use in the spin coloration of polyester, polypropylene, acrylonitrile, and polyamide fibers. Its use in paints has never approached the level of its use in plastics. In paints it provides a more bluish red than is seen when used in plastics. It is transparent and so provides attractive metallic shades, but its lightfastness is well below the level required in automotive finishes. Its lightfastness is rated as good to very good in full and deep shades, and good in pale reductions, although in strong shades it darkens on exposure to light. It is stable to 200 °C, and offers very good to excellent resistance to solvents, including aromatic hydrocarbons, ketones and esters. It is occasionally used in inks, such as those based on ketone or ester solvents, of the type that are used with vinyl and similar resins for printing on polymers or for wallcoverings.

53.5.1.2 CI Pigment Red 178 (3b)

This pigment offers a bluer shade of red compared with CI Pigment Red 149 (**3a**), finding use in both paints and plastics. It was introduced to the market by BASF and remains part of their pigment range for the paint and plastic industries. It is only offered as a relatively opaque grade. In plastics, it can be used in most polymers as it is very heat stable. It is recommended for the coloration of PVC, polyolefins, polystyrene and, with care, can be used in ABS, polyamides, and polycarbonate, offering heat stability up to 300 °C. It is somewhat less lightfast and weatherfast than CI Pigment Red 149. It has a serious detrimental effect on the dimensional stability of polyolefins that have been injection molded and on the effectiveness of HALS antioxidants. In paints, it can be used in general industrial finishes, including automotive original equipment in full and deep shades. It has good stability to aromatic hydrocarbon and alcoholic solvents but is not completely fast to overcoating.

53.5.1.3 CI Pigment Red 179 (3c)

This pigment, which also has the Generic Name CI Vat Red 23, can exist in two versions with quite different shades, either as a rather dull maroon shade or a much brighter mid-red, albeit somewhat bluer than CI Pigment Red 178 (**3b**). Even within these two pigment types, the shade can vary, thus offering formulators a versatile color range, all offering very good to excellent fastness properties with varying levels of opacity. The use of the maroon shade pigments is mainly limited to metallic and pearlescent finishes and therefore requires high transparency. CI Pigment Red 179 is probably the perylene pigment that is most used, with a usage of over 100 tons per annum registered for European REACH regulations. The pigment is used mainly in paints but also finds significant use in plastics. In 2019, the pigment was listed on the EU Community Rolling Action Plan (CoRAP), as the EU is challenging the solubility figures presented by the consortium responsible for its registration under REACH regulations [13].

It provides excellent fastness to light, in spite of the small particle size that is responsible for its transparency. It is specifically recommended for two coat metallic paints in original automotive finishes, with the pigment acquiring some additional protection from the second coat. It shows only slight fading after two years weathering in such systems and is stable to 250 °C, well in excess of heat stability requirements for this application. It has very good resistance to aromatic hydrocarbons, esters, and alcohol solvents, but its fastness to ketones is a little lower. It is recommended for most binder systems, although its high price and relative dullness in color precludes its use in most decorative paint systems. The larger particle size of the redder grade means that it is slightly less transparent, but it has similar fastness to light and weather. Its solvent fastness is slightly better, especially to ketones. Although the areas of use of this grade are broadly the same as the maroon grades, it is given only qualified approval in water-based systems. It is not recommended for systems requiring the use of amine curing agents as these can lead to degradation of the pigment due to chemical reduction. In plastics, the pigment has a wide range of applications, including in both plasticized and rigid PVC, and it shows excellent fastness to migration in all polyolefins, in which it is stable up to 300 °C. It has a quite severe tendency to cause warpage when used for injection molded articles and its combination with HALS stabilizers should be avoided. Its use in coloring polystyrene and styrenic copolymers should only be considered after careful testing, while it is not suitable for polycarbonates and polyamides. Its use in inks is limited to specialist applications.

53.5.1.4 CI Pigment Red 190 (3d)

This pigment has been used in coatings, where it produces rather dull yellowish red full shades. It offers excellent light fastness and very good solvent resistance and is stable to 200 °C. However, it is currently of much reduced importance. It is also designated as CI Vat Red 29.

53.5.1.5 CI Pigment Red 224 (3e)

This pigment was introduced by Ciba under the designation Irgazin Red BPT, manufactured in both transparent and opaque forms. It is no longer marketed by BASF who acquired the Ciba range of pigments. However, Sun Chemicals expanded their US production facility in 2019 to include this pigment. It is almost entirely used in the industrial paint industry, where it is used in original equipment automotive finishes. In such paints, it produces a very bright yellowish to mid-red shade in color and it has very good to excellent lightfastness in full shade, although becoming lower as the shade is reduced with titanium dioxide, thus limiting its value where highest fastness to light is demanded. Its poor fastness to alkali limits its use, especially in water-based finishes. However, this feature is less critical in two coat finishes as the colored film is protected by the clear topcoat. It is very fast to the solvents used in paint formulations and is heat stable to 200 °C.

53.5.1.6 CI Pigment Violet 29 (3f)

This pigment is structurally the simplest of the diimide perylene series and was the first perylene pigment to be registered in the Colour Index, originally as CI Pigment Brown 26. It was introduced commercially as PV Fast Bordeaux B by Hoechst, aimed at plastics applications. The pigment provides a dull, maroon shade, not a particularly attractive color, but in terms of fastness performance, it surpasses almost all organic pigments. It is highly heat stable so that it is suitable for most polymers used in plastics, withstanding high processing temperatures. Its fastness to light is excellent even in pale reductions with titanium dioxide. It is one of the few organic pigments that can be used for the spin coloration of polyester, capable of meeting the demands of the condensation process during which it is required to withstand temperatures of almost 300 °C for 5 hours or more. It remains recommended for paints, but its only use is as a transparent pigment for metallic and pearlescent finishes, where it offers excellent stability to overcoating. It is now manufactured mainly in the US and in the Far East.

53.5.1.7 CI Pigment Black 31 (3g)

This intensely bluish-black pigment was introduced to the market by BASF, more on account of its properties in the infrared region of the spectrum than in the visible region. The pigment absorbs most wavelengths of visible light, hence its black color, and reflects near infrared radiation. This has led to the use of the pigment in the so-called "cool pigment" technologies, which allow the formulation of dark colors which reflect near infrared radiation thereby limiting surface heating and in turn reducing energy costs. It is particularly useful in aircraft coatings, and in military camouflage.

53.5.1.8 CI Pigment Black 32 (3h)

This pigment is a successor to CI Pigment Black 31 (**3g**) and is promoted for the same applications. The pigment is stable to 200 °C. As it is intended for external use, it is important that it has very good weatherability in strong shades, although this property reduces significantly as the titanium dioxide level increases. Its resistance to aromatic hydrocarbon and ester solvents is very good, but only moderate to methyl ethyl ketone (MEK) and butanol. The pigment can be used in most binder systems, including water-based and amine-curing systems.

53.5.2 Perinone pigments

As described earlier, in the synthesis of the only perinone pigment of commercial importance, a mixture of the *cis* and *trans* isomers is initially formed. The *cis* isomer (**5**) is bordeaux in color, whereas the much more desirable *trans* isomer (**4**) is a brilliant orange color and has very good fastness properties. The commercial manufacture of the pigment is well-established. However, it is a challenging procedure, involving long lead times, estimated at well over a year, as some of the stages can only be carried out in individual locations, thus requiring shipping across continents. Consequently, the pigment is expensive, indeed one of the most expensive organic pigments on the market. The pigment is now manufactured in Asia but, whereas with many other organic pigments this has led to a crash in market price, the price in the case of CI Pigment Orange 43 (**4**) has dropped much less.

53.5.2.1 CI Pigment Orange 43 (4)

This pigment is also known as CI Vat Orange 7. It offers a very pure, reddish shade of orange. As a result of this, and because of its technical properties, CI Pigment Orange 43 is virtually impossible to substitute. Its spectral curve shows very low reflectance at wavelengths up to 550 nm and then a steep rise into the infrared region, which explains its unique purity of shade. It finds use throughout the range of pigment applications, although it is especially important in plastics and for spin coloration of fibers. In paints, it is suitable for all binder systems and is used across a wide variety of these. It is tinctorially strong, more typical of the strength of monoazo pigments and essentially double that of CI Pigment Red 168, a dibromoanthanthrone red pigment, and of quinacridone red pigments. It is semi-transparent, allowing its use in metallic finishes where it offers an interesting clean tone, although some of the metallic sparkle is lost. Its heat stability reaches 200 °C for 30 minutes exposure without affecting its shade. It shows unusual lightfastness properties. Whereas most pigment show reducing fastness to light as the depth of shade becomes paler, with CI Pigment Orange 43 it improves. In full shade it is rated good, rising to excellent at 1/25 standard depth when reduced with titanium dioxide. This phenomenon is due to the pigment darkening, which is very noticeable in full shade and compensates for fading at paler depths. Its resistance to solvents is very good to excellent, even in aromatic hydrocarbons, ketones and esters. It was used extensively for automotive finishes, but its tendency to darken and its lack of transparency limits it to the role of a tint for other pigments. Economic considerations limit its use in repair finishes. It is occasionally used in decorative (architectural) paints in pale reductions, where its purity of shade is unique. It is a very useful pigment in tinting systems, as it provides tints on the yellow side of mid red and on the red side of yellow, without compromising lightfastness. However, once its use has been accepted in such a system, it is virtually impossible to replace without seriously compromising the range of available shades [14]. In plastics it offers excellent heat stability in most polymers. In PVC it provides very good to excellent fastness to light and very good resistance to migration. It is suitable for cable sheathing. In polyolefins it is stable to 280 °C and offers excellent lightfastness. It is not recommended for injection molding where dimensional stability is required, as such moldings tend to warp. It is fully recommended for polystyrene and styrenic copolymers as well as for some engineering polymers such as polycarbonate. It is considered to have only limited suitability for nylon, but can be used for acrylonitrile fibers, including for use in tents and awnings, where it satisfies the lightfastness demands [15]. Its use in inks is limited by economics, but it can be used for specialty applications, for metal decorative inks up to 220 °C and for applications where steam sterilization is required [16]. It can be used for inkjet printing, where its strong, brilliant orange shade, together with very good heat resistance, lightfastness, and excellent fastness to weathering, is unique. It is used at very low concentrations to color Imperial Leather soap and other cosmetic applications. It is of growing importance in some new applications, for example, as an organic semiconductor.

53.5.2.2 CI Pigment Red 194 (5)

This pigment, the *cis* perinone isomer, is also known as CI Vat Red 15. The pigment provides a dull, brick red color. While the *trans* isomer (**4**) is obtained from the separation process in pure form, the *cis* isomer usually contains some *trans* isomer and, since the absorption characteristics of the two isomers are well separated in the spectrum, the mixture results in the dull red shade, as described in the Colour Index under the designation CI Pigment Red 194. The pigment is mainly used in paints, where it offers very good to excellent lightfastness. In contrast to the *trans* isomer, it does not darken and thus its fastness to light does not improve in paler depths. It is heat stable to 200 °C and it has almost as good solvent fastness properties as the *trans* isomer. Although unconfirmed, it is believed that the brick red colored Volkswagen Beetle car, beloved during the 1960–70s, used this pigment in the paint formulation employed at the time. Although never offered as a pigment for plastics, it offers excellent heat stability in polyolefins, in which it is stable to 270 °C. However, in more crystalline polymers, and even rigid PVC, it dissolves and acts as a temperature critical solvent soluble dye, offering a bright orange color, but in practice it is difficult to obtain consistent color. The equivalent vat dye, a mixture of the two isomers, used to be offered as CI Vat Red 74. This was often used in combination with carbon black to offer a range of brown shades. Traditional manufacturers have ceased offering this pigment, but Anshan HOP Technology company list it in the Colour Index.

References

1. Hunger K, Schmidt MU. Industrial organic pigments, 4th ed, Weinheim: Wiley-VCH Verlag GmbH, Ch 3 2019.
2. Thompson B. Perylene pigments. In: Faulkner EB, Schwartz RJ, editors. High performance pigments, 2nd ed. Weinheim: Wiley-VCH Verlag GmbH, 2009 Ch 16.
3. Christie RM. Colour chemistry, 2nd ed, London: RSC, Ch 9 2015.

4. Lewis P. Colored organic pigments. Appl Polym Sci Twenty-first Century. 2000;493–526.
5. Nagao Y. Synthesis and properties of perylene pigments. Prog Org Coat. 1997;31:43–9.
6. Klebe G, Graser F, Hadicke E, Berndt J. Crystallochromy as a solid-state effect; correlation of molecular conformation, crystal packing and color in perylene-3,4:9,10-bis(dicarboximide) pigments. Acta Crystallogr. 1989;B45:69–77.
7. Mizuguchi J. Electronic structure of perylene pigments as viewed from the crystal structure and excitonic interactions. J Phys Chem. 2002;106:767–72.
8. Christie RM. Fluorescent dyes. Rev Prog Color. 1993;23:1–18.
9. Christie RM. Fluorescent dyes. In: Clark M, editor. Handbook of textile and industrial dyeing: principles, processes and types of dyes. vol. 1, Cambridge: Woodhead Publishing, 2011 ch.17.
10. Mizuguchi J. Crystal structure and electronic characterization of trans-and cis-perinone pigments. J Phys Chem B. 2004;108:8926–30.
11. Merkle K, Schäfer H. In: Patton TC, editor. In pigment handbook, vol. 3, New York: John Wiley and Sons, 1973:157–67.
12. Tapmeyer L, Bolte M, Chierotti MR, Schmidt MU. Structure of the intermediates in the industrial separation of perinone isomers. Dyes Pigm. 2020;181:108442.
13. https://echa.europa.eu/documents/10162/c607549c-1c07-c5d6-d6e2-8d18bff91f3a. Accessed: 6 Apr 2021.
14. https://www.clariant.com/en/Solutions/Products/2014/03/18/17/10/Hostaperm-Orange-GR-for-Paints-and-Coatings. Accessed: 6 Mar 2021.
15. https://www.clariant.com/en/Solutions/Products/2014/03/18/17/14/PV-Fast-Orange-GRL. Accessed: 6 Mar 2021.
16. https://www.clariant.com/en/Solutions/Products/2014/10/14/15/12/Hostaperm-Orange-GR-for-Printing. Accessed: 6 Mar 2021.

Andrew Towns

54 Photochromic dyes

Abstract: This article describes the defining characteristics of photochromic dyes and highlights the subset of properties that are of greatest commercial importance. It outlines the history of the industrial exploitation of photochromic colorants before moving on to discuss current and potential applications. In doing so, a brief tour of key types of photochromic dye is provided.

Keywords: photochromism, photochromic, colorant, dye, functional

The subject of photochromic dyes encompasses a diverse array of chemistries and end-uses. The industrial exploitation of photochromism is, however, restricted to a select few structural classes with one application being commercially dominant. Nevertheless, many millions of dollars' worth of business across the globe depends upon the photochromism of dyes. In addition, they continue to attract much interest as the basis for a wide range of as yet unrealized technologies of enormous potential, from the digital to the medical to the 'nano'. While much research aimed at realizing these applications focuses on relatively new and esoteric dye classes, plenty of effort is also being directed at making use of the old too. This article does not just define photochromism: it delineates the most industrially important subset of the phenomenon. Consequently, colorant types of greatest commercial significance feature heavily in terms of their chemistry and behavior. A flavor is also given of the kinds of potential uses to which photochromic dyes might be put in future. It is not the intention of this article to go into great mechanistic detail concerning dye photochemistry or synthesis, but to point the reader in the direction of good sources of information on these subjects.

54.1 Fundamentals

54.1.1 Properties defining a photochromic dye

A photochromic dye is a colorant (i) that at some point during its application and use is in solution and (ii) which exhibits photochromism. The first part of this statement

This article has previously been published in the journal Physical Sciences Reviews. Please cite as: A. Towns, Photochromic Dyes *Physical Sciences Reviews* [Online] 2021, 6. DOI: 10.1515/psr-2020-0191

https://doi.org/10.1515/9783110587128-054

is the defining characteristic of all dyes [1]. The second part refers to a light-driven color-change phenomenon for which a standard definition appears in Box 54.1.

Box 54.1

Photochromism	*The reversible transformation of a molecular entity between two forms, A and B, having different absorption spectra, induced in one or both directions by absorption of electromagnetic radiation* [2]. *The spectral change produced is typically, but not necessarily, of visible color and is accompanied by differences in other physical properties.*

This definition is very broad, covering fully switchable appearance of color, disappearance of color, or alteration in hue. It could be argued that the definition is too general, because it does not preclude labelling as photochromic those changes in absorption properties that cannot be observed unaided by the typical human eye. The definition does not exclude instances in which the changes in spectra are restricted entirely to within the ultraviolet or that switch from and back to the original color very rapidly, such as on a sub-millisecond timescale. This article thus uses the term to signify a reversible change in the color of a system *that is visible to the naked human eye.*

Note use of the word of “reversible” in the definition. Color change in response to irradiation with light is not enough on its own to justify labelling something as photochromic. Just as crucial is the capability of its original color to become fully restored by removal of the light stimulus (or altering its wavelength appropriately). In other words, the change in color must be completely reversible. However, it is not uncommon to encounter incorrect terminology in scientific papers especially the oxymoron “irreversible photochromism” as well as the tautologies “reversibly photochromic” and “reversible photochromism”. A permanent one-way switch in color is not an example of photochromism: for example, a dyed substrate that fades irreversibly in sunlight through destruction of the colorant is undergoing photochemically driven change, but it is not photochromic.

All photochromic dyes necessarily share the property that their color is switched in one direction upon absorption of electromagnetic radiation. Transformations undergone by photochromic systems may be unimolecular or bimolecular, and occur by a variety of mechanisms [3–5], some of which are driven by simultaneous or stepwise absorption of two photons [6]. As we shall see, dyes currently being exploited commercially for their photochromism rely on reversible unimolecular pericyclic reactions, i.e. they depend on the breaking or formation of a ring structure within the dye, prompted by single photon absorption of an energy associated with the UV and/or visible light regions.

Photochromic colorants fall into one of two groups depending upon the means by which they revert to the state they were in prior to photoactivation, i.e. whether the color change is thermoreversible or triggered photochemically (see Box 54.2).

Box 54.2

T-type Photochromism — *The property of returning to the original colored state simply upon removing the stimulating radiation. This reversion is thermally driven, so maintenance of the color change requires constant irradiation: as soon as the illumination is removed, the system starts to turn back to its original color.*

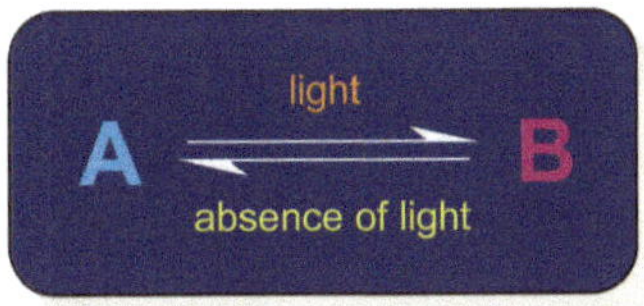

P-type Photochromism — *Behavior in which a color change is driven in the reverse direction only by photochemical means. Purely P-type colorants will remain indefinitely locked in one state once irradiated until they are flipped back to their original state by light of the appropriate wavelengths.*

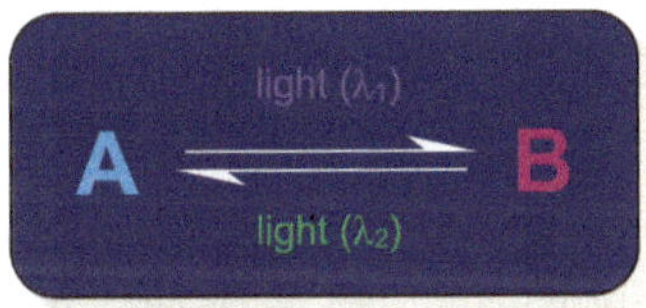

The behaviors of many commercially important photochromic dyes (as well as colorants that are under intense academic scrutiny for functional uses) are more complex than the two types defined in Box 54.2. Their photochromism often lies between these ideals, although a particular colorant will lean very heavily towards one of these characteristics. Real-world dyes described as P-type tend to show minor T-type behavior, although usually the former character is so dominant over the latter property that they can be exploited as switches: their light absorption properties are readily manipulated by exposure to different light sources. Consequently they have attracted great interest as functional colorants, but their potential has yet to be realized in a commercially significant technology – this aspect will be covered later. Many T-type dyes exhibit a small element of P-type behavior, which can be disadvantageous in applications that call for automatic and complete reversion to the original state after removal of illumination. Irrespective of this complication, they are of far less use as switches than P-type colorants, because of their tendency to revert thermally to their original state as soon as the source of photo-activating light is removed. Nevertheless, T-type dyes enjoy commercial dominance over P-types. The next section outlines their particular brand of photochromic properties: industrial photochromism.

54.1.2 Industrial photochromism

The definition in Box 54.1 is too broad to pin down the performance desired of photochromic dyes that are of greatest industrial importance. The term 'heliochromism' (see Box 54.3) is a good description of the subset of photochromism that covers their characteristics. Heliochromic dyes are the most commercially dominant variety of photochromic colorants. They account for the bulk of photochromic dyes consumed each year: over 90% of photochromic colorants end up used in the production of ophthalmic lenses that darken reversibly upon exposure to strong sunshine.

Box 54.3

Heliochromism	*Reversible photocoloration efficiently brought about by near UV radiation that forms part of unfiltered sunlight incident at the Earth's surface, which is accompanied by moderately efficient thermal fading at typical ambient temperatures and inefficient photobleaching by white light* [7, 8].

The definition in Box 54.3 describes photochromic dyes with strong T-type character that readily switch from a colorless to colored states in strong sunshine outdoors. Well formulated industrial dyes photoactivate to 90% of their peak darkness for a given intensity of irradiation within a minute. The activating radiation for commercial dyes is typically in the UV-A region of 320–400 nm, which forms a significant proportion of the solar radiation incident at the Earth's surface. Such dyes thus do not respond well to sunlight that has passed through architectural or automotive glass (since it largely filters UV-A wavelengths out), nor to light sources that lack a significant UV-A component such as incandescent bulbs. The range of activating wavelengths can extend out to the blue region of wavelengths as long as 420 nm for certain industrial dye types such as spiroindolinonaphthoxazines (SINO) [9]. While niche photochromic products that respond to sunlight filtered through glass have been commercialized using dyes sensitive to short wavelength visible light, e.g. sunglasses for drivers [10], most conventional photochromic ophthalmic lenses do not tend to work well when the wearer is indoors behind glass, or is illuminated with artificial light, owing to the lack of incident UV radiation.

The efficiency of thermal fading is a compromise, hence the word "moderate" in Box 54.3. A trade-off arises because a rapid thermal fade process translates to the equilibrium position attained at constant levels of irradiation and temperature ('photostationary state' – see Box 54.4) being one of a low concentration of colored dye species. As a consequence, the photocoloration will be perceived as weak. In the case of ophthalmic lenses, thermal fading must be sufficiently rapid that the dye perceptibly responds to light intensity changes in a matter of seconds. To avoid impairment of vision that is generally unacceptable to spectacle wearers, the dye should return to essentially a colorless state when no longer exposed to unfiltered sunlight after a few minutes. Consequently, a balance has to be found between the photocoloration intensity created upon irradiation by a light source and the rate at which it diminishes when the source is removed. Lens manufacturers thus carefully select T-type photochromic dyes and techniques of application to furnish products that develop adequately colored photostationary states in bright daylight which revert to acceptably colorless equilibrium states at low light levels in sufficiently short periods of time.

Box 54.4

Photostationary State *The equilibrium state eventually reached by a photochromic system at a constant intensity of irradiation. When this state is attained, the concentrations of the photo-activated and unactivated forms of dye do not alter. Consequently, the intensity of photocoloration stays unchanged. The equilibrium is dynamic: at any given moment, a proportion of the dye is incessantly switching between colored and colorless forms. Other than the magnitude of the flux of incident radiation and its spectral power distribution, the position of the equilibrium is dependent on numerous factors, including temperature and the nature of medium in which the dye is dissolved. As soon as one changes, the system is no longer in equilibrium and intensity of photocoloration shifts. The preceding description assumes that the rate of photodegradation of dye is negligible over the timescale of interest.*

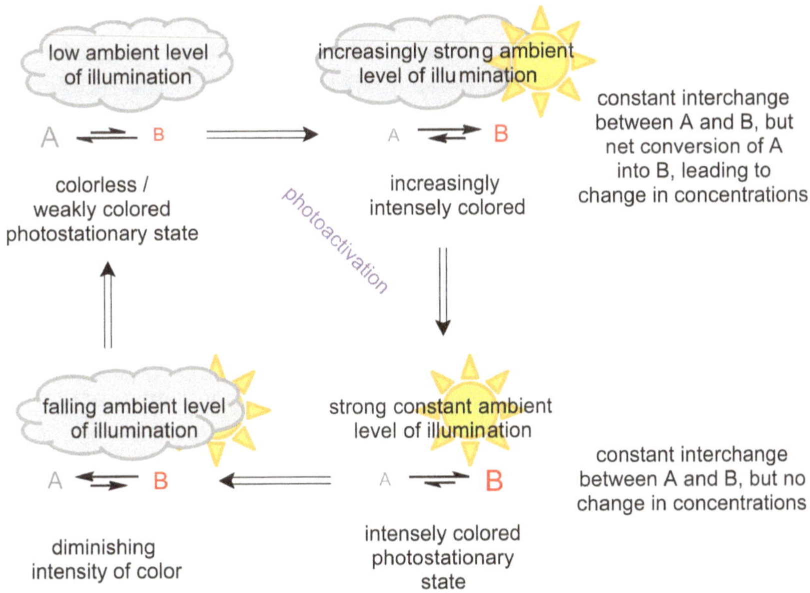

Note the type of arrows employed in Box 54.4: in T-type systems, switching between colored and colorless forms of dyes occurs constantly. Do not fall into the trap of thinking that no switching of dye occurs once a photostationary state has been reached.

The equilibrium is dynamic: while the concentrations of colored and colorless species remain constant, interconversion continues. The rate of photoactivation of dye is balanced by the rate of reversion to the dye's colorless state. When these rates are high, i.e. under irradiation by an intense UV light source, the impact of irreversible photochemical side reactions on the longevity of a dye can become very pronounced. Given the constant photoswitching that takes place, side-reactions with rates that are tiny compared to that of the desired reversible phototransformation become significant. (By way of illustration, consider the model system in which there is just one side-reaction upon photoactivation of "A" and none in the thermal or photochemical reversion of "B" to "A". Even if the quantum yield of the side-reaction is as low as 0.001, only 63% of the initial amount of "A" remains after one thousand cycles [11].) The concomitant irreversible formation of non-photochromic photoproducts, which may be colored or colored, is known as 'fatigue' because it leads to a reduction – and eventually complete loss – of photocoloration intensity. Fatigue is dependent on cumulative total exposure to light – it is likely to be greater in a photochromic system that has been irradiated for long periods on a few occasions in comparison to a scenario in which the system is switched multiple times but only by photoactivation for very short durations. Many academic papers show graphs in which photochromic dyes are cycled between colored and colorless states a few tens of times by brief interludes of exposure to radiation with little or no loss in photointensity each time: often their authors claim photostability is therefore good. However, these same systems actually fall apart readily under constant illumination by UV sources of only moderate intensity for a few hours, and in industrial terms show very high fatigue. One should thus be careful in adjudging robustness and thereby suitability for a particular application. While the kind of system just mentioned is not useful for typical current commercial outlets, they remain of interest for some of the potential uses mentioned later in this article which do not involve long periods of exposure to intense light sources.

Temperature has a significant influence over the position of a photostationary state: industrial T-type dyes work less well as temperature increases, because the rate of thermal fading rises. Consequently, in warmer environments, the depth of photocoloration produced by a particular intensity of sunlight falls. The phrase "typical ambient temperatures" in Box 54.3 as applied to the heliochromism of commercial T-type colorants is ~ 10–30°C. Their performance noticeably drops away in hot climates. In contrast, they develop especially strong photocoloration in cold regions, which tends to be slower to fade, owing to retardation of the thermal back reaction. Photochromic systems which show relative temperature independence thus offer a commercial advantage: one kind of T-type dye, the naphthopyran class [12], rose to prominence for ophthalmic lens production in part because its members minimize the likelihood of consumers noticing variations in performance. However, even in the most important T-type classes, such temperature dependence cannot be completely avoided.

The phrase “inefficient photobleaching” in Box 54.3 refers to a lack of P-type character rather than a tendency to be permanently destroyed, although as we shall see, the robustness of commercial photochromic dyes remains problematic. Industrial T-type colorants ideally display only very weak P-type behavior – if any at all – so that the visible wavelengths of sunlight do not tend to counteract the effect of the incident near-UV radiation and thus decrease the intensity of photocoloration.

It must be emphasized that the flavor of helichromism exhibited by the most commercially important T-type colorants is a transition from an essentially colorless state to one of intense color, dependent upon incident unfiltered sunlight intensity. The photocoloration is eventually lost upon complete removal of the irradiation. Such behavior is a subset of ‘positive photochromism’ (see Box 54.5).

Box 54.5

Positive photochromism	*Behavior in which a system’s color shifts bathochromically (i.e. its wavelengths of absorption lengthen) in a reversible manner upon irradiation with light. While this phenomenon might involve a colored to colored transition, industrial T-type dyes exclusively respond with shifts of their longest wavelength absorption bands from the UV into the visible region, leading to the appearance of color.*
Negative photochromism	*This phenomenon is the opposite of positive photochromism: the term does not signify the absence of a photochromic effect! Also referred to as ‘inverse photochromism’* [4]*, it has not been widely exploited commercially and features more rarely in the academic literature, but numerous examples are known* [13, 14]*. A T-type dye is said to exhibit ‘reversible photobleaching’ when irradiation with light leads to its coloration disappearing, and its original colored state returns by means of a thermal back reaction upon cessation of exposure.*

Industrially important photochromic dyes owe their usefulness to several key characteristics in addition to heliochromism, all of which are geared towards their application in ophthalmic lenses. These features translate to desired properties of the lens system:

(i) high transmittance in an unactivated state so that a lens has low ‘residual’ or ‘base’ color as a consequence of dyes not absorbing appreciable visible wavelengths before exposure to sunlight.

(ii) longevity of photochromism – even those industrial photochromic dyes which are the fruits of much research and development work are orders of magnitude less robust towards permanent photodegradation than conventional colorants. Through inclusion of additives to enhance dye photostability, dye lifetime can be extended so that performance drops off slowly enough for lenses to retain acceptably pronounced photochromism for a duration of around two years of everyday wear.

(iii) compatibility with the medium in which the dyes photocolor. Industrial T-type dyes only give a visible photoresponse when in solution – not as solids. They must therefore possess good solubility in their intended medium. (Note that this is not necessarily true of non-commercialized other kinds of photochromic colorant.)

The environment of a photochromic dye strongly influences the above characteristics: this aspect will be covered in more detail later in this article.

Industrial photochromism thus takes the form of an even narrower definition [15] to those already given. Of paramount importance to an industrial photochromic system is cost effectiveness – commercial success is of course dependent upon the economics being palatable to the market otherwise fulfilment of technical criteria counts for nothing. As well as meeting customer expectations concerning performance, longevity and cost, industrial photochromic products must not present unacceptable health and safety risks to users, nor infringe the claims of in-force granted patents held by third parties. A shortfall in any of these criteria makes the difference between a system exhibiting industrial photochromism and merely being photochromic! Box 54.6 captures the salient points in the form of a definition.

Box 54.6

Industrial Photochromism *Reversible coloration induced in a dye, which:*
- *(i) rapidly becomes intense on exposure to a source of strong UV light and quickly disappears when exposure to the source ceases;*
- *(ii) can be repeated over many cycles during a time period that is acceptable to the expectations of consumers;*
- *(iii) is cost-effective and safe to implement* [15]

Figure 54.1 depicts an attempt to show the relationship between the phenomena defined thus far. The representation is somewhat simplistic given previous mention that photochromic colorants do not always show pure T- or P-type behavior. Note that the size of the areas shown in the Figure bear no relation to their commercial importance, nor other measures such as the amount of current research or known instances of dyes. It is fair, however, to state that T-type classes and instances of positive photochromic dyes known to science, respectively, outnumber P-type classes and colorants exhibiting negative photochromism.

As indicated in Figure 54.1, industrial photochromism largely encompasses positive T-type photochromism, i.e. heliochromism, but does spread into P-type behavior for specialist applications. As will be discussed in Section 54.4.2, a great deal of R&D effort continues to be expended on attempts to expand practical use of functional photochromism further into the P-type region, which if commercially successful would lead to big changes to the representation in Figure 54.1. The exploitation of negative T-type photochromism remains of interest in certain fields, e.g. photopharmacology (see Section 54.4.2); again, developments here would require considerable re-drawing of Figure 54.1 as a map of photochromism use. The next section briefly explains how

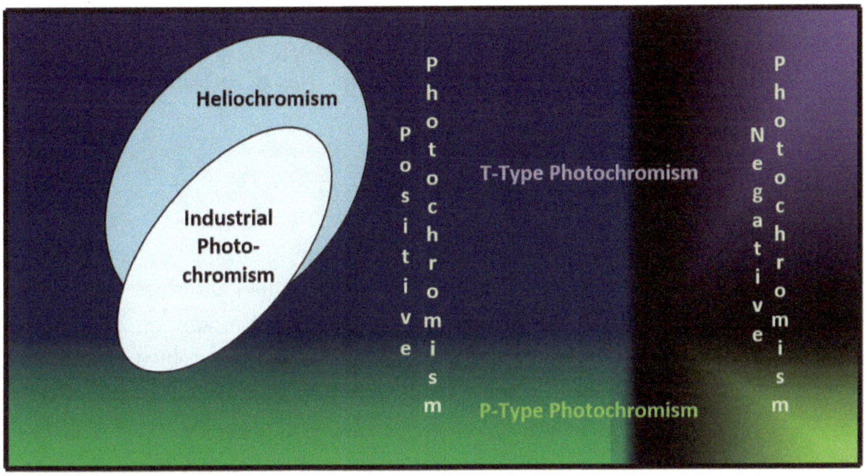

Figure 54.1: A representation of relationship between the behavior of industrially important photochromic dyes and types of photochromic phenomena.

photochromic dyes, now more commonly labelled with terms such as 'molecular photoswitches' in the academic literature, evolved to leave their status as scientific curiosities behind by enjoying some commercial success, and also sit poised to take real-world use of photochromism in significant new directions.

54.2 A brief history of industrial photochromism and photochromic colorants

Reports of organic materials that reversibly altered in color first appeared in the scientific literature in the second half of the nineteenth century at a time when chemists were only starting to make sense of the molecular structure of carbon-based compounds. Over a century passed before photochromic systems were exploited in a marketable application to significant commercial advantage.

Numerous sources (e.g. [3,4,5, 16]) incorrectly declare that the first scientific literature report of a phenomenon resembling photochromism was made in 1867 with an observation relating to tetracene. The content of this French language paper by the chemist Carl Julius Fritzsche [17] is not consistent with their claims. As succinctly summarized in [18], the article instead concerns the thermally reversible photodimerization by sunlight of anthracene [19], isolated during his studies on the components of coal tar. A German language communication authored by Fritzsche and published in the same year also reports on this effect of sunshine on anthracene, which led to precipitation of a less soluble substance that reverted to anthracene on melting [20]. The occurrence of photodimerization to dianthracene was confirmed by molecular

weight determination a quarter of a century later [21]. The transformation falls under the definition of photochromism given in Box 54.1 because:
(i) the absorption spectrum and fluorescence properties of anthracene are different from those of dianthracene;
(ii) sunlight initiates the conversion to the latter, and
(iii) heating of the photoproduct regenerates anthracene.

An earlier paper much less remembered by the photochromism literature, written by the same author in 1865 and published the next year [22], also describes a reversible phenomenon initiated by light. An orange-yellow solution of a component, labelled "chrysogen" by Fritzsche, of a heavy oil fraction from coal tar photobleached in sunlight, affording colorless crystals. When melted, this material became orange-yellow. Nearly seventy years later, a chromatographic investigation identified chrysogen from technical-grade anthracene as tetracene [23]. However, since the studies did not share the same hydrocarbon source and Fritzsche's data is equivocal by modern standards, it is not clear whether tetracene alone was responsible for the observation, nor what kind of photochemistry lay behind it. A speculative explanation put forward invokes photoinduced *endo*-peroxide formation of colored polycyclic aromatic hydrocarbons [18]. The lack of unambiguous characterization of the species involved in this reversible color change phenomenon renders Fritzsche's 1867 accounts of the behavior of anthracene – not tetracene contrary to what is written in much photochemical literature – as the earliest confirmed scientific reporting of photochromism exhibited by an organic compound.

However, the anthracene–dianthracene photochromic cycle described above sits far removed from the heliochromism of commercial photochromic dyes owing to:
(i) the absence of a significant change of color perceptible to the naked eye;
(ii) irradiation bringing about a phase change;
(iii) the need for exposure to temperatures substantially above that experienced by the system during irradiation (i.e. $>150°C$ higher), or short wavelength UV (<300 nm), to drive reversion at a significant rate.

A phenomenon more akin to industrial photochromism was described in 1876: the potassium salt of dinitroethane in solid form was discovered to redden in daylight and return to its original yellow color after being kept in the dark [24]. This material has been hailed as the first known example of a photochromic organic compound [8].

Perhaps as a consequence of misapprehension of a statement in the first paragraph of [25] [see Box 54.7(a)], numerous publications state that the English word 'photochromism' was first coined in 1950, citing a research paper published in French [26], which actually featured the term 'photochromie' in its title. Nevertheless, the term 'photochromism' started to appear with increasing frequency in English-language scientific literature during the early 1950s at a time of burgeoning understanding of the phenomenon. It eventually displaced the now-obsolete term 'phototropy' during the next two decades. This latter term is presently restricted to meaning 'growth of an organism in response to stimulation with light'; phototropy likely acquired the extra meaning of light-induced color change as a consequence of the German language term 'Phototropie' being proposed in 1899 to designate reversible changes of state encompassing color [27] [see Box 54.7(b)]. While 'Thermochromie' was first used in German over a century ago to refer to thermochromism [28], the precise definition of the word 'Photochromie' was still being settled during the mid-1960s [29] [see Box 54.7(c)] and is now the general German language term applied to the definition given in Box 54.1.

Reports occasionally surface in the academic literature, for example [18], [30], and [31], telling of a photochromic system employed by the ancient Greek army of Alexander the Great to time military action. These unsubstantiated accounts must not be taken seriously because they all originate from a play on words that appeared in a 1961 article of an American weekly magazine [32]. It tells a joke whose premise is that the ruler equipped his soldiers with " ... chemically treated cloth worn on the left forearm. Under the heat of the sun, the cloth changed colors each hour and provided Macedonian troopers with the world's first wristwatch ... " and which concludes with the punchline of "Among historians, it is known as 'Alexander's Rag Timeband'." – a homophonous reference to the title of Irving Berlin's much covered song and musical film "Alexander's Ragtime Band".

Inspired by the discovery of photochromic behavior in members of the now much-researched spiropyran (SP) dye class [33], EH Frei suggested in the mid-1950s [34] that photochromism could be exploited in producing high density recording of data. The proposal heralded considerable efforts within academia, industry and the military to make use of the phenomenon [35]. These attempts were ultimately unsuccessful owing to unsatisfactory technical properties of the dyes, usually a lack of photostability [35, 36]. Consequently, the first successful industrialization of photochromism to reach the mass market took the form of light-responsive eyewear reliant on inorganic materials rather than dyes: silver halide in mineral glass [37]. These photochromic lenses were launched in the 1960s. Their success fueled the desire to develop more lightweight and robust photosensitive lenses based on polymers rather than glass. However the former media cannot host silver halide-based systems: organic rather than inorganic materials, specifically photochromic dyes, were required for polymeric media.

Box 54.7

(a) *The phenomenon of reversible formation of colored modifications of compounds related to dianthrone by ultraviolet irradiation at low temperatures was first reported by Y. Hirschberg in 1950* [Y. Hirshberg, Compt. rend. **231**, 903 (1950)] *and named photochromism.* (Hirshberg and Fischer, 1955 [25])

(b) ...*Man kennt nun Stoffe, welche unter der Einwirkung der Lichtstrahlen eine Zustandsänderung erleiden, welche jedenfalls von der Polymerisation, wie sie etwa beim Übergang des gelben in roten Phosphor, des Aldehyds in Paraldehyd u. s. w. erfolgt, völlig verschieden ist, und welche auf Umlagerung der Atome im Molekül nicht beruht. Diese Lichtwirkungen sind dadurch charakterisiert, dass die Zustandsänderung nach der Belichtung schneller oder langsamer wieder verschwindet. Dahin gehört z. B. die Phosphorescenz, die Erscheinung, dass Selen nach der Belichtung seine Leitfähigkeit für Elektrizität ändert und, wenn wir den Begriff der „Lichtwirkung“ weiter fassen, vielleicht auch die Färbung gewisser Krystalle durch Kathodenstrahlen. Für solche Zustandsänderungen durch Lichtwirkung schlage ich die Bezeichnung „Phototropie“ vor. Dazu werde ich durch die Entdeckung einer eigenartigen Lichtwirkung veranlasst, welche im folgenden beschrieben werden soll…* (Marckwald, 1899 [27])

(Translation: "…We now know substances which under the action of light rays undergo a change of state, which is completely different from polymerization, such as conversion of yellow into red phosphorus, aldehyde into paraldehyde, and so on, and which are not based on the rearrangement of atoms in molecules. These light effects are characterized by the fact that the change in state disappears more quickly or more slowly after exposure. To these belong, for example, phosphorescence, the phenomenon that selenium changes its electrical conductivity after exposure and, if we take the term "light effect" more broadly, perhaps also the coloration of certain crystals by cathode rays [electron beams]. For such changes of state due to the effect of light I propose the term 'phototropy'. I am prompted to do this by the discovery of a peculiar light-driven effect [reversible color changes caused by light], which will be described in the following…")

(c) *Der Begriff Phototropie [**] kann nach unseren Kenntnissen wie folgt definiert werden: Unter Phototropie versteht man durch Lichtabsorption erzeugte reversible Ubergange in quantenmechanisch stabile, thermodynamisch instabile Zustande.* (Luck and Sand, 1964 [29])

***In der Biologie werden unter „Phototropismus“ Wachstumsbewegungen unter Lichteinfluss verstanden. Es ware daher guenstiger – wie im deutschen Sprachgebrauch oft ueblich – die reversiblen photochemischen Vorgange als Photochromie zu bezeichnen. In der vorliegenden Darstellung schliessen wir uns der angelsaechsischen Bezeichnung „Phototropie“ an. Phototropie und Photochromie werden gleichrangig nebeneinander gebraucht. Man konnte daran denken, „Phototropie" als Oberbegriff zu verwenden und „Photochromie" als Unterbegriff fur Falle sichtbaren Farbwechsels zu reservieren.*

(Translation: "To our knowledge, the term "phototropy" ["Phototropie"] [**] can be defined as follows: Phototropy is understood to be reversible transitions, produced by the absorption of light, into quantum-mechanically stable, thermodynamically unstable states.

**In biology, "phototropism" ['Phototropismus'] refers to the movement of growth under the influence of light. It would therefore be more favorable – as is often common in German usage – to designate reversible photochemical processes as photochromism ['Photochromie']. In the present account we associate it with the Anglo-Saxon term "phototropy" ['Phototropie']. Phototropy and photochromism are used alongside one another on an equal footing. One could think of using "phototropy" as a "generic term and reserving "photochromism" as a sub-term for cases of visible color change.")

Organic colorants with the requisite color, kinetics and longevity to generate sustained market pull were not developed until the late 1980s and early 1990s. Their industrialization was the culmination of efforts to translate the recognized potential of photochromic organic compounds into a commercial reality. This development was made possible by exploiting the discovery that SINO dyes could furnish reversible coloration with greater robustness than hitherto known types (e.g. SPs) [9] and the realization that members of the previously overlooked naphthopyran class could be designed to produce suitable heliochromism [12]. Other important aspects included tailoring the polymer medium and dye application to get the best out of the colorants as well as finding ways to stabilize dyes with additives to prolong the lifetime of the photochromic effect to a commercially acceptable duration; Section 54.4 covers this aspect in a little more depth. Enhancements to the technology now mean that the manufacture and sale of ophthalmic photochromic lenses relying on T-type dyes has grown into a global business worth a combined annual turnover of hundreds of millions of dollars.

The invention of strongly P-type photochromic dyes during the 1980s and further refinement in their properties sparked a great renewal of interest in embedding photochromism into a wide range of nascent technologies that required controlled two-way photoswitching to make them work. Many of them have the potential to disrupt and transform certain sectors, such as communications, information technology and computing, as well as extend others, for example in medicine, imaging, material science and nanotechnology. Section 54.4 furnishes more detail. The field has attracted very able minds and spawned ingenious approaches, with some work being Nobel prize-winning [38]. Despite the past few decades of endeavor, technological breakthroughs with P-type dyes have yet to reach the mainstream. Should any of these of these avenues become successful, then it is likely they would quickly supplant the industrial dominance of T-type colorants. At least in the short-term, the commercial picture is one of continued importance of photochromic units exhibiting thermal bleaching rather than fully photoswitched color changes. The following section gives a brief tour of dye classes of each camp.

54.3 Classes of photochromic dyes

We have already met important means of classifying photochromic dyes: the nature of reversion from the metastable photoproduct (photochemical/thermal, i.e. P-/T-type) and the color change (positive and negative photochromism). Plenty of other distinctions exist by which colorants can be categorized, such as grouping them by the kinds of mechanism behind their light responsiveness, e.g. geometric isomerism, tautomerism, dissociation, etc. However, this section will look at the gamut of known dyes from the perspective of chemical type since this approach allows generalizations to be made about the collective properties of individual classes of colorants in a reasonable manner. Given the article's emphasis on those kinds of photochromic dye which hold greatest industrial utility, the discussion will start with them, but move on to others that have attracted considerable attention and which may go on to be of importance. (It must be remembered that the literature on photochromism is vast, spanning roughly a century and a half. During this time, the numbers of researchers working with photochromic systems has increased enormously. Consequently, even in the case of some of the least explored types of photochromic species, considerable numbers of publications reporting investigatory studies exist.)

The most commercially important families exhibit T-type behavior. Table 54.1 lists in alphabetical order just a few of them: those tabulated carry the greatest historical, industrial or potential significance. Of the numerous kinds which thermally revert from metastable states, just three are currently of much note from a commercial perspective: in order of rapidly increasing importance, these are the SP [33], SO [9] and naphthopyran [12] classes of dyes. For their color changes, all rely on isomerization by ring-opening and -closure via pericyclic reactions following heterolytic cleavage of a σ-bond. Much has been learnt about how to tailor dye structure within these dye classes to produce colorants exhibiting industrial photochromism. Since each of the aforementioned sets of dyes forms the subject of its own chapter, nothing further will be said of them in this section.

Table 54.1: Some examples of photochromic dye families with predominantly T-type character.

Chemical class	Mechanism	Example
Anil	Tautomerism/ hydrogen transfer	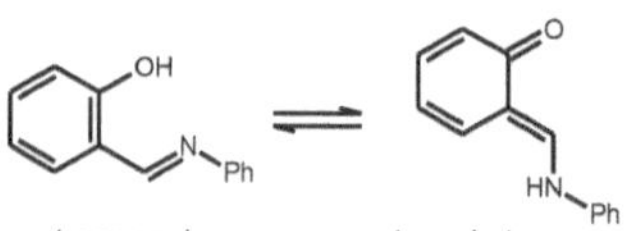 *trans*-enol ⇌ *trans*-keto
Azo/diazene	Geometric isomerization	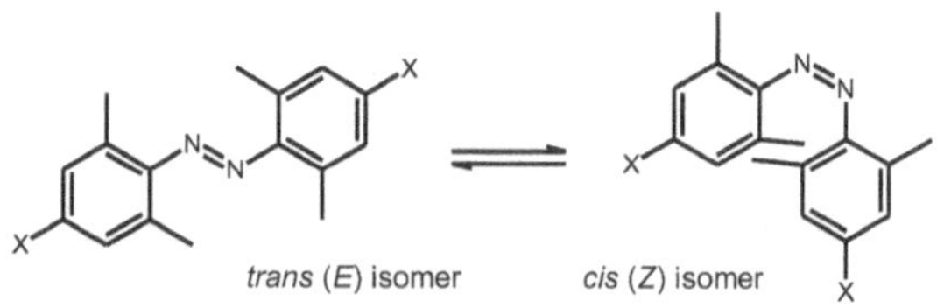 *trans* (*E*) isomer ⇌ *cis* (*Z*) isomer
Dihydroindolizine	Electrocyclization	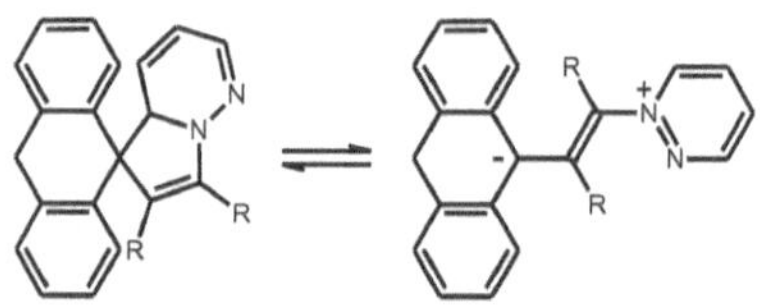
Hexaarylbisimidazole (HABI)	Photocleavage / dissociation	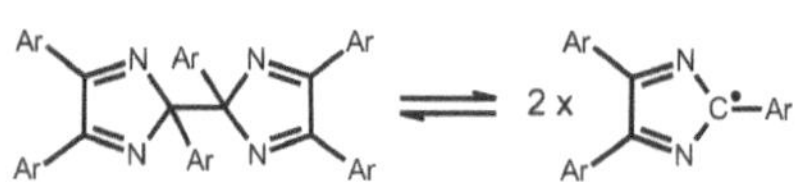 lophyl radical
Naphthopyran	Pericyclic reaction / ring-opening	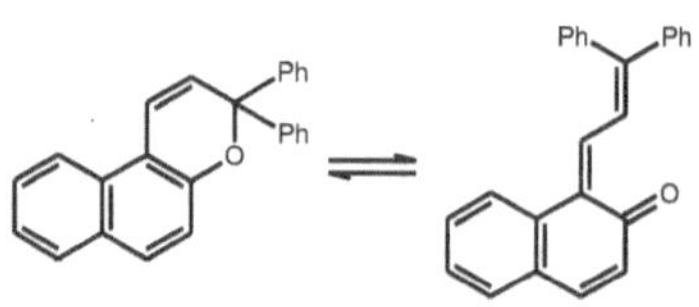

Table 54.1 (continued)

Chemical class	Mechanism	Example
Spirooxazine (SO) Spiropyran (SP)	Pericyclic reaction / ring-opening	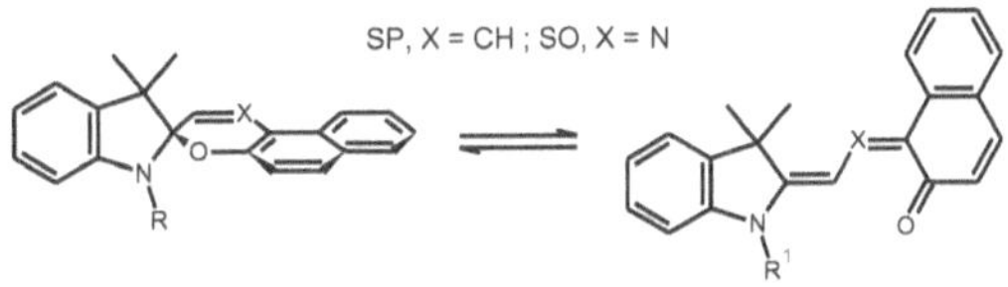
Donor–acceptor Stenhouse adduct (DASA)	Geometric isomerization & 4π-electro-cyclization	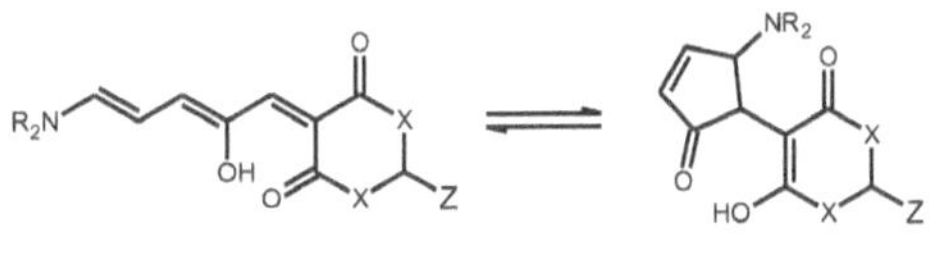 *triene*-DASA *cyclo*-DASA
Triarylmethane	Photoionization/ dissociation	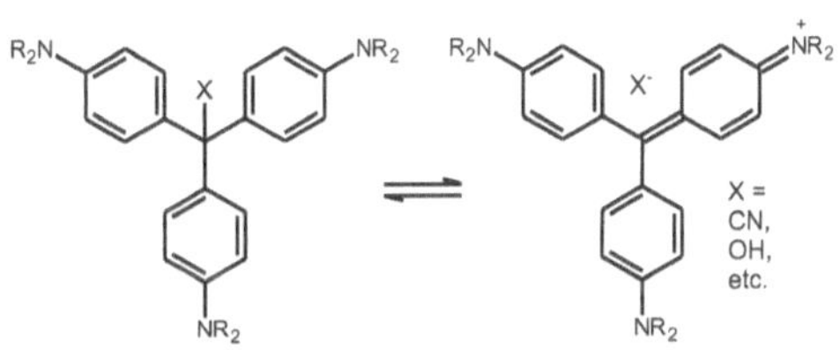

A variety of alternative mechanisms behind the photochromism exhibited by organic molecules is known. Table 54.1 exemplifies a few in greatly simplified form. Although much has been published concerning these other systems, none has seen extended commercial use, for example, the 1,5-electrocyclization of dihydroindolizines [39, 40], and the photo-induced tautomerism of anils [16, 41, 42]. In the latter instance, a hydrogen shift from a *trans*-enol to a *cis*-keto tautomer precedes isomerization to an intensely colored *trans*-keto form that thermally reverts to the enol [43].

Thermally reversible photodissociation is another mode of T-type change: homolytic or heterolytic bond cleavage occurs, for example, in hexaarylbisimidazoles (HABI) [44, 45] and substituted triarylmethanes [46], respectively, generating strongly colored lophyl radicals or triarylmethine cations. While HABIs see niche commercial use as photoinitators in free radical polymerization [47], most T-type families do not enjoy widespread application as industrial colorants because their photochromism, stability and/or economics fall short of requirements.

Note that some of the T-type classes shown in Table 54.1 also exhibit an element of P-type behavior to varying degrees. This aspect is undesirable in the case of lens grade naphthopyrans [12], since it leads to unwanted persistent residual photocoloration following the usual thermal decay after a drop in ambient light level; research aimed at eliminating this problem continues. However, the mild P-type character of SP dyes is exploited in many explorations of their functional uses to expedite reversion from photoactivated states [33]. The chromism of certain azobenzenoid dyes was originally viewed as a nuisance by dyers, and thus azo colorants (see Table 54.1) were designed to obviate such shifts in shade. These same dye users would likely regard as ironic the considerable research effort being devoted to the manipulation of structure to enhance the P-type character of azobenzene (AB) chromophores. For example, tetrasubstitution *ortho* to the azo bridge [48], as illustrated by the azo structure depicted in Table 54.1, slows the thermal decay of *E*-isomers, as does replacing one of the phenyl rings with a pyrazole system [49], creating dyes with greater photocontrol in both directions of isomerization. This kind of work is at the heart of many of the substantial strands of endeavor in progress that are seeking to exploit photochromism through the creation of light-responsive functional materials.

The appeal of azobenzenoid units as photochromic materials lies in: (i) their synthetic accessibility and robustness; (ii) the efficiency of photoisomerization and its relative freedom from competing photochemical side-reactions; (iii) the large changes in geometry and electronic properties that occur upon isomerization. The chromophore length shrinks by roughly half, while dipole moment increases significantly, upon transitioning from AB *E*-isomer to *Z*-isomer. For an excellent introductory overview of AB-based photoswitching, see [50]. Given that azo chemistry dates back to near the dawn of the synthetic dye (and chemical) industry [51], it is fascinating that the old azo dog is being taught to perform new tricks. Some will be briefly outlined in Section 54.4.

Today's commercial applications rely on use of dyes in their capacity as photoresponsive compounds of relatively low molecular mass. However, much research work to develop novel technologies involves linking photochromic dye motifs with functional compounds or macromolecules to enable photomodulation of their properties. A large proportion of this effort is devoted to colorants displaying either pure or substantially P-type photochromism rather than T-type colorants because the switching of the former is in principle more controllable. Colorants that exhibit purely P-type behavior are rare beasts and have yet to make the leap to the substantial commercial importance achieved by their T-type counterparts. The only two dye families that act as sources of pure P-type switches [11] are the dihetarylethene and fulgide-related classes (see Table 54.2). Consequently, they have attracted much attention during the past few decades. Since both of these classes receive their own chapters [52, 53], attention will be turned to the other family listed in Table 54.2: indigoid colorants.

While human use of indigo stretches back into antiquity [1] and the related hemiindigo (HI; see Table 54.2) has been known for over a century [54], the latter's photochromism was only explored just before the turn of the millennium [55]. The analogous hemithioindigo dyes (HTI; see Table 54.2) gained attention as functional photoswitches only recently too [56]. Both HI and HTI depend on geometric isomerization for their color changes, but as with azobenzenoid units, it is the switching of their other properties through action of light that drives interest in them. Features which have proved attractive include the controlled manner by which HTI isomerization can be effected in

Table 54.2: Some examples of photochromic dye families with predominantly P-type character.

Chemical class	Mechanism	Example
Dihetarylethene	6π-electro-cyclization	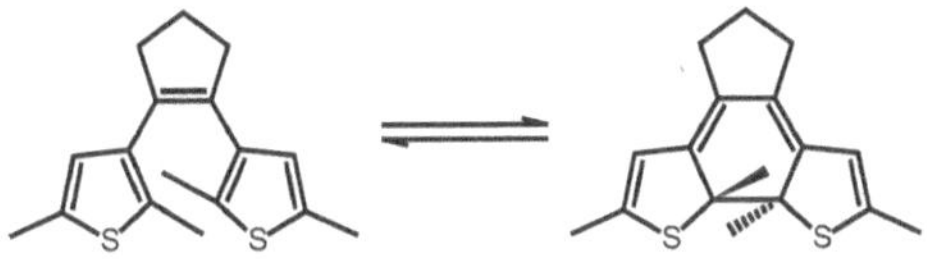
Fulgide Fulgimide	6π-electro-cyclization	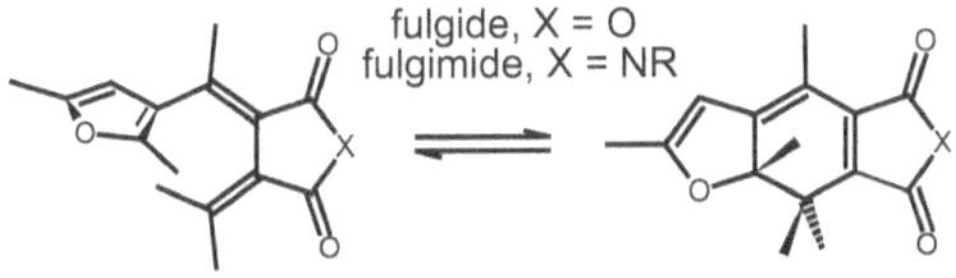
Hemiindigo (HI) Hemithioindigo (HTI)	Geometric isomerization	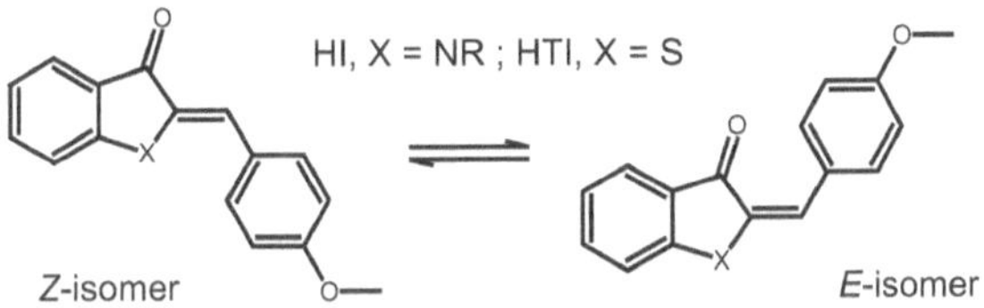

either direction with visible light only (no UV needed), causing relatively little fatigue. The absorption bands of the *Z*- and metastable *E*-isomers of simple derivatives, e.g. the HTI example of Table 54.2, lie sufficiently far apart (λ_{max} values differ typically by 20–30 nm) that blue and green light, respectively, drives *Z*→*E* and *E*→*Z* isomerization. (HTIs also exhibit T-type character in that *E* *Z* isomerization may be driven thermally.)

Another class which has come to the fore is that of the negatively photochromic 'donor–acceptor Stenhouse adducts' (DASA) [57, 58] (see Table 54.1). Their rise to prominence since they were reported in 2014 [59] stems not just from the visible light activation of T-type switching, but also the large differences in geometry, physicochemical properties and absorbance exhibited between 'triene' and 'cyclo' isomers. These hemioxonol dyes [60] consist of an electron donor (amine group) linked to a heterocycle (acceptor) by a hydroxytriene bridge. Simple DASA examples are violet to blue and cyclize upon absorption of white light to colorless cyclopentenones. These relatively hydrophilic cyclo forms thermally revert to the intensely colored hydrophobic trienes on cessation of irradiation. Resistance to fatigue of the most robust examples is claimed to be as good as that of members of other classes.

Photochromic dyes are manufactured by speciality organics producers either for offer to the open market or on an exclusive confidential basis for particular end-users. Gram quantities of examples of the more major colorant classes are available from suppliers of fine chemicals to laboratories [61]. Whereas commercially significant textile dyes are produced in the order of tens or hundreds of tonnes per year, the corresponding unit for 'high-volume' T-type dyes is the kilogram! Usage is orders of magnitude lower, while cost is orders of magnitude higher: pricing tends to be framed in units of dollars per gram rather than dollars per kilogram as per conventional non-photochromic colorants [62]. Part of this high cost is dictated by the expense of low-volume manufacture, although the necessity of multi-step syntheses from commercially available raw materials is a contributory factor in many cases.

The array of photochromic dye classes spans the very old to the very new. Several are drawn from chromophoric types that have been known for more than fifty years, e.g. azo, indigoid, diarylethene, fulgide and naphthopyran. However, they only came into their own as photochromic colorants of interest to technologists during the 1980s or much later as a consequence of breakthroughs in manipulation of photochromism though dye molecule design. Despite the palette of photoresponsive colorants available, their limitations and the specialized requirements of potential uses means that the search for new dyes of improved or more suitable behaviors goes on [63].

54.4 Applications of photochromic dyes

Care is needed when working with photochromic dyes to create light-responsive products of the desired performance and longevity. Numerous aspects must be borne in mind when trying to meet these targets. The dye is just one piece of the jigsaw puzzle. Detail is king. All of this holds true for the most important commercial outlet of photochromism: ophthalmic lenses. Well over 90% of the commercial value of photochromic dye manufacture lies in the production of T-type colorants for this sector. This section will therefore continue with general considerations for the application of such dyes, before moving on to cover specifics of their incorporation into lenses. It closes with a look at the potential uses of P-type colorants that are under intense current scrutiny.

54.4.1 Applications of T-type dyes

Naphthopyran colorants predominate usage in lenses [15, 62]. This application dwarfs industrial consumption within other sectors in which SO dyes, and to a lesser extent SPs, have an increased share of the market. In comparison to non-photochromic industrial colorants, each of these commercial T-type dyes tends to be less chemically and photochemically robust. Their delicate nature, either in part or whole, means that photochromic colorants cannot often be applied by the usual coloration techniques employed for many non-chromic dyes and pigments. This aspect is often not appreciated by those unused to handling photochromic dyes. To get the best performance out of them, they must be incorporated into suitable media using appropriate means. The dye, however, is not the whole story. The choice of substrate is crucial to achieving the targeted photochromism: if a medium excessively restricts the reversible structural changes in T-type molecules associated with photochromism, then the effect will be stifled. Selection of the application method can make all the difference too ... The colorant must be placed into a final state of solution (since commercial T-type dyes do not furnish visible color change in pure solid form) that provides a favourable environment without exposing it to conditions that degrade the dyes. In addition, measures must often be taken to enhance the useful working lifetime of the photochromic system by incorporation of additives or physical barriers; they are essential to achieving acceptable performance in certain applications, particularly ophthalmic lenses. When considering the responsiveness and resilience of photochromic effects, it is thus important not to consider just the dye, but the system as a whole. The next two subsections illustrate this point, as well as the others raised above, by looking at how T-type photochromism is put into practice industrially.

54.4.1.1 Non-lens applications

Photochromism is a striking effect. It has captured the imaginations of many. Consequently, a large number of uses for T-type colorants has been proposed in the commercial, academic and patent literature. Despite the investment sunk into development efforts, particularly in the latter half of the twentieth century [35, 36], most have not come to pass. In such cases, the unsurmountable obstacles may have included:

(i) *Unsatisfactory dye–substrate technical properties.* For example, all known photochromic dyes exhibit inadequate photostability for long-term application in architectural glazing and exterior coatings.
(ii) *Legislative barriers.* Depending on jurisdiction, regulatory restrictions may forbid cosmetic use and/or entail prohibitive expense by industry to demonstrate safety.
(iii) *Cost expectations.* Even in those applications calling for only very low inclusion percentages of photochromic dye, the expense of the colorants coupled with margin requirements fully or partly prevents the displacement, or adoption of, non-photochromic technologies in very price sensitive sectors, typified by marking of fuel and security printing.

Nevertheless, non-lens products created with T-type dyes that undergo eye-catching heliochromic changes have made it to market, albeit in low volumes, e.g. toys, inks, novelty plastic items, textile prints, etc. In the vast majority of cases, the colorants are applied in the manner of water-insoluble solvent dyes, thus ending up dissolved in hydrophobic liquids or polymers. Commercial T-type colorants also lack water solubility, exhibiting sufficient robustness under certain circumstances for them to be incorporated into thermoplastic polymeric materials by means of injection molding. Known as 'mass coloration', this technique involves the mixing of colorant as a neat powder or a masterbatch (i.e. plastic containing a high concentration of dye) with polymer and the combination heated to create a solution of dye in molten thermoplastic. The temperatures employed are dependent on the polymer, but the majority of commercial T-type colorants must not be exposed even briefly to temperatures of greater than 250°C to avoid dye decomposition, permanent discoloration and loss of photochromism. The dye must dissolve to give a solution because conventional T-type colorants do not exhibit observable photochromic effects as neat solids. The restrictive nature of their crystal lattices in terms of rigidity and low free volume inhibits the reversible changes in molecular structure associated with T-type dye photochromism from taking place. The end result must be a uniform mono-molecular distribution of colorant in the polymer matrix, i.e. a solid solution of dye throughout the whole mass of the polymer rather than a suspension of fine solid particles in the manner of a pigment. Just as the chemical and physical nature of the colorant dictate its photochromic properties, the rigidity and free volume of the polymer matrix strongly influence the behavior of the system in terms of the kinetics of the photochromism and the longevity of the effect [64]. Commercially important T-type colorants perform well in polyolefins because the hydrophobic matrices of these polymers consist of flexible chains that solubilize the dyes and do not suppress changes in molecular shape during the interconversion of colored and colorless forms.

Owing to the intense absorption of their activated forms, only relatively low loadings (0.05–0.1%w/w or less) of industrial dyes produce striking effects. A common mistake in attempting to remedy a weak or absent photochromic response is to increase greatly colorant concentration, which can be counter-productive. The excess dye may lead to a strong background color and limit the extent of changes. Instead, attention may need to be paid to other aspects of the system, e.g. the influence of the matrix.

One means of circumventing suppression of photochromism by the polymer medium that is employed industrially relies upon microencapsulation of a solution of dye in a high-boiling water-immiscible liquid. The procedure results in micron-scale spheres of composite consisting of photochromic solvent trapped within a shell of non-photochromic polymeric wall material, as described for example in [65]. These composites can be dispersed like pigment particles during mass coloration of polymers. Their photochromic response is akin to that of the dye dissolved in a solvent – it is independent of the polymer matrix surrounding the microcapsules. The approach also makes possible incorporation of water-insoluble T-type dyes into aqueous surface coatings, particularly inks, whilst preserving their photochromism. Aqueous dispersions of microencapsulate are commercially available which suit this purpose. Photochromic colorants also find use in solvent-based surface coatings where encapsulation is not required, for example, simple dissolution of dye in solvent-rich gloss varnish formulations for application to fingernails.

Photochromic dyes have been applied in commercial settings to textiles for many years, but such usage remains very small. This status is unlikely to change because their photostability is poor as numerous academic studies merely confirm [66, 67]. Photochromic textiles are only suited to markets where customer expectations for product lifetime are low. Screen printing is the usual mode of dye application, whereby a thermally curable liquid formulation containing photochromic colorant, typically in microencapsulated form, is coated onto the surface of a garment. Photochromic detail may also be woven into a textile by means of commercially available thread consisting of poly(alkene) fiber melt-spun after doping with photochromic dye, i.e. mass coloration through extrusion, rather than molding, of dye dissolved in molten polymer. Aqueous dyeing methods do not suit textile coloration with conventional T-type colorants. The dyes are not substantive to natural fibers and are insoluble in water. Lack of stability to aqueous exhaustion dyeing conditions, coupled typically with suppression by the polymeric matrix of photochromism, precludes successful application by this means to commercially important synthetic hydrophobic fibers, like polyester [68]. Despite various academic investigations into dye design, use of non-aqueous media, and novel methods of application [15], this sector will remain niche. Although numerous academic reports describe light-responsive textiles produced from T-type dyes as being "smart", "intelligent", or "chameleonic" (e.g [69].), it is arguable that labelling of their passive photochromism in this way is not appropriate. Use of current T-type technology to produce effects that are worthy of these descriptions and which address needs calling for genuinely controllable and adaptive manipulation of color (such as camouflage design in military uniforms) has long been fanciful, irrespective of photostability considerations.

54.4.1.2 Lens applications

Production of polymer-based ophthalmic lenses remain the primary outlet for photochromic dyes; a significant proportion (5–20%) of corrective spectacles prescribed in North America, Europe and Asia contain such colorants [70]. Successful introduction of the technology to market during the late 1980s and 1990s depended upon overcoming numerous technical challenges relating to dye design and lens construction [71]. While there may be some latitude in adjusting the polymerizable formulation from which the bulk of the lens is produced to optimize photochromic performance, options are restricted owing to demands on the optical properties (e.g. refractive index, clarity) and physical characteristics (e.g. impact resistance). Consequently, production technique has proved to be a key determinant of success in addition to the constitution of lens matrices: see Box 54.8 for a summary of methods.

Box 54.8

– *In mass* Useful only when the lens medium permits satisfactory photochromism, a commonly used technique is to dissolve dye in lens pre-polymer, e.g. acrylate mixtures, which is then cured thermally or by UV irradiation into a semi-finished lens. Although relatively simple and flexible, the technique consumes dye in greater amounts than other techniques, despite concentrations of only around 300–500ppm typically being employed, because the colorant is distributed throughout the lens and part of the lens body is removed during grinding to the finished lens.

- *Coating* A layer of photochromic resin solution is spin- or dip-coated onto the front or rear of lenses enabling the use of matrices for the lens body that would otherwise inhibit the photochromism of conventional T-type dyes, e.g. polycarbonate. Despite the concentration of colorant in the coating being relatively high in order to achieve serviceable photochromism, less colorant is consumed per lens because the coating is thin. The elevated concentration of dye, held in only a thin layer call for good formulations and appropriate dye design to minimize absorption by the dye in non-photoactivated form (i.e. 'base color') and maximize photostability.
- *Lamination* A photochromic film is bonded to the surface of a lens and then sandwiched behind another lens layer. The approach offers similar technical advantages to those associated with coating.
- *Imbibition* Originally proprietary to Transitions Optical [70], this process takes advantage of the environment of the polymer near its surface being more forgiving in its impact on photochromic performance. After application of dye by spin-coating, the lens is baked, which results in dye diffusing from the coating into the surface of the lens. The coating is washed off to leave the dye trapped as a ca. 150–200 μm thick photochromic layer.

The requirements of lens photochromism are exacting [15]. It must endure for at least two years of everyday wear with performance dropping away at a sufficiently low rate during this time for it to still meet consumer expectation. Minimizing fatigue sufficiently entails trial-and-error dye selection as well as the identification of effective formulations of stabilizing additives. The latter typically include UV absorbers to filter out damaging high energy radiation and thus protect dyes from photodegradation, but without impeding photocoloration by excessive removal of those UV wavelengths that efficiently lead to photoisomerization. Another commercialized means of improving resilience of dyes in polymeric matrices, as well as increasing rate of activation and decay without significant detriment to peak intensity, is to attach silicon- or carbon-based oligomeric chains to them [15].

Although established for several decades, the position of T-type dyes in this market is not necessarily secure in the long-term from rival technologies that do not depend on photochromic organic molecules. For example, proof of concept for neutral lens coloration originating from electrochromic materials has been demonstrated [72], potentially offering active and more responsive control of light transmission. Numerous factors will dictate whether the approach becomes a genuine threat to photochromic dye use, e.g. cost, development of sufficiently lightweight power source, fatigue resistance, etc.

54.4.2 Applications of P-type dyes

Table 54.3 lists the main current industrial outlets of photochromic dyes, already covered in Section 54.4.1, as well as some of their potential uses. The former depend largely on T-type dyes, By contrast, the latter applications tend to rely on photoswitching with predominant or complete P-type character. They are the focus of much research and development within the realms of information technology, optoelectronics, and nanotechnology, amongst others [56, 73,74,75,76,77] as exemplified by entries in Table 54.3. All of these applications exploit photochromic colorants as functional dyes [1]: their color is of little or no importance. It is the change in the system's spectroscopic properties, or the modulation of another of its characteristics (see Table 54.4), in response to irradiation that is key. (Such differences may serve as the basis for altering interactions with other materials, e.g. reversibly photomodulating the strength of adhesion to create 'debonding-on-demand' systems of commercial interest in areas like semiconductor manufacture and wound treatment [78].) The technical requirements of dyes demanded by such outlets are largely dissimilar to the set of criteria fulfilled in current commercial uses. Not least is the need for tightly controlled reversible switching between two (or more) states that do not decay. T-types are not suited to this kind of task owing to their thermal fading, hence the predominance of the P-type designation in Table 54.3 [79]. They act as molecular switches: these dyes permit controlled and reversible transitions between at least two states by means of a non-invasive non-contact stimulus, light, which can often be directed spatially and temporally with high resolution. A bistable switch consists of two states which occupy local energy minima. Typically one is stable and the other metastable, separated by an energy barrier that is essentially unsurmountable thermally at ambient system temperature. The two states of the material comprising the switch may differ significantly not only in color, but also – as shown in Table 54.4 – in terms of molecular geometry as well as other optical and electrical properties. The attributes tabulated number among those most heavily investigated. Examples of other switchable characteristics and related applications may be found in some of the chapters on individual dye classes [33, 52, 53].

Interest in the old idea of utilizing photochromism to store data at high densities lives on, for example [135], but using photoswitchable properties other than absorbance to record bits of data. Doing so enables the information to be read without resorting to irradiation by wavelengths of light that would otherwise disturb the state of the photochromic material (i.e. to read the data non-destructively). Although fluorescence, chirality and conductivity are highlighted in this regard in Table 54.3, numerous colorant classes and other switched properties have also been examined for this purpose over the decades. Nothing of commercial significance has yet to emerge from this effort.

Photomechanics constitutes another significant field of exploration involving photochromic colorants but in the capacity of creating materials that convert light

Table 54.3: Some current commercial and potential applications of photochromic dyes.

Sector	Photochromism type	Comment
ophthalmic	T	The first commercially significant mass-market application of photochromic dyes and still the most dominant one [15, 62].
thermoplastics	T	Restricted to niche applications (e.g. novelty products, toys, fishing line, etc.) in which dye is usually applied either neat or as microencapsulate by mass coloration to polymers such as polyalkylenes [15].
textile [66, 67, 69]	T	Poor robustness and cost of dyes severely limit usage; suited to mass coloration or screen printing only [15]
actinometry	P	Specialized technical application to monitor exposure to UV or light by measurement of coloration or decoloration, respectively [80]
security printing	T, P	Covert and overt marking for authentication through visual inspection or machine-readable response [81], for example, by means of photosensitive ink; cost-conscious secretive industry
windows [82]	T	Widespread commercialization in architectural and automotive glazing requires greater photostability than can be provided by current dye technologies; rival technologies like electrochromism show more promise [83,84,85]
agricultural film	T	For passive protection of crops from excessive sunlight; fatigue-resistance of current commercial dyes unlikely to be improved sufficiently for application to be realized
data storage	P	As binary (and even multi-state) memory elements at the heart of next generation storage systems that theoretically offer higher data densities than extant technologies, e.g. [86]
information technology	P	Light-controlled processing of electrical and optical signals, respectively, by e.g. optically-responsive organic field-effect transistors [87] and photomodulated liquid crystalline materials [88, 89].
photonics	P	In creation of optical equivalents to electronic components, such as switches and logic gates, which are essential building blocks for photonic circuity that offers speedier, higher bandwidth, and less power hungry data handling than conventional electronics [90,91,92].

Table 54.3 (continued)

Sector	Photochromism type	Comment
photomechanics	P	Relates to the construction of light-responsive materials for use as components in machinery at the macro [93, 94] and nano scale [63, 94], e.g. actuators to convert energy to motion [95, 96]
biomedicine [97]	P,T	Remote and precise control of bioactivity [98–100], e.g. photopharmacological chemotherapy [101–104], photoswitchable receptors [105], drug delivery [106–111], high resolution imaging [112, 113], non-invasive photoswitchable probes [114]

Table 54.4: Some property changes accompanying photochromic effects that are of interest in regard to example functional dye applications.

Reversible property	Example(s)
absorbance	visualizing regions of UV spectrum with naphthopyran and SO dyes [115]; resists for optical nanolithography with vacuum UV [116]
chirality [117, 118]	overcrowded alkenes as non-destructive readout memory elements [119] and unidirectional molecular rotors [120]
conductivity [121, 122]	non-destructive readout in memory [123] and nanometer-scale patterning of conductivity in polymer films for organic electronic devices [124] based on diarylethene photochromic units
fluorescence [125]	non-destructive readout in memory [126]; super-resolution microscopy [112]
geometry	light-powered macroscopic motor driven by belt constructed from photomobile azo-based polymer [93]; unidirectional molecular rotors derived from HTI chromophore [127]
refractive index [128]	diarylethene optical switch in photonic circuitry [129]
state of matter [77]	AB-based systems furnishing either light-triggered thermal release of stored energy through phase change [130] or photocontrollable adhesion [131]
wettability [132–134]	separation of materials; controlled transportation of liquids in microfluidic devices

energy to motion and thus mechanical work. Reversible changes in geometry at the molecular scale, arising through photoisomerization, act in concert to produce macroscopic effects [136]. Azo-based polymeric materials have been particularly intensively researched for this purpose [137]. Their incorporation in an appropriate manner into polymeric materials enables the large geometry change associated with interconversion between *Z*- and *E*-isomeric forms to be translated into bending and twisting motions at the macroscale for potential use as light-responsive actuators in devices [94, 96, 138]. Success requires the following: sufficient light penetration into the material; an environment that does not inhibit isomerization; light-triggered molecular scale transformation translating to a macroscopic effect [139]. Research directed at creating materials that move in a continuous motion through conversion of light energy to a mechanical response has already produced some amazing prototypes, for example light-driven belts of photochromic azo-based polymer [93], but with respect to industrial exploitation it remains in an embryonic state [140, 141]. Machines scaled to micro- and nano-dimensions that are capable of performing useful tasks and which take their instructions and power from light [142] lie in the realm of science fiction for now. The potential of photo-fuelled unidirectional molecular motors – the subject in part of Nobel Prize-winning work [38] – remains some distance away from practical application. Despite numerous academic demonstrations of concept, for example those based on HTI [127], considerable challenges on the road to commercializing such soft robotic technologies have yet to be surmounted: these include attaining satisfactory degrees of speed of actuation, controllability, and of course fatigue [143] so that new avenues become economic and technically viable, either in isolation or in competition with rival technologies.

An important theme of recent academic work is the creation of safer photochromism-based technologies through the development of chromophores that may be switched with visible or near infrared (NIR) light. It is of particular relevance in the promising, and intensely researched, field of photocontrolled biological systems [97]. UV irradiation, which is required for the photoactivation of most conventional photochromic dyes, presents hazards and poses technical difficulties [106]; its carcinogenicity is of particular concern in the context of potential biological applications of photochromic systems. In addition, UV light penetrates biological tissue to a much lesser extent than red and NIR light, whilst its propensity to trigger degradative side-reactions is much greater. Various strategies to obviate use of UV with traditional dye classes have been demonstrated, such as inclusion of additives to transfer energy to (and thereby activate) photoswitches. Others center on two-photon absorption [144] and/or take on the considerable challenge of designing colorants that may be switched with nothing other than visible or NIR wavelengths [139]. Some of the effort devoted to dye structure modification was touched on in Section 54.3: (i) T-type negative photochromic systems and (ii) P-type switches that color up and photobleach following the absorption of light frequencies of lower energies than those of UV. These approaches are of major interest to researchers attempting to put photochromism to work in biomedicine [145]. Photo-controllable guest–host systems [106–109] constitute one instance of how it might be employed therapeutically. Examples of this approach include micelles formed from polymer incorporating AB units: these supramolecular structures disassemble, releasing their biologically active payloads, only where commanded in living tissue by localized irradiation with light through a photo-triggered increase in hydrophilicity [106]. Long-wavelength responsive guest–host technologies [109, 111] are in principle an exciting tool of considerable benefit, but their development remains in its infancy.

Another new field of medicinal chemistry in which photochromism plays a central role is photopharmacology. At the heart of this novel discipline is the concept of localized reversible photoswitching of inactive pro-drugs into active drugs that deliver therapeutic effects only where they are needed [101–103]. The approach utilizes photochromic motifs to regulate biomolecule pharmacokinetic or phamacodynamic properties in a reversible manner. They may be built into small molecule analogues of known therapeutic agents, or used in the guise of ligands or linkers within macromolecules: photoisomerization, respectively, alters molecular geometry sufficiently to alter binding affinity with biomolecules or it distorts the shape of the macromolecule into which it is incorporated [104]. AB-based photochromic units with a mix of T- and P-type character have dominated photopharmacological work to date [104, 146]. The bending and straightening associated with isomerization about the azo bridge causes substantial changes in small molecule shape and distortions in macromolecule conformations [147]. The (non-photochromic) azo compound Prontosil proved to be a landmark discovery in antibiotics as an antibacterial pro-drug, earning a Nobel Prize: photopharmacology has the potential to revolutionize the treatment of various diseases, perhaps in the shape of azo derivatives also as pro-drugs but with reversible activity. (T-type antibiotics have even been proposed to combat the growing problem of increased microbial resistance [148]. A functional photochromic dye is irradiated to generate a metastable antibiotic form, which does its work and then thermally reverts to its inactive state. As the latter, it is not considered likely to enable bacteria to acquire resistance). Nothing worth doing is easy, which holds true for the development of photopharmaceuticals: the involvement of photochromic species complicates the already difficult job of identifying safe and effective medicinal compounds. When designing photochromic dyes for lens use, compromises must be made between aspects such as rate of thermal decay and intensity of photocoloration to achieve balanced but acceptable performance [15, 149]. The identification of successful photopharmaceutical candidates must resolve obstacles of even greater complexity. Their structures will not only have to gain regulatory approval through achieving the usual metrics of medicinal chemistry (target potency, selectivity, pharmokinetics, toxicology, etc.), but also possess appropriate photoswitching performance [146]. A further regulatory wrinkle may be demonstrating the safety of two species, not just one, given that use of azobenzenoid systems will entail exposure to both *Z*- and *E*-photoisomers during treatment (with other photochromic classes also facing an analogous situation [150]). Photopharmacology represents a new, rapidly expanding, and potentially very rewarding direction of research into photochromic colorants. Some candidates have shown promise in preclinical development [151]. Nevertheless, getting therapies based upon them to, and through, clinical human trials will be no mean feat. Their introduction remains years away [144].

Another potential medical use of photochromic AB derivatives is the treatment of degenerative retinal diseases [152]. Introduced to the eyes of animal models, they have been found to improve or restore vision by acting as photosensitizing surrogates for lost rods and cones, providing signals to retinal neurons. Example structures of some of the azobenzenoid small molecules examined so far are shown in Figure 54.2. Although exploiting these kinds of functional dye as sight-restorative drugs is a fascinating possibility, they are "still far from being used in the clinic" [152].

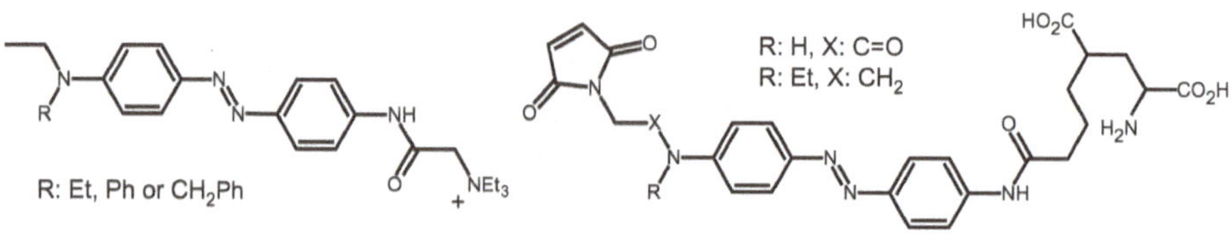

Figure 54.2: Some photochromic ABs that have been explored as therapeutic agents for treatment of degenerative retinal diseases.

The academic literature describing efforts to exploit photochromism in new technologies by building light-responsive units into materials is so enormous that there has only been space to scratch its surface. This article does not pretend to cover everything that has, and is, being attempted with photochromic dyes, but does provide an overview of what has been achieved commercially, and furnishes a glimpse of what might become a reality. For more detail, please see the chapters on individual photochromic dye classes. In the case of those colorant types not covered by them, the references cited in the previous few paragraphs, as well as Tables 54.3 and 54.4, provide examples of how they have been – and might yet be – employed.

54.5 Summary

Numerous types of photochromic dye are known. The most commercially important colorants possess positive T-type character, i.e. thermal decay of a metastable colored state generated upon absorption of UV light by dye in colorless form. A more precise label for their behavior is heliochromism, while the term industrial photochromism captures their traits from both a technical and commercial perspective. Ophthalmic lens manufacture is by far the largest outlet for such colorants: it has been appositely called the first "killer app" of photochromic dyes [153]. The search for the next one continues, despite the roots of many strands of research and development stretching back decades. Most focus on functional uses lying within fields such as optoelectronics, nanoengineering, and therapeutics. The dyes of principal interest within a large proportion of these areas exhibit predominantly P-type character. However, for certain applications, particularly those relating to biological systems, negative T-type colorants (as well as P-types switchable entirely by visible wavelengths of light) have recently attracted increasing attention as photoswitches, because they avoid drawbacks associated with irradiation by UV. Much recent research aimed at exploiting functional photochromism entails covalently bonding photochromic units to other materials rather than utilizing dyes as light-responsive small molecules. Many such avenues promise to be genuinely disruptive (and at least two of them have been connected with the work of scientists who were awarded the Nobel Prize [38, 113] within the last decade), yet most are still embryonic. Their commercialization would lead to existing T-type dyes being supplanted as the most industrially important organic photochromic substances, but realization of this prospect lies many years into the future.

References

1. Towns A. Colorants: general survey. Phys Sci Rev. 2019;4. DOI:10.1515/psr-2019-0008.
2. International union of pure and applied chemistry organic and biomolecular chemistry division. Glossary of terms used in photochemistry, 3rd edition (IUPAC Recommendations 2006). Pure Appl Chem. 2007, 79, 293.
3. International union of pure and applied chemistry organic chemistry division commission on photochemistry. Organic photochromism (IUPAC Technical Report). Pure Appl Chem. 2001, 71, 639.
4. Dürr H. In: Photochromism molecules and Systems. Dürr H, Bouas-Laurent H, editors. Amsterdam: Elsevier, 2003 Chapter 1 ("General Introduction") 1–14.
5. Samat A, Guglielmetti R. Chromogenic materials, photochromic. In: Kirk-Othmer enyclopedia of chemical technology. 5th ed. vol. 6, Wiley, 2004:587–606.
6. Kobayashi Y, Mutoh K, Abe J. Stepwise two-photon absorption processes utilizing photochromic reactions. J Photochem Photobiol C: Photochem Rev. 2018;34:2–28.
7. Heller HG, Oliver SN, Whittal J, Johncock W, Darcy PJ, Trundle C. (The Plessey Company plc). UK Patent GB2146327B 1986.
8. Heller HG. In: Miller LS, Mullin JB, editors. Electronic materials: from silicon to organics. New York: Springer, 1991 Chapter 31 ("Photochromics for the future") 471–83.
9. Towns A. Spirooxazine dyes. Phys Sci Rev. 2020;5. DOI:10.1515/psr-2020-0013.
10. Eppig T, Speck A, Gillner M, Nagengast D, Langenbucher A. Photochromic dynamics of ophthalmic lenses. Appl Opt. 2012;51:133.
11. Irie M. Molecular switches. Feringa BL, editor. Weinheim: Wiley-VCH, 2003 Chapter 2 ("Photoswitchable molecular systems based on diarylethenes") 37–62.
12. Towns A. Naphthopyran dyes. Phys Sci Rev. 2020;5. DOI:10.1515/psr-2019-0085.
13. Aiken S, Edgar RJ, Gabbutt CD, Heron BM, Hobson PA. Negatively photochromic organic compounds. Dyes Pigm. 2018;149:92.

14. Barachevsky VA. Negative photochromism in organic systems. Rev J Chem. 2017;7:334.
15. Towns AD. In: Bergamini G, Silvi S, editors. Applied photochemistry. Switzerland: Springer, 2016 Chapter 5 ("Industrial Photochromism") 227–79.
16. Rawat MS, Mal S, Singh P. Photochromism in anils – A review. Open Chem J. 2015;2:7–19.
17. Fritzsche J. Note sur les carbures d'hydrogène solides, tirés du goudron de houille. Compt Rend Hebd Séances Acad Sci. 1867;64:1035–7.
18. Roth HD. The beginnings of organic photochemistry. Angew Chem Int Ed Engl. 1989;28: 1193–207.
19. Becker H-D. Unimolecular photochemistry of anthracenes. Chem Rev. 1993;93:145–72.
20. Fritzsche J. Bericht an die physikalisch-mathematische Classe der Kaiserlichen Akademie der Wissenschaften vom Akademiker Fritzsche über seine Arbeit, betreffend die festen Kohlenwasserstoffe des Steinkohlentheers. Bull Acad Imp Sci St Pétersbourg. 1867;11:385–97.
21. Elbs K. Ueber Paranthracen. J Prakt Chem. 1891;44:467–9.
22. Fritzsche J. Über die festen Kohlenwasserstoffe des Steinkohlentheers. Bull Acad Imp Sci St -Pétersbourg. 1866;9:406–19.
23. Winterstein A, Schön K, Vetter HV. Mitteilung: Anthracen, Chrysen, Pyren. Z Physiol Chem. 1934;230:158–69.
24. Ter Meer E. Ueber Dinitroverbindung der Fettreihe. Justus Liebig's Ann. Chemie. 1876;181:1.
25. Hirschberg Y, Fischer E. Thermochromism and Photochromism. J Chem Phys. 1955;23:1723.
26. Hirschberg Y. Photochromie dans la série de la bianthrone. Compt Rend. 1950;231:903.
27. Marckwald W. Ueber Phototropie. Z Phys Chem. 1899;30:140.
28. Stobbe H. Die farbigen Anhydride der Butadiën-β, χ-dicarbonsäuren; ihr Verhalten gegen Licht und Wärme. (II. Abhandlung über Butadiënverbindungen.). Ber Deutschen Chemischen Ges. 1904;37:2236–40.
29. Luck W, Sand H. Über Phototropie. Angew Chem. 1964;76:463–73.
30. Dessauer R, Paris JP. In: Noyes Jr WA, Hammon GS, Pitts Jr JN, editors. Advances in photochemistry, vol. 1, John Wiley, 1963 "Photochromism" 275–321.
31. Nakatani K, Piard J, Yu P, Métivier R. In: Photochromic materials: preparation, properties and applications. Tian H, Zhang J, editors . Weinheim: Wiley-VCH, 2016 Chapter 1 ("Introduction: Organic Photochromic Molecules"):1–45.
32. Beatty JM. Trade winds. Saturday review. 5 Aug 1961. 5–6 (www.unz.com/print/SaturdayRev-1961aug05-00005/). Accessed: 19 Apr 2021.
33. Towns A. Spiropyran dyes. Phys Sci Rev. in press. DOI:10.1515/psr-2020-0197.
34. Hirschberg Y. The photochemical memory. New Scientist. 1960;7:1423–5. https://books.google.co.uk/books?id=RRVRuUO5vFIC&pg=PA1379&source=gbs_toc&cad=2#v=onepage&q&f=false. Accessed: 19 Apr 2021.
35. Bertelson RC. Reminiscences about organic photochromics. Mol Cryst Liq Cryst. 1994;246:1–8.
36. Bertelson RC. Chapter 10 ("Applications of photochromism"). In: Brown GH, editor. Photochromism – techniques of chemistry. vol. III. London: Wiley, 1971.:733–840.
37. Hoffman HJ. In: Photochromism molecules and systems. Dürr H, Bouas-Laurent H, editors. Amsterdam: Elsevier, 2003 Chapter 22 ("The Use of Silver Salts for Photochromic Glasses") 822–54.
38. Feringa BL. The art of building small: from molecular switches to motors (Nobel Lecture). Angew Chem Int Ed. 2017;56:11060–78.
39. Dürr H. In: Crano JC, Guglielmetti RJ, editors. Organic photochromic and thermochromic compounds vol. 1: main photochromic families. New York: Plenum, 1999 Chapter 6 ("Photochromism of Dihydroindolizines and Related Systems") 223–66.

40. Dürr H. In: Dürr H, Bouas-Laurent H, editors. Photochromism molecules and systems. Amsterdam: Elsevier, 2003 Chapter 6 ("4n+2 Systems Based on 1,5-Electrocyclization") 210–69.
41. Hajoudis E. In: Dürr H, Bouas-Laurent H, editors. Photochromism molecules and systems. Amsterdam: Elsevier, 2003 Chapter 17 ("Tautomerism by Hydrogen Transfer in Anils, Aci-Nitro and Related Compounds") 685–712.
42. Hadjoudis E, Mavridis IM. Photochromism and thermochromism of Schiff bases in the solid state: structural aspects. Chem Soc Rev. 2004;33:579–88.
43. Sliwa M, Mouton N, Ruckebusch C, Poisson L, Idrissi A, Aloïse S, et al. Investigation of ultrafast photoinduced processes for salicylidene aniline in solution and gas phase: towards a general photo-dynamical scheme. Photochem Photobiol Sci. 2010;9:661–9.
44. Eigenmann G. Chapter 4 ("Photochromic Processes By Homolytic Cleavage"). In: Brown GH, editor. Photochromism – techniques of chemistry. vol. III. London: Wiley, 1971:433–69.
45. Aldag R, In: Dürr H, Bouas-Laurent H, editors. Photochromism molecules and systems. Amsterdam: Elsevier, 2003 Chapter 18 ("Photochromism Based on Dissociation Processes") 713–37.
46. Bertelson RC. Chapter 3 ("Photochromic Processes Involving Heterolytic Cleavage"). In: Brown GH, editor. Photochromism – techniques of chemistry. vol. III. London: Wiley, 1971:45–431.
47. Green WA. Industrial photoinitiators: a technical guide. Boca Raton: CRC Press, 2010.
48. Rullo A, Reiner A, Reiter A, Trauner D, Isacoff EY, Woolley GA. Long wavelength optical control of glutamate receptor ion channels using a tetra-ortho-substituted azobenzene derivative. Chem Commun. 2014;50:14613–15.
49. Weston CE, Richardson RD, Haycock PR, White AJ, Fuchter MJ. Arylazopyrazoles: azoheteroarene photoswitches offering quantitative isomerization and long thermal half-lives. J Am Chem Soc. 2014;136:11878–81.
50. Yager KG, Barrett CJ. In: Zhao Y, Ikeda T, editors. Smart light responsive materials: azobenzene-containing polymers and liquid crystals. Hoboken: Wiley, 2009 Chapter 1 ("Azobenzene polymers for photonic applications") 1–46.
51. Cliffe WH. The Life and Times of Peter Griess. J Soc Dyers Colourists. 1959;75:278–85.
52. Towns A. Diarylethene dyes. Phys Sci Rev. 2020;5. DOI:10.1515/psr-2019-0146.
53. Towns A. Fulgide Dyes. Phys Sci Rev. in press. DOI:10.1515/psr-2020-0173.
54. Baeyer A. Ueber die Verbindungen der Indigogruppe. Ber Deutsch Chem Ges. 1883;16:2188–204.
55. Arai T, Ikegamai M. Novel photochromic dye based on hydrogen bonding. Chem Lett. 1999;28:965–6.
56. Pianowski ZL. Recent implementation of molecular photoswitches into smart materials and biological systems. Chem Eur J. 2019;25:5128–44.
57. Mallo N, Foley ED, Iranmanesh H, Kennedy AD, Luis ET, Ho J, et al. Structure–function relationships of donor–acceptor Stenhouse adduct photochromic switches. Chem Sci. 2018;9:8242–52.
58. Lerch MM, Symański W, Feringa BL. The (photo)chemistry of Stenhouse photoswitches: guiding principles and system design. Chem Soc Rev. 2018;47:1910–37.
59. Helmy S, Leibfarth FA, Oh S, Poelma JE, Hawker CJ, De Alaniz JR. Photoswitching using visible light: A new class of organic photochromic molecules. J Am Chem Soc. 2014;136:8169–72.
60. Mustroph H. Hemioxonol dyes. Phys Sci Rev. in press. DOI:10.1515/psr-2020-0175.
61. For example: Tokyo Chemical Industry. Photochromic dyes. www.tcichemicals.com/eshop/en/gb/category_index/12989/ Accessed: 19 Apr 2021.

62. Corns SN, Partington SM, Towns AD. Industrial organic photochromic dyes. Color Technol. 2009;125:249.
63. Harris JD, Moran MJ, Aprahamian I. New molecular switch architectures. PNAS. 2018;115:9414.
64. Krongauz VA. Chapter 21 ("Environmental Effects on Organic Photochromic Systems"). In: Dürr H, Bouas-Laurent H, editors. Photochromism molecules and systems, revised edition. Amsterdam: Elsevier, 2003:793–821.
65. Vázquez-Mera N, Roscini C, Hernando J, Ruiz-Molina D. Liquid-filled capsules as fast responsive photochromic materials. Adv Optical Mater. 2013;1:631–6.
66. Parhizkar M, Zhao Y, Lin T. Photochromic fibers and fabrics. In: Tao X, editor. Handbook of smart textiles. Singapore: Springer. 2014. DOI: https://doi.org/10.1007/978-981-4451-68-0_7-1.
67. Periyasamy AP, Vikova M, Vik M. A review of photochromism in textiles and its measurement. Textile Prog. 2017;49:53–136.
68. Reduwan Billah SM, Christie RM, Shamey R. Direct coloration of textiles with photochromic dyes. Part 1: Application of spiroindolinonaphthoxazines as disperse dyes to polyester, nylon and acrylic fabrics. Color Technol. 2008;124:223–8.
69. Ramlow H, Andrade KL, Immich APS. Smart textiles: an overview of recent progress on chromic textiles. J Textile Inst. 2021;112:152–71.
70. Meslin D. Materials and treatments. Paris: Essilor Academy Europe, 2010.
71. Crano JC, Kwak WS, Welch CN. In: McArdle CB, editors, Applied photochromic polymer systems. Glasgow: Blackie, 1992.
72. Österholm AM, Shen DE, Kerszulis JA, Bulloch RH, Kuepfert M, Dyer AL, et al. Four shades of brown: tuning of electrochromic polymer blends toward high-contrast eyewear. ACS Appl Mater Interfaces. 2015;7:1413–21.
73. Feringa BL, Browne WR, editors. Molecular switches. 2nd ed. Weinheim: Wiley-VCH, 2011.
74. Dong H, Zhu H, Meng Q, Gong X, Hu W. Organic photoresponse materials and devices. Chem Soc Rev. 2012;41:1754.
75. Zhang Q, Qu D-H, Tian H. Photo-regulated supramolecular polymers: shining beyond disassembly and reassembly. Adv Optical Mater. 2019;7:1900033.
76. Pantuso E, de Filpo G, Nicoletta FP. Light-responsive polymer membranes. Adv Optical Mater. 2019;7:1900252.
77. Goulet-Hanssens A, Eisenreich F, Hecht S. Enlightening materials with photoswitches. Adv Mater. 2020;1905966.
78. Hohl DK, Weder C. (De)bonding on demand with optically switchable adhesives. Adv Optical Mater. 2019;7:1900230.
79. Irie M, Yokoyama Y, Seki T, editors. New frontiers in photochromism. Tokyo: Springer, 2013.
80. Gauglitz G. In: Dürr H, Bouas-Laurent H, editors. Photochromism molecules and systems. Amsterdam: Elsevier, 2003 Chapter 25 ("Actinometry") 883–903.
81. Van Renesse RL. In: Van Renesse RL, editor. Optical document security. 2nd ed, London: Artech House, 1998 Chapter 9 201–24.
82. Ke Y, Chen J, Lin G, Wang S, Zhou Y, Yin J, et al. Smart Windows: Electro-, Thermo-, Mechano-, Photochromics, and Beyond. Adv Energy Mater. 2019;19:1902066.
83. Mortimer RJ, Rosseinsky DR, Monk PM, editors. Electrochromic materials and devices. Weinheim: Wiley-VCH, 2015.
84. Casini M. Active dynamic windows for buildings: A review. Renewable Energy. 2018;119:923–34.
85. Tällberg R, Jelle BP, Loonen R, Gao T, Hamdy M. Comparison of the energy saving potential of adaptive and controllable smart windows: a state-of-the-art review and simulation studies of thermochromic, photochromic and electrochromic technologies. Solar Energy Mater Solar Cells. 2019;200:109828.

86. Tsujioka T, Matsui N. Electrical characterization of photochromic diarylethene films consisting of extraordinarily large crystallites. J Mater Chem C. 2014;2:3589.
87. Fu L-N, Leng B, Li Y-S, Gao X-K. Photoresponsive organic field-effect transistors involving photochromic molecules. Chin Chem Lett. 2016;27:1319–29.
88. Bisoyi HK, Li Q. Light-driven liquid crystalline materials: from photo-induced phase transitions and property modulations to applications. Chem Rev. 2016;116:15089–166.
89. Kim Y, Tamaoki N. Photoresponsive chiral dopants: light-driven helicity manipulation in cholesteric liquid crystals for optical and mechanical functions. ChemPhotoChem. 2019;3:284–303.
90. Andréasson J, Pischel U. Storage and processing of information using molecules: the All-Photonic approach with simple and Multi-Photochromic switches. Israel J Chem. 2013;53:236.
91. Fang Y, Sun M. Nanoplasmonic waveguides: towards applications in integrated nanophotonic circuits. Light Sci Appl. 2015;4:e294.
92. Zhao P, Wang D, Gao H, Zhang J, Xing Y, Yang Z, et al. Third-order nonlinear optical properties of the "clicked" closed-ring spiropyrans. Dyes Pigm. 2019;162:451.
93. Yamada M, Kondo M, Mamiya J-I, Yu Y, Kinoshita M, Barrett CJ, et al. Photomobile polymer materials: towards light-driven plastic motors. Angew Chem Intl Ed. 2008;47:4986–8.
94. Dattler D, Fuks G, Heiser J, Moulin E, Perrot A, Yao X, et al. Design of collective motions from synthetic molecular switches, rotors, and motors. Chem Rev. 2020;120:310–433.
95. Manrique-Juárez MD, Rat S, Salmon L, Molnár G, Quintero CM, Nicu L, et al. Switchable molecule-based materials for micro- and nanoscale actuating applications: achievements and prospects. Coord Chem Rev. 2016;308:395.
96. Han -D-D, Zhang Y-L, Ma J-N, Liu Y-Q, Han B, Sun H-B. Light-Mediated manufacture and manipulation of actuators. Adv Mater. 2016;28:8328–43.
97. Szymánski W, Beierle JM, Kistemaker HAV, Velema WA, Feringa BL. Reversible photocontrol of biological systems by the incorporation of molecular photoswitches. Chem Rev. 2013;113:6114–78.
98. Zhang J, Wang J, Tian H. Taking orders from light: progress in photochromic bio-materials. Mater Horiz. 2014;1:169–84.
99. Fleming C, Remón P, Li S, Simeth NA, König B, Grøtli M, et al. On the use of diarylmaleimide derivatives in biological contexts: an investigation of the photochromic properties in aqueous solution. Dyes Pigm. 2017;137:410.
100. Cheng H, Yoon J, Tian H. Recent advances in the use of photochromic dyes for photocontrol in biomedicine. Coord Chem Rev. 2018;372:66–84.
101. Velema WA, Szymanski W, Feringa BL. Photopharmacology: beyond proof of principle. J Am Chem Soc. 2014;136:2178–91.
102. Lerch MM, Hansen MJ, Van Dam GM, Szymanski W, Feringa BL. Emerging targets in photopharmacology. Angew Chem Int Ed. 2016;55:10978–99.
103. Arrue L, Ratjen L. Internal targeting and external control: phototriggered targeting in nanomedicine. ChemMedChem. 2017;12:1908–16.
104. Hüll K, Morstein J, Trauner D. *In Vivo* photopharmacology. Chem Rev. 2018;118:10710–47.
105. Paramonov SV, Lokshin V, Fedorova OA. Spiropyran, chromenes or spirooxaxine ligands: insights into mutual relations between complexing and photochromic properties. J Photochem Photobiol C: Photochem Rev. 2011;12:209.
106. Lv G, Chen L, Lan H, Yin T. In: Tian H, Zhang J, editors. Photochromic materials: preparation, properties and applications, 1st ed, Wiley-VCH: Weinheim, 2016 Chapter 4 ("Photoswitchable Supramolecular Systems") 109–66.

107. Hao Y, Meng J, Wang S. Photo-responsive polymer materials for biological applications. Chin Chem Lett. 2017;28:2085.
108. Basílio N, García-Río L. Photoswitchable vesicles. Curr Opinion Colloid Interfac Sci. 2017;32:29–38.
109. Moncelsi G, Ballester P. Photoswitchable host-guest systems incorporating hemithioindigo and spiropyran units. ChemPhotoChem. 2019;3:304–17.
110. Jia S, Fong W-K, Graham B, Boyd BJ. Photoswitchable molecules in long-wavelength light-responsive drug delivery: from molecular design to applications. Chem Mater. 2018;30:2873–87.
111. Weinstain R, Slanina T, Kand D, Klán P. Visible-to-NIR-Light activated release: from small molecules to nanomaterials. Chem Rev. 2020;120:13135–272.
112. Chozinski TJ, Gagnon LA, Vaughan JC. Twinkle twinkle little star: photoswitchable fluorophores for super-resolution imaging. FEBS Lett. 2014;588:3603–12.
113. Hell SW. Nanoscopy with focused light (Nobel Lecture). Angew Chem Int Ed. 2015;54:8054–66.
114. Pei JV, Heng S, De Ieso ML, Sylvia G, Kourghi M, Nourmohammadi S, et al. Development of a photoswitchable lithium-selective probe to analyse nonselective cation channel activity in migrating cancer cells. Mol Pharmacol. 2019;95:573.
115. Gentili PL, Rightler AL, Heron BM, Gabbutt CD. Extending human perception of electromagnetic radiation to the UV region through biologically inspired photochromic fuzzy logic (BIPFUL) systems. Chem Commun. 2016;52:1474–7.
116. Ayt A, Barachevsky V, Kobeleva O, Valova T, Gorelik A, Krayushkin M, et al. Masking photochromic films for nanolithography technology. Phys Status Solidi C. 2011;8:2866–9.
117. Nakagawa T, Ubukata T, Yokoyama Y. Chirality and stereoselectivity in photochromic reactions. J Photochem Photobiol C: Photochem Rev. 2018;34:152–91.
118. Browne WR, Feringa BL. In: Browne WR, Feringa BL, editors. Molecular switches, 2nd ed., Weinheim: Wiley-VCH, 2011 Chapter 5 ("Chiroptical Molecular Switches") 121–79.
119. Feringa BL, Van Delden RA, Koumura N, Geertsema EM. Chiroptical molecular switches. Chem Rev. 2000;100:1789–816.
120. Feringa BL. In control of motion: from molecular switches to molecular motors. Acc Chem Res. 2001;34:504–13.
121. Tsujioka T, Irie M. Electrical functions of photochromic molecules. J Photochem Photobiol C: Photochem Rev. 2010;11:1–14.
122. Van der Molen SJ, Liljeroth P. In: Browne WR, Feringa BL, editors. Molecular switches, 2nd ed, Weinheim: Wiley-VCH, 2011 Chapter 20 ("Conductance Properties of Switchable Molecules") 719–77.
123. Tsujioka T, Hamada Y, Shibata K. Nondestructive readout of photochromic optical memory using photocurrent detection. Appl Phys Lett. 2001;78:2282.
124. Kim Y, Kim E. Conductive polymer patterning on a photoswitching polymer layer. Macromol Res. 2006;14:584–7.
125. Kim D, Park SY. Multicolor fluorescence photoswitching: color-correlated versus color-specific switching. Adv Optical Mater. 2018;6:1800678.
126. Yildiz I, Deniz E, Raymo FM. Fluorescence modulation with photochromic switches in nanostructured constructs. Chem Soc Rev. 2009;38:1859–67.
127. Schildhauer M, Rott F, Thumser S, Mayer P, de Vivie-Riedle R, Dube H, et al. A prospective ultrafast hemithioindigo molecular motor. ChemPhotoChem. 2019;3:365–71.
128. Bertarelli C, Bianco A, Castagna R, Pariani G. Photochromism in optics. Opportunities to develop light-triggered optical elements. J Photochem Photobiol C: Photochem Rev. 2011;12:106–25.

129. Majumder A, Mondol M, Andrew TL, Menon R. Nanophotonic Optical Switching Using Digital Metamaterials and Photochromism. 2019. arXiv:1904.00837 (arxiv.org/abs/1904.00837). Accessed: 19 Apr 2021.
130. Han GGD, Li H, Grossman J. Optically-controlled long-term storage and release of thermal energy in phase changed materials. Nature Commun. 2017;8:1446.
131. Norikane Y, Uchida E, Tanaka S, Fujiwara K, Nagai H, Akiyama H. Photoinduced phase transitions in rod-shaped azobenzene with different alkyl chain length. J Photopolym Sci Technol. 2016;29:149–57.
132. Wagner N, Theato P. Light-induced wettability changes on polymer surfaces. Polymer. 2014;55:3436–53.
133. Guo F, Guo Z. Inspired smart materials with external stimuli responsive wettability: a review. RSC Adv. 2016;6:36623–41.
134. Zhang T, Wu LYL, Wang Z. Smart UV/Visible light responsive polymer surface switching reversibly between superhydrophobic and superhydrophilic. Surf Coatings Technol. 2017;320:304–10.
135. Li N-Y, Chen J-M, Tang X-Y, Zhang G-P, Liu D. Reversible single-crystal-to-single-crystal conversion of a photoreactive coordination network for rewritable optical memory storage. Chem Commun. 2020;56:1984–7.
136. Bossi ML, Aramendía PF. Photomodulation of macroscopic properties. J Photochem Photobiol C: Photochem Rev. 2011;12:154–66.
137. Mendonça CR, Balogh DT, De Boni L, Dos Santos Jr DS, Zucolotto V, Oliveira Jr ON. In: Browne WR, Feringa BL, editors. Molecular switches, 2nd ed, Weinheim: Wiley-VCH, 2011 Chapter 12 ("Optically Induced Processes in Azo Polymers") 399–422.
138. Kuenstler AS, Hayward RC. Light-induced shape morphing of thin films. Curr Opinion Colloid Interfac Sci. 2019;40:70–86.
139. Boelke J, Hecht S. Designing molecular switches for soft materials applications. Adv Optical Mater. 2019;7:1900404.
140. Bléger D. Orchestrating molecular motion with light – from single (macro)molecules to materials. Macromol Chem Phys. 2016;217:189–98.
141. Kageyama Y. Light-power self-sustainable macroscopic motion for the active locomotion of materials. ChemPhotoChem. 2019;3:327–36.
142. Nocentini S, Parmeggiani C, Martella D, Wiersma DS. Optically driven soft micro robotics. Adv Optical Mater. 2018;6:1800207.
143. Stoychev G, Kirillova A, Ionov L. Light-responsive shape-changing polymers. Adv Optical Mater. 2019;7:1900067.
144. Kellner S, Berlin S. Two-photon excitation of azobenzene photoswitches for synthetic optogenetics. Appl Sci. 2020;10:805.
145. Wilson D, Branda NR. In: Tian H, Zhang J, editors. Photochromic materials: preparation, properties and applications, 1st edn. Weinheim: Wiley-VCH, 2016 Chapter 9 ("Photochromic Materials in Biochemistry") 361–92.
146. Fuchter MJ. On the promise of photopharmacology using photoswitches: a medicinal chemist's perspective. J Med Chem. 2020;63:11436–47.
147. Yu Z, Hecht S. Remote control over folding by light. Chem Commun. 2016;52:6639–53.
148. Wegener M, Hansen MJ, Driessen AJM, Szymanski W, Feringa BL. Photocontrol of antibacterial activity: shifting from UV to red light activation. J Am Chem Soc. 2017;139:17979–86.
149. Towns A. Olympian colour chemistry. Chem Ind. 2012;32–5.

150. Babii O, Afonin S, Schober T, Garmanchuk LV, Ostapchenko LI, Yurchenko V, et al. Peptide drugs for photopharmacology: how much of a safety advantage can be gained by photocontrol? Future Drug Discov. 2020;2:FDD28.
151. Morstein J, Trauner D. New players in phototherapy: photopharmacology and bio-integrated optoelectronics. Curr Opinion Chem Biol. 2019;50:145–51.
152. Tochitsky I, Kienzler MA, Isacoff E, Kramer RH. Restoring vision to the blind with chemical photoswitches. Chem Rev. 2018;118:10748–73.
153. Van Gemert B. Plenary lecture. Int Symp Photochromism Vancouver. 2007.

Robert Christie and Adrian Abel

55 Phthalocyanine blue pigments

Abstract: Copper phthalocyanine is the dominant blue organic pigment by far, used extensively in printing ink, paint, plastics, and a range of other applications. A historical perspective of the development of phthalocyanine pigments, from their original serendipitous discovery, through the characterization of their molecular structures, to their development as pigments, is discussed in this chapter and in the separate chapter entitled Phthalocyanines: General Principles. Copper phthalocyanine exhibits polymorphism. The α- and β-forms are the most important crystal phases used as pigments, while the ε-form has only minor significance. Although structurally complex, the synthesis of copper phthalocyanine is relatively straightforward, involving readily available commodity starting materials to provide the products in high yield. After-treatments are required not only to convert the crude pigment into an appropriate pigmentary physical form, but also to provide stability towards crystal phase change and flocculation in application.

Keywords: phthalocyanine pigments, phthalocyanine blue, copper phthalocyanines, metal-free phthalocyanines, Scottish Dyes, Heliogen, Monastral, polymorphism, phthalic anhydride route, phthalonitrile route, CI Pigment Blue 15, CI Pigment Blue 15:1, CI Pigment Blue 15:2, CI Pigment Blue 15:3, CI Pigment Blue 15:4, CI Pigment Blue 15:5, CI Pigment Blue 16

55.1 Fundamentals

The most important organic blue pigment by far is copper phthalocyanine (**1**) (Figure 55.1) (CuPc, CI Pigment Blue 15). The pigment provides a brilliant intense blue color, combined with exceptional technical performance, and all this at fairly modest cost [1–6]. It finds almost universal use in the range of organic pigment applications. It should be noted that the classical yellow, orange, and red organic pigments, dominated by azo pigments, include a multitude of separate pigments with varying color and technical performance, from which end-users can make their selection based on the application requirements. This contrasts with the blue shade area, in which CuPc is adaptable to the vast majority of organic pigment applications. Indeed, while generally considered as a classical organic pigment, it may equally be categorized as a high-performance pigment. Although phthalocyanine complexes have been prepared from

This article has previously been published in the journal Physical Sciences Reviews. Please cite as: R. Christie, A. Abel, Phthalocyanine Blue Pigments *Physical Sciences Reviews* [Online] 2021, 6. DOI: 10.1515/psr-2020-0192

https://doi.org/10.1515/9783110587128-055

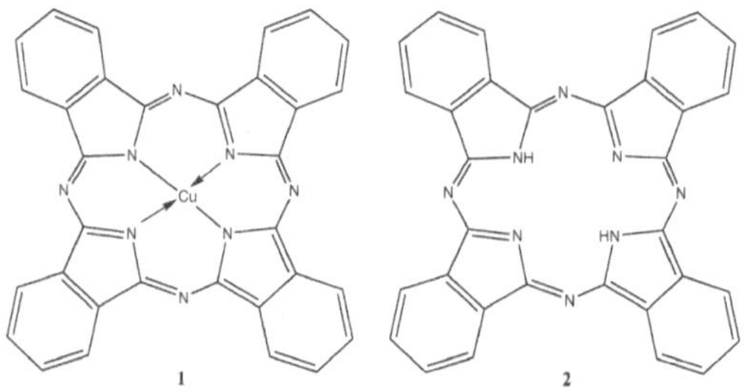

Figure 55.1: The structures of copper phthalocyanine (**1**) and metal-free phthalocyanine (**2**).

most metallic elements in the periodic table, only the copper complex is of significant commercial importance as a pigment, simply because it provides the optimum combination of color and technical properties. However, metal-free phthalocyanine (**2**) finds some use as a greenish-blue pigment of high stability.

55.2 History

The discovery and development of pigments based on the phthalocyanine chemical class, the most important chromophoric system developed during the twentieth century, ushered a new era in organic pigments. The historical elements of the accidental discovery of the phthalocyanines through to the characterization of the molecular structures is discussed in the chapter on Phthalocyanines: General Principles. In 1928, at Scottish Dyes, later to form part of ICI, it was observed that, in the industrial manufacture of phthalimide from the reaction of phthalic anhydride with ammonia, a blue contaminant formed in certain production batches. The impurity, iron phthalocyanine (FePc), was isolated as a dark blue, insoluble, crystalline substance. The source of iron was identified as the reactor wall, exposed due to a flaw in its lining. A series of academic investigations led by Linstead characterized the unique molecular structures of the phthalocyanines, which were confirmed by Robertson using x-ray crystal structure determination. The phthalocyanines emerged as one of the most extensively studied classes of organic compound, and the color manufacturing industry recognized, and ultimately exploited, the commercial potential of copper phthalocyanines as pigments. Following these studies, it was quickly established that the copper complex (**1**) offered the best products, and ICI made a major effort to ensure that they had adequate patent protection with a long list of patents submitted. The serendipitous discovery of copper phthalocyanine by de Diesbach and von der Weid in 1927 (although they did not establish its chemical structure) appears to have been unknown to the groups involved in the early years of the research [7]. However, when this earlier discovery came to light, it meant that ICI could not patent the compound, as there was prior knowledge. Thus, it left them with no option but to attempt to patent their manufacturing process. The company initiated a development program towards the commercial production of the new pigment, in which scaling up the manufacturing process presented many problems. Ultimately, serious production trials started in 1933 and this was followed by the launching of Monastral Fast Blue BS onto the market in 1935 [8]. However, competition to the ICI discovery was close behind. In 1934, at the Ludwigshafen works of IG Farben (which would later form BASF), a different way of manufacturing copper phthalocyanine had been developed that circumvented the ICI patents. This led to their release of Heliogen Blue B in 1935. Although a blow to ICI, as it had so quickly lost its monopoly, the ICI team recognized that IG Farben had found an excellent manufacturing route, which had advantages over the ICI processes. Eventually ICI and IG Farben came to a cross licensing agreement which included a technical information exchange. In 1937, IG Farben proceeded to release the first significantly greener metal free phthalocyanine (**2**), CI Pigment Blue 16, under the designation Heliogen Blue G. In 1936/37 DuPont introduced their version of copper phthalocyanine from their facility at Deepwater Point, New Jersey, US. In 1940, recognizing that the pigment exhibited polymorphism, they introduced the first product based on the greenish blue shade β-crystal form, commonly designated Blue G, which they described as Peacock Blue. This product opened the possibilities for three- and four-color printing, as it was close to the standard cyan color which had been difficult to achieve until then. This paved the way for greatly improved three- and four-color printing processes. The gap between the pigment first being described and a commercial pigment entering the market was almost 30 years, initially because it was not recognized as a potential pigment and then later because of difficulties in converting the crude product into usable pigment [9].

The earliest classical organic pigments were derived by modifying established textile dyes while, in contrast, phthalocyanines were developed as pigments and later converted into dyes. However, while textile dyes based on phthalocyanines are of relatively limited importance, the phthalocyanines provide by far the most important blue and green organic pigments. CuPc (**1**), CI Pigment Blue 15 types, are by far the most important blue pigments finding almost universal use as a colorant in a wide range of paint, printing ink and plastics applications. In fact, it may be argued that it is the most important of all organic pigments. It owes this dominant position to its intense brilliant blue color and its exceptional stability to light, heat, solvents, alkalis, acids and other chemicals. Among the features that demonstrate this stability are the ability to sublime unchanged at temperatures above 500 °C, and to dissolve without decomposition in concentrated sulfuric acid. In addition, despite its structural complexity, CuPc is a relatively low-cost product, because its manufacture is straightforward giving high yields from inexpensive, commodity starting materials.

55.3 Structures and properties

The elucidation of the structure of the metal phthalocyanines constitutes one of the classical applications of single crystal X-ray crystallography carried out in the 1930s [10], although it is interesting that the crystal structure of CuPc (**1**) (in its *β*-crystal form) was not reported until 1968 [11]. The molecules are planar with an internal 16-membered central ring where nitrogen atoms form a square planar complex with the copper atom. CuPc is extensively resonance stabilized with structure (**1**) as only one of the many resonance forms that contribute to the molecular structure. The phthalocyanines are aromatic molecules, attributed to the 18 π-electrons in the perimeter of the molecules.

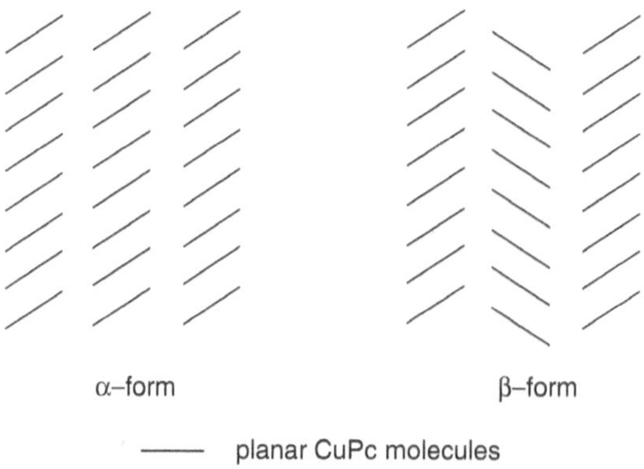

Figure 55.2: Representations of the *α*- and *β*-crystal forms of CuPc.

Copper phthalocyanine is one of the many organic pigments that exhibit polymorphism. Six different crystal phases have been established, each with unique physical and chemical properties. The *α*- and *β*-forms are the most important crystal phases used as pigments, with the ε-form having only minor significance. The *α*- and *β*-forms exhibit different hues, the α-form being reddish-blue while the *β*-form is greenish blue. The ε-form is very reddish blue. *β*-CuPc is of particular importance as the cyan pigment for printing inks, while the α-form is more important in paint and plastics applications. The *β*-form is the more stable phase thermodynamically, particularly towards organic solvents. A structural comparison between the *α*- and *β*-phases is shown in simplified form in Figure 55.2. In all crystal phases, the planar CuPc molecules form a close-packed stacking arrangement in one dimension. It has been suggested that in the "herring-bone" arrangement in the crystal structure of the *β*-form, the copper atom, which is coordinated to the four central nitrogen atoms, is also coordinated to nitrogen atoms in adjacent molecules, thereby forming a distorted octahedron, a coordination geometry that is particularly favored in copper complexes. This feature may contribute to the higher stability of this polymorph. No such coordination is possible in the parallel arrangement of molecules in the crystal structure of α-CuPc [12, 13]. The α-form may convert to the β-form in the presence of certain organic solvents with a corresponding change in shade unless stabilized. The designation CI Pigment Blue 15 refers to the α-phase in its unstabilized form. However, this phase may be stabilized towards phase change by the incorporation of a small amount of chlorine substitution (averaging around 0.5 chlorine atoms per molecule). The resulting phase-stabilized α-CuPc is designated as CI Pigment Blue 15:1. It has been suggested that the chlorine atom sterically prevents conversion to the β-form. It is possible to incorporate additional chlorine into the molecules, but this adds cost, reduces the tinctorial strength and provides an increasingly greener shade. A group of pigments were introduced that used this principle, but the market was limited, mainly to producing unique effects for automotive metallic paints.

CuPc pigments may also require stabilization against flocculation for certain applications, especially in paints. Flocculation involves a loose re-aggregation of dispersed pigment particles. If the paint is applied by brushing, the shearing forces can de-flocculate the pigment, leading to increased color strength. The issues that might result may be exemplified by a blue and yellow blend used to produce a green paint. If the blue de-flocculates, the color would shift towards a bluer shade of green. Since the degree of deflocculation depends on how much brushing is used, especially as the paint is drying and the shear increases, an uneven color may result even from paint from the same can. It has been suggested that flocculation is associated with the non-polarity of certain CuPc crystal surfaces where the H-atoms on the exterior of the molecules dominate. To address this feature, CuPc derivatives with outer ring substituents containing basic amino groups as (e. g., $-CH_2NR^1R^2$) have been developed as additives. The CuPc unit in these additives attaches to the crystal surfaces, and the substituents provide electronic and steric protection at the surface, essentially acting as a bridge between the pigment and the binder, and thus minimizing flocculation. The treatment adds cost but also value to the pigment and therefore the treated pigments tend to be used in applications of high quality where the higher price can be justified. The stabilized pigment also tends to provide a purer shade. CI Pigment Blue 15:2 refers to flocculation-resistant (also phase-stabilized) α-CuPc. β-CuPc is CI Pigment Blue 15:3, and the flocculation-resistant β-phase is designated as CI Pigment Blue 15:4. CI Pigment Blue 15:6 is the ε-form.

55.4 Synthesis and manufacture

Although the structure of copper phthalocyanine may be complex, its synthesis is relatively straightforward. The relatively low cost of the pigment may be explained by the fact that it is prepared in high yield from readily available, low cost starting materials. Two chemically related methods, the phthalic anhydride and phthalonitrile routes, are most used. Both involve simultaneous synthesis of the ligand and metal complex formation in a template procedure.

(a) Phthalic anhydride route

In the most encountered version of this method, as illustrated in Figure 55.3, phthalic anhydride (**3**) is heated with urea, copper (I) or (II) chloride and a catalytic amount of ammonium molybdate either in a high boiling solvent, such as trichlorobenzene, nitrobenzene or kerosene. A reaction temperature of 200–300 °C is used with a reaction time of 2–3 hours. The solvent is removed from the product by filtration or centrifugation. Alternatively, a "dry bake" process may be employed, in which a mixture of the solid reactants and catalyst are heated at around 200 °C. There are difficulties in this process including the formation of by-products and in overcoming mixing issues. This was the original method of CuPc manufacture. Over the years the method declined in favor of the more efficient solvent-based process, although more recent years have seen some renewed interest for a combination of economic and ecological reasons, for example the use of undesirable solvents. The solvent process is easier to conduct and affords a product of higher purity. On the other hand, a significant disadvantage is the need for solvent recovery. In the phthalic anhydride process, the urea acts as a source of nitrogen, and the carbonyl group in the urea is lost ultimately as carbon dioxide. Mechanistic schemes have been proposed to explain the course of this synthetic sequence although much of the detail remains to be established unequivocally. In broad terms, phthalic anhydride reacts with urea or products of its decomposition or polymerization, resulting in progressive replacement of the oxygen atoms with nitrogen and the formation of the key intermediate 1-amino-3-iminoisoindoline (**4**) (Figure 55.3). The presence of ammonium molybdate is essential to catalyze these parts of the synthetic sequence. This intermediate undergoes a tetramerization with cyclization aided by coordination with the copper ion to form crude copper phthalocyanine.

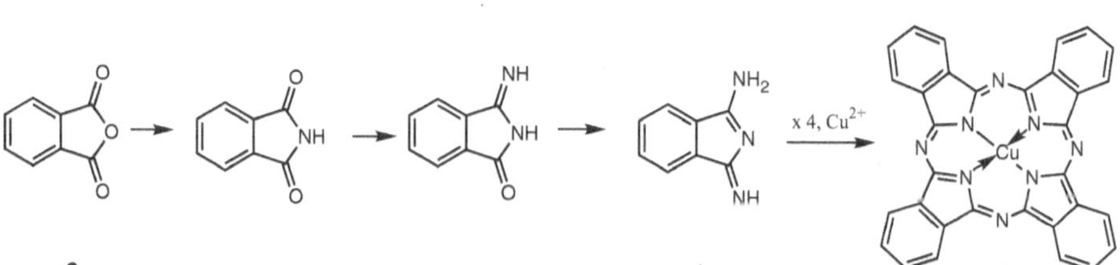

3 4 1

Figure 55.3: The phthalic anhydride/urea synthetic route to CuPc.

In the most encountered version of this method, as illustrated in Figure 55.3, phthalic anhydride **(3)** is heated with urea, copper (I) or (II) chloride and a catalytic amount of ammonium molybdate either in a high boiling solvent, such as trichlorobenzene, nitrobenzene or kerosene. A reaction temperature of 200–300 °C is used with a reaction time of 2–3 hours. The solvent is removed from the product by filtration or centrifugation. Alternatively, a 'dry bake' process may be employed, in which a mixture of the solid reactants and catalyst are heated at around 200 °C. There are difficulties in this process including the formation of by-products and in overcoming mixing issues. This was the original method of CuPc manufacture. Over the years the method declined in favor of the more efficient solvent-based process, although more recent years have seen some renewed interest for a combination of economic and ecological reasons, for example the use of undesirable solvents. The solvent process is easier to conduct and affords a product of higher purity. On the other hand, a significant disadvantage is the need for solvent recovery. In the phthalic anhydride process, the urea acts as a source of nitrogen, and the carbonyl group in the urea is lost ultimately as carbon dioxide. Mechanistic schemes have been proposed to explain the course of this synthetic sequence although much of the detail remains to be established unequivocally. In broad terms, phthalic anhydride reacts with urea or products of its decomposition or polymerization, resulting in progressive replacement of the oxygen atoms with nitrogen and the formation of the key intermediate 1-amino-3-iminoisoindoline **(4)** (Figure 55.3). The presence of ammonium molybdate is essential to catalyze these parts of the synthetic sequence. This intermediate undergoes a tetramerization with cyclization aided by coordination with the copper ion to form crude copper phthalocyanine.

(b) Phthalonitrile route

In this process, phthalonitrile (**5**) is heated to around 200 °C with copper metal or a copper salt, with or without a high boiling solvent. In this case, there is no requirement for an ammonia-donating ingredient, such as urea, as phthalonitrile contains sufficient nitrogen for the process. When copper (I) chloride is used, a small percentage of chlorine substitution occurs, so that this can be used as a simple one-step route to the phase-stabilized *α*-CuPc. For manufacture of the β-form by this method, a small amount of urea may be added, acting as a chlorine-trap. A speculative but plausible mechanism for the phthalonitrile route to copper phthalocyanine has been proposed as illustrated in Figure 55.4 [14]. It is suggested that reaction is initiated by attack by a nucleophile (Y^-), for example the counteranion associated with the Cu^{2+} ion, at a cyano group activated by coordination with the Cu^{2+} ion, followed by cyclization to isoindoline derivative (**6**). Attack by this intermediate on another molecule of phthalonitrile occurs and, following a series of similar reactions with cyclization, facilitated by the coordinating role of the Cu^{2+}, intermediate (**7**) is formed. When copper metal is used as the reactant, it is proposed that two electrons are transferred from the metal, allowing elimination of Y^- to form copper phthalocyanine [route (i)]. Consequently, the Cu(0) is oxidized to Cu(II) as required to participate further in the reaction. When a copper (II) salt is used, it is suggested that Y^+ (the chloronium ion in the case of $CuCl_2$) is eliminated to form CuPc [route (ii)]. The product in this case is a monochloro derivative, formed by electrophilic attack of Cl^+ on the CuPc formed initially.

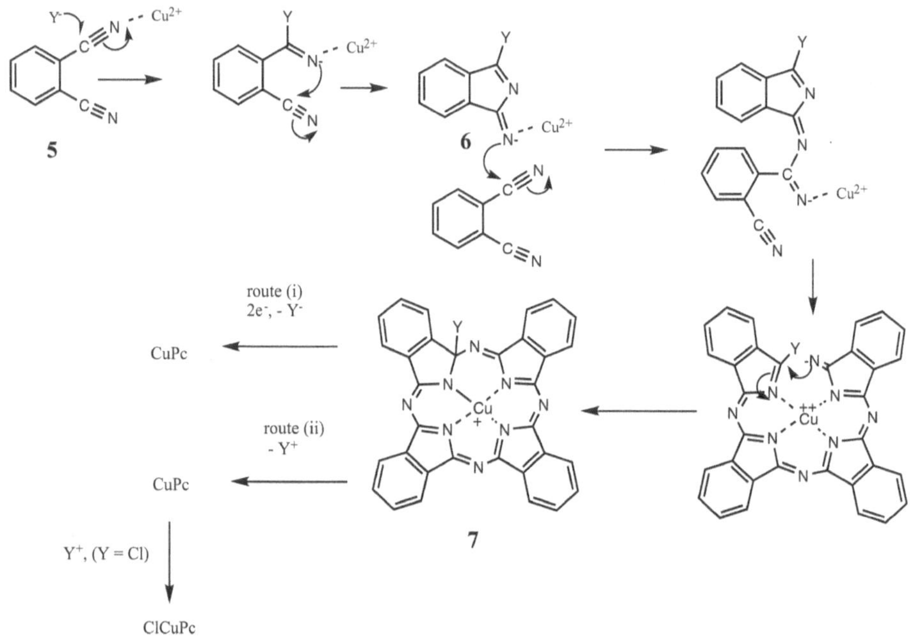

Figure 55.4: Mechanism of the phthalonitrile route to CuPc.

Both the phthalic anhydride and phthalonitrile routes produce a "crude" blue product, which is of far too large a particle size to be of use as a pigment, unable to provide the coloristic properties for which CuPc is known. Thus, post-synthesis conditioning must be carried out, to break down the coarse particles to an optimum particle size distribution and form the desired crystal modification. While the general principles of conditioning of crude CuPc are well known, the exact processes conducted by individual manufacturers are a guarded commercial secret as they can add significant value to their products. In some cases, a preliminary purification is carried out involving a wash with dilute acid or alkali, or both. An original method developed for particle size reduction used was acid pasting, which involves dissolving or swelling the crude product in concentrated sulfuric acid, followed by reprecipitation with water. This method provides the α-modification of the pigment in a fine particle size form. Mechanical grinding or milling of the crude blue product in the presence of inorganic salts, such as sodium chloride or sulfate, or calcium chloride, often produces a mixture of α- and β-CuPc, which may be converted to pure pigmentary β-CuPc by careful treatment with certain organic solvents, such as aromatic hydrocarbons or alcohols. Alternatively, grinding the crude material with inorganic salts in the presence of small amounts of "crystallization-stimulating" organic solvents can lead directly to β-CuPc in a fine particle size form.

Metal-free phthalocyanine (**2**) is conveniently prepared by subjecting certain labile metal phthalocyanine derivatives, such as those of sodium or lithium, to acidic conditions.

55.5 Applications

Commercially, phthalocyanine blue pigments account for about 25% of all organic pigment manufactured, with phthalocyanine green raising this total to over 30% [15]. A large proportion of phthalocyanine pigments is now manufactured in Asia, and so accurate production statistics are not available, but as it was estimated to be around 80,000 tonnes at the turn of this century, the total may now be approaching 100,000 tonnes. Phthalocyanine blue pigments are used in virtually all coloring processes. It has been estimated that 42% is used in printing inks, 30% in paint, 20% in plastics with the remaining 8% in a plethora of other applications such as textile printing, stationery, artists' materials, agricultural applications. In the description of the fastness properties of the different pigment grades that follows, there is monotonous use of the term excellent. However, the individual commercial grades may be either suitable or unsuitable for specific applications, depending on the treatments to which they have been subjected. The technical information given here for each Colour Index designation assumes that each pigment has been assessed technically and recommended by a reputable manufacturer for the intended application.

CI Pigment Blue 15

This pigment is the unstabilized α-modification of CuPc. A particular deficiency of this pigment is its relative instability to solvents, especially aromatic types, a feature exacerbated by heat, causing it to convert to the β-modification and consequently lose much of its tinctorial strength. It is also susceptible to flocculation in certain applications. However, it is the most economical grade providing high tinctorial strength and the reddest shade, apart from the ε-phase products. Its fastness to light and weather, and resistance to acids and alkalis are excellent. In printing inks, it is too red in hue for use in process inks. It is employed to some extent in water-based inks, where it is often used in the form of an aqueous dispersion, as its hydrophobic surface character means that it is difficult to wet out. It can also be used to color PVC in which it is fast to migration and has excellent lightfastness although it can bleed slightly in certain plasticizers. Its fastness to weathering is distinctly inferior to the stabilized grades. In polyolefins, its use is limited by heat stability, as it starts to convert to the β-modification above 220 °C, with loss of tinctorial strength and purity of shade, in addition to becoming greener. It is rarely used in paints, even in water-based paints, in which it can be affected by the coalescing organic solvents used in their formulation.

CI Pigment Blue 15:1

The low chlorine content of this α-form pigment is sufficient to prevent conversion to the β-form. It has excellent fastness to almost all solvents, excellent lightfastness and very good weather fastness. Current pigments of this type offer higher tinctorial strength and purity of shade than could be obtained in earlier grades, so that it has largely displaced CI Pigment Blue 15 for many applications. It offers around 20% higher tinctorial strength than CI Pigment Blue 15:3 types. In inks, it is too red to be used in process colors, but it is favored for applications where its higher strength offers greater economy. In plastics, it is widely used for coloring virtually all polymers, including styrenic copolymers, through to polyamides and other engineering polymers, a feature that is almost unique for organic pigments. Most grades easily meet the requirement for stability up to 300 °C in polyolefins. However, it can cause quite severe warping when used in injection moldings. Caution is necessary in the coloration of rubber and rubber latex. Rubber is sensitive to free copper, which causes it to age prematurely and can affect the vulcanization process, so many manufacturers offer special grades with no free copper. In paints it can be used for almost all applications, offering excellent lightfastness and resistance to solvents and chemicals. It is used in automotive finishes, although generally specialist grades are preferred, especially for metallic finishes. Some of these may contain additional chlorine substituents to provide a greener shade.

CI Pigment Blue 15:2

This pigment is described as a flocculation-stable α-modification CuPc. The Color Index does not generally recognize differences in after treatments that do not involve a change in constitution or in crystal modification. However, manufacturers sought to differentiate the pigments that had been treated to provide additional stability to flocculation and so this classification was adopted. No detailed specification was ever issued, so it is essentially up to the manufacturer to decide which classification to adopt, an approach that appears to have been respected. As CI Pigment Blue 15:1 may be given some treatment to improve stability to flocculation, there is no precise dividing line between these two grades. It is mainly used in the paint industry, where flocculation is a more significant problem, certainly than in plastics. This pigment generally has a smaller particle size distribution and is easier to disperse. It may therefore offer higher transparency, and so is often preferred for metallic and pearlescent finishes. It may be the preferred pigment for tinting systems on account of its higher stability with respect to shade and color strength, important considerations when used in combination with other tints to achieve consistent colors. The use of the additives to reduce flocculation can sometimes cause slight bleed under certain conditions.

CI Pigment Blue 15:3

This pigment is the β-modification of CuPc, the most stable crystal form, offering a very pure greenish blue shade that is almost the perfect hue for the cyan primary in three and four process color printing. This is the main application for the pigment, although paints and plastics are also large outlets. In inks, as well its color so close to the cyan standard, it offers other technical advantages, such as excellent light fastness, heat stability, fastness to solvents and chemicals. Unlike many other process colors, this pigment does not recrystallize in modern bead mills that can generate high temperatures during the dispersion process. In the US, the pigment is often supplied as a flush paste. While flush pastes are increasingly used in Europe (mainly by subsidiaries of US companies), powder pigments still dominate. Special grades of the pigment have been introduced for the toners used in laser printers and color photocopiers. As well as offering the appropriate coloristic properties, they comfortably satisfy the requirements for heat stability and lightfastness. However, they also require optimum triboelectric properties, which need careful control as chemically identical pigments can offer different performance in this respect. In plastics it can be used in most polymers, from rigid PVC, through plasticized PVC, polyolefins, polystyrene and styrenic copolymers, and even many engineering polymers. It generally has a considerable effect on dimensional stability in polyolefins and other crystalline polymers, leading to severe warpage in injection molded articles. It is used for the spin coloration of most fibers, including polyester and acrylonitrile. Special pigment grades with low free copper content are required when used to color rubber and latex to prevent premature aging. In paints, it is used for all types of finish, but it tends to flocculate, so that care must be taken when there is a processing change such as letting down with solvents or resins after the dispersion process. This is particularly problematic with tinting systems. The β-form is often preferred when aiming to optimize the number of shades that can be achieved with a limited number of pigments on account of its greater purity of shade.

CI Pigment Blue 15:4

This flocculation-stable β-modification can be used for virtually all applications, with the same levels of fastness as provided by CI Pigment Blue 15:3, with the minor reservation that some surface treatments used to provide resistance to flocculation may adversely affect solvent fastness, as is the case with the flocculation-resistant α form. As the treatment provides improved dispersibility, the dispersed particle size is reduced, resulting in higher color strength and a purer shade. This CuPc pigment grade is therefore increasingly used in place of CI Pigment Blue 15:3, mainly in paints but also in some inks. In inks, its main use is in liquid inks, including publication gravure, for which special grades are recommended that provide much lower abrasion, thus causing less damage to the expensive gravure printing plates. It also provides better flow properties in application. In paints, flocculation is often a problem in stoving paints, so that this grade offers some protection, if it retains good solvent fastness. Although flocculation is less of an issue in plastics, this pigment grade is often easier to disperse and offers a purer shade. At least one manufacturer claims that it can be used in injection moldings based on polyolefins, where it is much less likely to produce warpage than conventional grades. Flocculation-stable grades carry a significant price premium over untreated grades, so that the higher cost needs to be justified through improved performance. The stabilized grades are commonly around 20% more expensive, although there is some variation among suppliers.

CI Pigment Blue 15:5

This designation represented a modification that was never confirmed.

CI Pigment Blue 15:6

This is the ε-modification of the pigment and is the reddest form of CuPc, producing a purer tone in this shade than can be obtained using combinations of other CuPc grades with dioxazine violet pigments, or with the high value anthraquinonoid pigment, Indanthrone Blue (CI Pigment Blue 60). Initially, grades of this pigment were unstable, and so its use has started only relatively recently. Its high price limits its application to specialist paint and printing ink applications, including UV-curing inks. In paints, its main application is in metallic and pearlescent finishes, as it has even better weatherfastness than the other modifications when used in thermosetting acrylic binders. Its alkali stability is good but inferior to other grades.

CI Pigment Blue 16

This pigment is metal free phthalocyanine (**2**). At first it held an important position in applications as it was the greenest form of the unchlorinated phthalocyanines. However, it has been displaced by the much superior β-CuPc. It must be used with great care as it is polymorphic and less stable than the copper complexes.

It is unstable towards many solvents that cause the pigment to flocculate, losing color strength, and its stability to acids is less than the CuPc pigments. It is, however, transparent and is specifically recommended for water-based paints, including metallic and pearlescent finishes. It can also be used in inks, usually water- and solvent-based packaging inks.

References

1. Moser FH, Thomas AL. The phthalocyanines vols I and II. Boca Raton, Florida: CRC Press Inc, 1983.
2. Leznoff CC, Lever AB. Phthalocyanines: properties and applications. Weinheim: VCH, 1989.
3. McKeown NB. Phthalocyanine materials: synthesis, structure and function. Cambridge: Cambridge University Press, 1998.
4. Zollinger H. Color chemistry: syntheses, properties and applications of organic dyes and pigments, 3rd ed ch. 5. Weinheim: Wiley-VCH, 2003.
5. Christie RM. Colour chemistry, 2nd ed, London: RSC, Ch 5: 2015.
6. Hunger K, Schmidt MU. Industrial organic pigments, 4th ed, Weinheim: Wiley-VCH Verlag GmbH, Ch 3; 2019.
7. de Diesbach H, Von Der Weid E. Quelques sels complexes des *o*-dinitriles avec le cuivre et la pyridine. Helv Chim Acta. 1927;10:886.
8. Fox M R. Dyemakers of Great Britain, 1856 – 1976. Manchester: Imperial Chemical Industries PLC; 1987.
9. Eastaugh N, Walsh V, Chaplin T, Siddall R. Pigment compendium. Abingdon: Routledge 2008.
10. Robertson JM. An X-ray study of the structure of the phthalocyanines. Part I. The metal-free; nickel, copper and platinum compounds. J Chem Soc. 1935;613.
11. Brown CJ. Crystal structure of β-copper phthalocyanine. J. Chem. Soc (A). 1968:2488.
12. Honigman B, Lenne HU, Schrodel R. Beziehungen zwischen den Stukturen der Modifikation des Platin- und Kupferphthalocyanins und einiger Chlorderivate. Z Krystallog. 1965:122; 185–205.
13. Fryer JR, McKay RB, Mather RR, Sing KS. The technological importance of the crystallographic and surface properties of copper phthalocyanine pigments. J Chem Tech Biotechnol. 1981:31;371.
14. Christie RM, Deans D. An investigation into the mechanism of the phthalonitrile route to copper phthalocyanines using differential scanning calorimetry. J Chem Soc Perkin. 1989:2; 193–8.
15. Endriβ H. Brilliant solutions: Heliogen – the copper phthalocyanine pigments of BASF, 2nd ed. EFO 9906e, 2001.

Robert Christie and Adrian Abel

56 Phthalocyanine green pigments

Abstract: The dominant organic green pigments are polyhalogenated copper phthalocyanines (CuPcs). These pigments exhibit similar excellent coloristic and technical performance to CuPc blue pigments and are produced worldwide in large quantities. The green pigments are halogenated CuPcs. There are two pigments with CI designations. CI Pigment Green 7 is a bluish green product, in which the 16 ring hydrogen atoms of the CuPc molecule are replaced virtually completely by chlorine. Products designated as CI Pigment Green 36 incorporate a range of bromo- and bromochlorocopper phthalocyanines and are yellowish-green pigments. CuPc green pigments are manufactured by direct exhaustive halogenation of crude CuPc blue with chlorine or bromine or an appropriate mixture of the two halogens, depending on the composition of the product required. The synthesis is carried out at elevated temperatures in a suitable solvent, commonly a molten $AlCl_3/NaCl$ eutectic mixture.

Keywords: phthalocyanine pigments, phthalocyanine green, copper phthalocyanines, CI Pigment Green 7, CI Pigment Green 36, bromocopper phthalocyanines, bromochlorocopper phthalocyanines

56.1 Fundamentals

In pigment applications, green is an important color, although of much less importance than blue. In the same way as copper phthalocyanine (CuPc) blue pigments dominate the blue shade for most applications, polyhalogenated CuPcs are the dominant organic green pigments. These pigments exhibit similar excellent coloristic and technical performance to their blue counterparts and are also produced worldwide in large quantities. Most large organic pigment manufacturers offer the two phthalocyanine pigments that completely dominate the sector [1–6].

56.2 History

A historical background to the discovery of the phthalocyanines and the commercial establishment of CuPc blue pigment are discussed in two separate chapters, entitled Phthalocyanine Pigments: General Principles, and Phthalocyanine Blue Pigments.

This article has previously been published in the journal Physical Sciences Reviews. Please cite as: R. Christie, A. Abel, Phthalocyanine Green Pigments *Physical Sciences Reviews* [Online] 2021, 6. DOI: 10.1515/psr-2020-0193

https://doi.org/10.1515/9783110587128-056

The history surrounding the blue pigments provides an essential historical background that is relevant to the green pigments, because they are manufactured from the blue pigments. Industrial manufacture of CuPc blue was first introduced in 1935 by ICI in the UK, followed later in the same year by IG Farben in Germany, and in 1938 by DuPont in the US, and they subsequently developed as the dominant high performance blue pigments. An early green pigment was copper tetraphenylphthalocyanine. This product has long been superseded by the range of polyhalogenated CuPc s which have become the dominant green organic pigments. Chlorination of CuPc was found to lead to the formation of a bluish green pigment, that was investigated and patented by Linstead. However, it is BASF who claim to have been the first company to release it as a commercial pigment in 1937, under the designation Heliogen Green G (CI Pigment Green 7). Further investigations resulted in the discovery and development of a version containing both chlorine and bromine (CI Pigment Green 36), which has a brilliant emerald green shade.

56.3 Structures and properties

A general structure of the CuPc pigments is displayed in Figure 56.1. CuPc blue is the unsubstituted derivative (**1**, $X = H$). The green pigments are halogenated CuPc pigments. CI Pigment Green 7, in which the 16 ring hydrogen atoms of the CuPc molecule are replaced virtually completely by chlorine (**1**, $X = \mathrm{Cl}$), is a bluish green product. However, it is reported to be very difficult to achieve absolutely complete chlorination. The products designated as CI Pigment Green 36, which incorporate a range of bromo- and bromochlorocopper phthalocyanines (**1**, $X = \mathrm{Br,Cl,H}$), are yellowish-green pigments. The hue of these pigments becomes progressively yellower with increasing bromine content. Attempts to produce even yellower shade greens, using iodine, do not appear to have had any success and no such pigments have been registered. The phthalocyanine greens exhibit the same outstanding coloristic and technical performance as the blue pigments from which they are derived and find equally widespread use in the coloration of paints, printing inks, and plastics. Unlike the blue pigments, phthalocyanine greens do not appear to exhibit polymorphism, so that phase stability is not an issue. Their crystal form is similar to that of α-CuPc.

56.4 Synthesis and manufacture

The chlorination and bromination reactions involved in the synthesis of CuPc green pigments are examples of electrophilic substitution on the outer benzene rings of the CuPc molecule, reflecting its aromatic character. Synthesis of the phthalocyanine green pigments involves the direct exhaustive halogenation of crude CuPc blue with chlorine or bromine or an appropriate mixture of the two halogens, depending on the

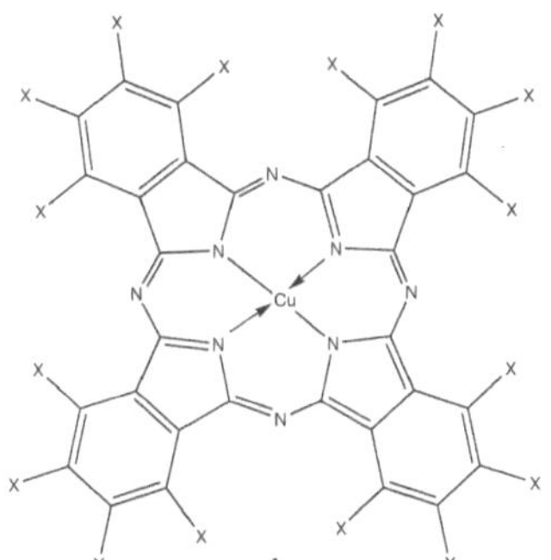

1

Figure 56.1: The structure of copper phthalocyanines (**1**).

composition of the product required. The synthesis is carried out at elevated temperatures (180–200 °C) in a suitable solvent, commonly a molten $AlCl_3$/NaCl eutectic mixture, in the presence of a catalyst such as $FeCl_3$. The green products arising directly from the synthesis, after isolation by filtration and washing, are of an appropriate primary particle size but are generally highly aggregated. The crude products require further conditioning by thermal treatment either with suitable organic solvents or with aqueous surfactant solutions. These processes effect a de-aggregation of the product that gives rise to a more uniform particle size distribution and increase the crystallinity of the products. Both provide a dramatic beneficial effect on the performance of the pigments.

56.5 Applications

In the past, it was economical to achieve green colors by mixing yellow organic pigments with CuPc blue. However, as the manufacture of CuPc green pigments has expanded, both geographically and in the number of companies engaged in its production, prices have fallen. Consequently, it is now usually less expensive to use phthalocyanine greens, with the added advantage that they generally provide superior fastness properties than most blue/yellow blends.

CI Pigment Green 7

This pigment is used in inks, coatings, and most plastics, offering an excellent level of fastness properties. It provides a clean bluish green shade with high tinctorial strength and is quite transparent. It is highly insoluble, and thus shows no tendency to migrate or bleed in application. It is the standard green pigment used in printing inks and for the packaging industry, having mostly replaced some green cationic (basic) dye complex pigments. There is only one crystal form of this pigment, which resembles α-CuPc blue, and thus shows no tendency to change crystal form on heating or in contact with solvents. In inks, green is an important spot color, especially for packaging, but it does not match the high color strength of CuPc blue pigments, having only around half the tinctorial strength of CI Pigment Blue 15:1, so this leads to some competition with blue/yellow blends. The excellent lightfastness, heat stability, and resistance to solvents mean that it is suitable for most demands, including metal decoration, where it is stable to at least 200 °C, wallcoverings and laminate printing. However, when used in melamine resins on paper, the prints change color in the dark, and return to their original shade when exposed to light, i.e., they are photochromic. They are thus usually avoided for such applications. The largest market for this pigment is in paints, in which the pigment offers excellent technical performance. Modern grades are treated to reduce their tendency to flocculate. In plastics, it is fast to migration in both plasticized and rigid PVC, with excellent fastness to light and weather. In polyolefins, it is stable to at least 300 °C, and gives the highest levels of fastness to light and weather. The pigment has a serious effect on injection moldings, where it can cause warpage. In other polymers, it is recommended for coloring polystyrene and ABS. It may be used under certain conditions for coloring polyamide and is highly recommended for the spin coloration of polypropylene, polyester and nylon fibers. The pigments are often quite difficult to disperse in polymers, and so it is often used in the form of pigment preparations, especially master batches.

CI Pigment Green 36

CI Pigment Green 36 is a brilliant emerald green color, in contrast to the blue shade green of CI Pigment Green 7. The exact shade of the pigment is dependent on the extent to which chlorine atoms are replaced by the heavier bromine atoms. The higher the bromine content, the yellower and purer the shade, although this is accompanied by reduction in tinctorial strength. In typical pigments, bromine can represent around 57% by weight of the molecule, a feature that explains the lower tinctorial strength. The pigment is mainly used for paints, where its excellent fastness to light, heat stability, and resistance to solvents mean that it can meet virtually all the demands in application. It is used mainly for industrial paints, including automotive original finishes, both solid and metallic shades. It also finds some use for coloring plastics. It is entirely fast to migration in PVC, with excellent lightfastness and very good to excellent fastness to weathering. In polyolefins it is stable to 300 °C, with excellent lightfastness, but can have an adverse effect on the dimensional stability of injection moldings, leading to warpage. It is recommended for use in polystyrene, and in some circumstances can be used to pigment ABS and nylon. It is less stable than CI Pigment Green 7 when used in polypropylene, polyester and nylon fibers.

References

1. Moser FH, Thomas AL. The phthalocyanines vols I and II. Boca Raton, Florida: CRC Press Inc; 1983.
2. Leznoff CC, Lever AB. Phthalocyanines: properties and applications. Weinheim: VCH; 1989.
3. McKeown NB. Phthalocyanine materials: synthesis, structure, and function. Cambridge: Cambridge University Press; 1998.
4. Zollinger H. Color chemistry: syntheses, properties and applications of organic dyes and pigments, 3rd ed. ch. 5 Weinheim: Wiley-VCH; 2003
5. Christie RM. Colour chemistry, 2nd ed. London: RSC, Ch 5; 2015.
6. Hunger K, Schmidt MU. Industrial organic pigments, 4th ed. Weinheim: Wiley-VCH Verlag GmbH, Ch 3; 2019.

Robert Christie and Adrian Abel

57 Phthalocyanine pigments: general principles

Abstract: This chapter describes some of the fundamental chemical and structural principles underlying the phthalocyanine system. Historically, phthalocyanines had been isolated as insoluble blue products in the early twentieth century, although the structures were not established at the time. After the serendipitous re-discovery of metal complex phthalocyanines by industry, and the elucidation of their structures, copper phthalocyanine (CuPc) pigments were introduced industrially in the 1930s to become, and remain, by far the most important blue and green organic pigments, finding almost universal use as colorants for paints, printing inks, plastics, and a wide range of other applications. The phthalocyanines have become one of the most extensively studied classes of organic compounds, because of their unique molecular structures, light absorption properties that produce strong, bright colors, and their exceptionally high stability. While their dominant use is as colorants, they are also of interest for a range of other applications, for example in electronic devices, biological applications, and as catalysts.

Keywords: phthalocyanine pigments, copper phthalocyanine, metal-free phthalocyanine, Scottish Dyes, iron phthalocyanine, chlorophyll, 16-membered ring, phthalic anhydride, phthalimide, phthalonitrile, resonance-stabilization, π-functional materials

57.1 Fundamentals

Phthalocyanines represent the most important new chromophoric system developed during the twentieth century [1–6]. Products based on phthalocyanines have provided by far the most important blue and green organic pigments for more than 85 years. In contrast, textile dyes based on the phthalocyanine system are of some, but rather limited, importance. The phthalocyanines readily form metal complexes, which have been prepared from most of the common metals in the periodic table, although it is only the copper derivatives that are used as pigments. Copper phthalocyanine (CuPc) (**1**), CI Pigment Blue 15, the structure of which is illustrated in Figure 57.1, is by far the most important blue organic pigment, finding almost universal use as a colorant for paints, printing inks, plastics, and a wide range of other applications. Similarly, its halogenated derivatives provide the most important green organic pigments. Since their serendipitous discovery, the phthalocyanines have become one of the most extensively studied classes of organic compounds, because of their strong, bright colors, their high

This article has previously been published in the journal Physical Sciences Reviews. Please cite as: R. Christie, A. Abel, Phthalocyanine Pigments: General Principles *Physical Sciences Reviews* [Online] 2021, 6. DOI: 10.1515/psr-2020-0194

https://doi.org/10.1515/9783110587128-057

stability, and their unique molecular structure. While their dominant use is as colorants, they are also of interest in a range of other applications (and potential applications), for example in electronic devices, such as displays and organic solar cells, in biological applications, such as photodynamic cancer therapy, and as catalysts [7–11].

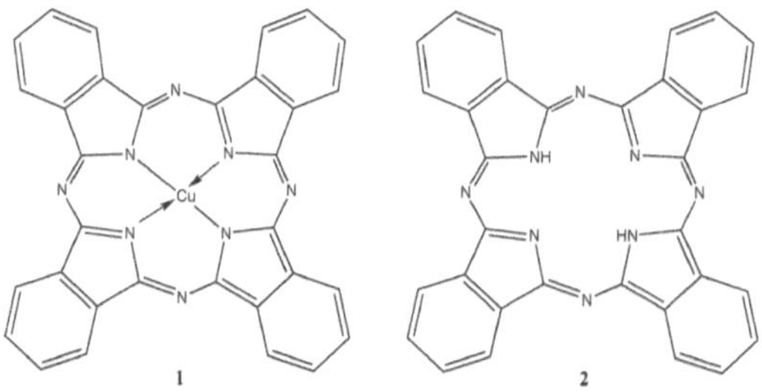

Figure 57.1: The molecular structures of copper phthalocyanine (**1**) and metal-free phthalocyanine (**2**).

57.2 History

The discovery of the phthalocyanine system and its subsequent development into commercial pigments was the start of a new era in organic pigments. Before its discovery, blue was a difficult color for organic pigments. Apart from Indanthrone Blue (CI Pigment Blue 60), an anthraquinonoid pigment, almost all other blue organic pigments had inferior solvent fastness and faded badly when exposed to light. Two inorganic blue pigments were available. However, Ultramarine (CI Pigment Blue 29) lacks color strength and is sensitive to acids, while Prussian Blue (CI Pigment Blue 27) lacks brightness of color and is sensitive to alkaline conditions, such as in paints to be applied on plaster. The earliest report of the formation of a phthalocyanine was in 1907 when Braun and Tcherniac of the South Metropolitan Gas Company, London, reported the formation of a blue substance when investigating the reaction of phthalimide with acetic anhydride, although they did not follow up their discovery [12]. This product was most probably metal free phthalocyanine (**2**). In 1927, de Diesbach and von der Weid of the University of Fribourg, Switzerland, reported that when 1,2-dibromobenzene was treated with copper (I) cyanide in pyridine at 200°C in a sealed tube, a blue product was obtained [13]. They obtained its molecular formula from elemental analysis and observed its remarkable stability to alkali, concentrated acids and to high temperatures. They proposed that it was a copper complex involving phthalonitrile and pyridine in its structure. This product was almost certainly copper phthalocyanine (CuPc) (**1**). The real significance of these early observations was not fully recognized at the time.

The most important event in the industrial history of the phthalocyanines was their accidental discovery by a dye manufacturing company in Scotland. In 1928, a group led by Dandridge at Scottish Dyes Ltd, Grangemouth, observed the formation of a dull, bluish impurity in certain production batches of phthalimide, an intermediate in dye manufacture, which was synthesized by passing ammonia gas into molten phthalic anhydride in a porcelain lined iron vessel. The impurity was isolated as a dark blue insoluble crystalline substance and was found to have remarkable stability. Ultimately, the compound proved to be iron phthalocyanine (FePc), the source of the iron being the wall of the reactor which had become exposed due to a flaw in the lining of the reaction vessel. An independent synthesis involving passing ammonia gas through molten phthalic anhydride in the presence of iron filings confirmed the findings. Without fully appreciating the commercial potential, they pursued their interest and reproduced the blue color using nickel and copper. The discovery was patented in 1928 [14]. However, at this stage, the structure of the compounds remained unknown. Around this time, Scottish Dyes was acquired by the British Dyestuff Corporation, the company that eventually became ICI, with its headquarters in Blackley, Manchester. The elucidation of the structure of the phthalocyanines followed from a pioneering program of research into the chemistry of the system by Linstead of Imperial College, University of London. The structure that we now recognize was first proposed in a 1934 publication based on the results of analysis of a number of derivatives, which confirmed the molecular formula, combined with an investigation of the products obtained from degradation studies [15]. At the time, prior to the introduction of the modern analytical instrumentation that is available now, this was the standard strategic approach to structure elucidation. Ultimately, Robertson confirmed the structure as a result of one of the classical applications of single crystal x-ray crystallography [16]. Linstead gave the name phthalocyanine to the new chromogen.

There is an interesting three part, very old, and low-quality movie about the discovery of CuPc and Linstead's subsequent efforts to identify the structures, available on the internet [17]. However, its content is not recommended for training in laboratory safety procedures! Further discussion of the development of the phthalocyanines that has led to the industrial dominance of the copper complexes as pigments in their characteristic shade areas is presented in the separate chapters on phthalocyanine blue and green pigments.

57.3 Structure and properties

The phthalocyanine system, as illustrated in Figure 57.1 for copper phthalocyanine (1) and the metal-free derivative (**2**) may be regarded as a tetraaza derivative of tetrabenzoporphin. The molecular structures are planar and rigid, consisting of four isoindole units connected by nitrogen atoms, thus forming an internal 16-membered ring of alternate carbon and nitrogen atoms. In the metal complexes, the metal atom is coordinated to the phthalocyanine ring system in a square planar arrangement. The phthalocyanines are related structurally to certain natural pigments, for example the familiar green pigment, chlorophyll (**3**), a magnesium complex of a porphyrin derivative, as shown in Figure 57.2. However, while the natural colorants generally show poor stability, the phthalocyanines are amongst the most stable of all synthetic organic colorants. Copper phthalocyanine, used here as a representative example, is usually illustrated as a single structure (**1**), which contains three outer benzenoid rings and one that is *o*-quinonoid. However, the molecule is centrosymmetric, so that structure (**1**) is only one of a large number of resonance forms that contribute to the molecular structure. The extensive resonance-stabilization of the phthalocyanines is a major factor in ensuring their exceptional stability.

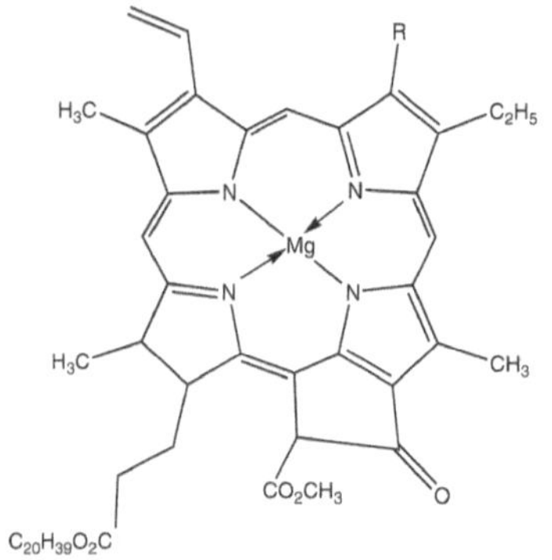

3

R = CH_3 : chlorophyll a
R = CHO: chlorophyll b

Figure 57.2: The structure of chlorophyll (**3**).

Metal phthalocyanines generally provide brilliant, intense colors. The UV/visible spectrum of metal-free phthalocyanine (**2**) in solution shows two absorption bands of similar intensity at 664 and 699 nm. The corresponding spectrum of copper phthalocyanine exhibits a single narrow major absorption band. This feature has been explained by the higher symmetry of the complexed species compared with the metal-free species. These absorption band characteristics explain the brilliance and intensity of the color of CuPc, compared with the less bright color of the metal-free compound (**2**). The exact position of the absorption band of metal phthalocyanines varies with the nature of the central metal ion, the substituent pattern on the outer rings, and the degree of ring annelation [18]. In the complexes of the first transition metal series, iron, cobalt, nickel, copper, and zinc, the color is affected little by the nature of the central metal ion, with λ_{max} values in the range 670–685 nm. In the wider range of complexes derived from the metals throughout the periodic table, the most hypsochromic compound is PtPc (λ_{max} 652 nm) while the most bathochromic is PbPc (λ_{max} 714 nm). Substituents on the outer aromatic rings invariably shift the absorption band to longer wavelengths. Copper hexadecachlorophthalocyanine, for example, absorbs at 720 nm in solution, giving rise to its green color. The colors of classical phthalocyanine dyes and pigments are restricted to blues and greens. However, recent decades have seen investigations of several derivatives, for example phthalocyanine complexes with multiple arylthio substituents and certain naphthalocyanines, whose absorption extends into the near infrared region of the spectrum.

57.4 Synthesis of phthalocyanine pigments

Despite the structural complexity of the metal phthalocyanines, their synthesis can be relatively straightforward. The processes involve simultaneous synthesis of the ligand and metal complex formation in a template procedure. Essentially three components are required: a phthalic acid derivative (phthalic anhydride, phthalimide, phthalonitrile or *o*-cyanobenzamide, a source of nitrogen (in cases where the phthalic acid derivative does not itself contain sufficient nitrogen), and an appropriate metal derivative. Commonly the reaction requires high temperatures carried out in a high boiling organic solvent or as a "dry bake" process, and often requires the use of catalysts. Using appropriate starting materials and reaction conditions, virtually the entire range of metal phthalocyanines may be synthesized. Substituted phthalocyanines may be synthesized either using an appropriately substituted phthalic acid derivative as starting material, or by means of substitution reactions conducted on unsubstituted phthalocyanines. Details of the methods of synthesis and manufacture of copper phthalocyanine blue and green pigments are discussed in the separate chapters that focus on these products.

57.5 Applications

The principal applications of copper phthalocyanines as the dominant organic blue and green pigments are discussed in detail in the two separate chapters on these particular groups of pigments. The phthalocyanine system has in more recent years created much interest well outside traditional applications as colorants, encompassed by the term "functional dye or pigment" [15]. The properties required of these materials are provided by compounds with extended conjugated π-electron systems, and hence they are alternatively referred to as "π-functional materials". These applications may involve color properties, including their use in color filters, in liquid crystal display devices, and inkjet printing applications. Phthalocyanines have also been developed for applications based on their absorption characteristics, for example where are the absorption maximum is shifted towards longer wavelengths in the near infrared region. A use for these materials is as radiation absorbers in CD-R devices. They may play a role in semiconductors, catalysts, lasers and lubricants, as well as being used as deodorizers for textiles. The medical profession has also been attracted to phthalocyanines as photosensitizers for malignant tumor treatment in photodynamic therapy [9–11].

References

1. Moser FH, Thomas AL. The phthalocyanines vols I and II. Boca Raton, Florida: CRC Press Inc; 1983.
2. Leznoff CC, Lever AB. Phthalocyanines: properties and applications. Weinheim: VCH; 1989.
3. McKeown NB. Phthalocyanine materials: synthesis, structure and function. Cambridge: Cambridge University Press; 1998.
4. H. Zollinger, Color Chemistry: Syntheses, Properties and Applications of Organic Dyes and Pigments, 3rd ed., ch. 5, Wiley-VCH, Weinheim, 2003.
5. Christie RM. Colour chemistry, 2nd ed. London: RSC, Ch 5: 2015.
6. Hunger K, Schmidt MU. Industrial organic pigments, 4th ed. Weinheim: Wiley-VCH Verlag GmbH, Ch 3; 2019.
7. De La Torre G, Bottari G, Hahn U, Torres T. Functional Phthalocyanines. Berlin: Springer-Verlag; 2010.
8. Tanaka M. Phthalocyanines – high performance pigments with new applications. In: Smith H, editor. High performance pigments. Wiley VCH, 2002; Ch. 17.
9. Wainwright M. Photodynamic therapy - from dyestuffs to high-tech clinical practice. Rev Prog Color. 2004:34;95–109.
10. Wainwright M. Therapeutic applications of near-infrared dyes. Color Technol. 2010;26:115–26.
11. Nyokonga T, Gledhill I. The use of phthalocyanines in cancer therapy. Proc AIP Conf. 2013:49;1517.
12. Braun A, Tcherniak J. Über die Produkte der Einwirkung von Acetanhydrid auf Phthalamid. Ber Dtsch Chem Ges. 1907:40;2709.
13. de Diesbach H, Von Der Weid E. Quelques sels complexes des o-dinitriles avec le cuivre et la pyridine. Helv Chim Acta. 1927:10;886.
14. Dandridge AG, Drescher HA, Thomas J. GB 322,169. 16 May 1928.
15. Bourne GT, Linstead RP, Lowe AR. Phthalocyanines. Part II. The preparation of phthalocyanine and some metallic derivatives from o-cyanobenzamide and phthalimide. J. Chem. Soc. 1934:1017.
16. Robertson JM. An X-ray study of the structure of the phthalocyanines. Part I. The metal-free, nickel, copper and platinum compounds. J Chem Soc. 1935:613.
17. The discovery of a new pigment – the story of monastral blue by imperial chemical industries. http://www.colorantshistory.org/PhthaloDiscovery.html. Accessed: 14 Feb 2021.
18. Christie RM. Colour & constitution relationships in organic pigments, part 5, the influence of solvents, the central metal atom and substituents on the electronic spectra of phthalocyanines. Dyes Pigm. 1995:27;35–43.

Uwe Hempelmann

58 Pigments: general aspects

Abstract: The chapter defines the terms pigments and fillers according to international standards and gives a short overview over the history and economic aspects and uses. The general common chemical and physical properties are outlined and basic methods for characterization of the pigments and their behavior in binders are described.

Keywords: pigment, filler, binder, characterization of pigments, particle size distribution, primary particles, aggregates, agglomerates

58.1 Introduction

The word pigment as used in the following is, defined in DIN EN ISO 18451–1 "a colorant consisting of particles, which are insoluble in the application medium (e.g., coating material or plastic)" [1]. Pigments can be classified according to their chemical composition or functional use (coloring, corrosion protection or magnetic storage [2]).

The word "pigment" has its origins in Latin ("pigmentum"), where it described generally coloring matter. So-called biological pigments which are produced by living organism to color skin, eyes, feathers or furs are not covered here.

Today, a distinction is made between two types of colorants – dyes and pigments. Dyes are by definition soluble in their application system. In contrast, pigments are insoluble in their application medium. As they are not soluble, the pigment particles are present in the liquid and the cured application medium and play an integral part in the properties of the pigmented system: Absorption and scattering of the pigment particles are responsible for the color of the medium.

The diameter of the particles must be between 50 and 1000 nm, putting them in the range of the wavelength of the visible light. At the same time pigments are also in some cases nanomaterials. As their use is confirmed for more than 100,000 years they easily qualify as the oldest nanomaterial in use [3].

To be an effective scatterer, pigments are characterized by a high index of refraction. Only material with this characteristic can be used as pigments. Relevant in practice is the relation between the index of refraction of the pigment and the medium. This means an effective scatterer in ambient air may be dull in a co ating.

This article has previously been published in the journal Physical Sciences Reviews. Please cite as: U. Hempelmann, Pigments - General Aspects *Physical Sciences Reviews* [Online] 2021, 6. DOI: 10.1515/psr-2020-0212

https://doi.org/10.1515/9783110587128-058

The index of refraction of air is about 1.0 while that of a coating is about 1.5. A comparable effect is the near invisibility of wet chalk on a board compared to the dry.

Good overviews about the causes of color of substances as well as the scattering of particles can be found in [4–6].

The scattering relative to the binder is also the main difference between pigments and extenders (fillers). These are also particular materials that are insoluble in the application medium, but they are used to modify or influence physical properties [1]. The difference is given by the purpose the material is selected for, so the question whether a given substance is called a pigment or a filler depends on the use.

Generally, a substance with an index of refraction below 1.7 can be regarded as a filler when used in a coating.

The properties of a pigment can be described broadly by three different categories:
- the chemistry of the solid substances as such
- the particle size and its distribution as well as the particle shape
- the chemistry of the surface responsible for the interaction with a binder [7].

The chemistry of the solid particle determines the chemical solubility and the complex index of refraction. These properties govern together with particle size (incl. its distribution and the morphology) the optical properties of the pigment. The chemistry also determines the crystal structure of pigment.

The chemistry of the phase interface is important in the respective application: whether or not a pigment is wettable and easily dispersible, whether it succumbs to flocculation or flooding, etc., is determined here.

Inorganic pigments are mainly oxides, sulfides, oxide hydroxides, silicates, sulfate or carbonates. Generally, they consist of a single component, e.g., α-Fe_2O_3, TiO_2 or Cr_2O_3. More often than not, pigments are used in mixtures. It must be taken care to mix them properly and to avoid any separation during the use.

In some cases, an aftertreatment is necessary to make the pigments more amenable for use or more stable against environmental influences. Typically, the aftertreatment is applied by a precipitation reaction on the pigment particle, so that a thin film of an organic or inorganic substance is formed on the surface of the particle. Alternatives are adsorption of suitable substances from solutions (usually aqueous), or by steam hydrolysis.

In case of titanium dioxide this is necessary to suppress the photocatalytic activity of the substance; otherwise, the use as a pigment would be severely compromised. Additionally, the dispersibility of the pigment can be enhanced by giving them specific hydrophilic or hydrophobic surfaces.

One characteristic of inorganic pigments is that they are well crystallized. Typical crystal lattices that can be found are zinc blend (cubic, e.g., CdS), spinel (cubic, e.g., Fe_3O_4, $CoAl_2O_4$), rutile (tetragonal, e.g., TiO_2, SnO_2), goethite (rhombic, e.g., α-FeOOH), corundum (hexagonal, e.g., α-Fe_2O_3, Cr_2O_3), and monazite (monoclinic, e.g., $PbCrO_4$).

The chemical base structure of organic pigments is often based on the corresponding dyes. The pigments can be obtained from the dyes by excluding the solubilizing groups, by formation of insoluble salts, by metal complex formation or by the introduction of groupings which reduce the solubility.

Organic pigments can be classified according to their chemical structure. According to DIN EN 18451, Part 1 [1], there are as follows:
- azo pigments
- polycyclic pigments: copper phthalocyanines
- polycyclic pigments: diketopyrrolopyrroles
- polycyclic pigments: quinacridones
- polycyclic pigments: isoindolines and isoindolinones
- polycyclic pigments: perylenes and perinones
- other polycyclic pigments.

It is important to note that the differentiation according to the chemical structure is no indication of the economic importance of the respective class.

58.2 History

Natural occurring pigments were used since the beginning of humanity. The use of iron oxides as ochre is known for 100,000 years for homo sapiens and also for the homo sapiens neanderthalensis [3, 8]. In neolithic ages there is already evidence that iron oxide yellow was calcined to maghemite or hematite on purpose [9, 10]. This was probably the first chemical reaction utilized on purpose for synthesizing a pigment.

In antiquity [11] the palette was enlarged by the discovery of arsenic sulfide and lead antimonate (Naples yellow) as yellow pigments. Egyptian blue and cobalt aluminum spinel were employed as blue pigments together with ultramarine (lapis lazuli). Egyptian blue is a calcium copper silicate, also called artificial lapis lazuli. If fell from use at the end of the Roman period its synthesis being rediscovered in modern times.

Synthetic copper hydroxychloride together with terra verte and malachite were green pigments. Lead oxide was used as white pigment albeit rarely due to its high price. Chalk and calcites are more common. Antimony sulfide and lead sulfide (galena) were utilized as black pigments. Cinnabar (mercury sulfide) was used as red pigment.

The end of the Roman empire and the age of migration led to a decline of knowledge about the synthesis and application of pigments. New development, reinventions came at the end of the Middle Ages and the Renaissance.

The start of the industrial pigment production dates back to the eighteenth century. Iron blue, cobalt blue, Scheele's green and chrome yellow are among the first pigments to be produced on an industrial scale. In the nineteenth century, synthetic ultramarine, cadmium yellow pigments and iron oxide pigments were developed [12].

In the twentieth century, titanium dioxide was introduced as the white pigment of choice due to its scattering efficiency. Organic pigments as well as inorganic mixed metal oxide pigments were introduced [13].

In the last decades pearlescent and interference pigments with their angle dependent color introduced new way of coloring [14].

In the last decades there are quite few new substances which are introduced as new pigments although from time to time the discovery of new chromophore is published. An example is the new blue pigment based on three valent manganese in a trigonal bipyramidal coordination [15, 16]. It remains to be seen whether this and other development will ever enter the market.

The main development in the field of pigments and extenders is dedicated to ease the handling and application. Reducing the time and energy necessary to obtain a proper dispersion by aftertreatment and additives is one direction. New form of delivery as granules and slurries is another way forward.

58.3 Economic aspects and uses

The most important applications and uses of pigments and fillers are of course paints, varnishes, plastics and artist's colors. Others are printing inks for paper, textiles and ceramics, leather decoration, building materials, leather imitates, floor coverings rubber, and cosmetics.

Pigments and fillers form an important presence in everybody's daily life. Besides some protective effects within a coating, they are also part of creating a livable environment as part of the 17 UN sustainability goals.

The worldwide production of pigments of all kind was estimated to be about 11 million tons in 2019; more than the half of this is titanium dioxide. The rest consists of carbon black and inorganic and organic pigments [17].

Iron oxide pigments form together with carbon black and zinc oxide the next group with a high production volume of about one million ton. In both cases it is difficult to give more exact figures as the majority of the carbon black and the zinc oxide production as well is not used as a pigment and in case of iron oxide pigments there are both natural and synthetic iron oxide pigments which have to be taken into account.

Organic pigments of all classes are estimated to be about 350,000 t/a. Chromium oxide pigments, bismuth vanadate pigments, mixed metal oxide pigments, ultramarine pigments are thought to be below 400,000 t; transparent effect pigments and metal effect pigments together are estimated to be about 170,000 t/a.

58.4 General chemical and physical properties

The paint industry demands high-quality pigments where a uniform particle size is important and necessary to obtain the optimal hiding power, tinting strength and gloss. Thus, the paint film can be as thin as possible.

The following properties are important beside the color of a pigment and of a filler:

- Stability behavior: lightfastness, weather resistance, thermal and chemical resistance, anticorrosive properties.
- Behavior in binders, plastics and construction materials: dispersibility,compatibility with the binder.
- General physical and chemical properties: chemical composition, moisture and salt content, content of soluble matter in water or acid, particle size, density and hardness.

Most of the methods used to determine the characteristics of pigments and fillers are available as international ISO or EN standards allowing a common ground for communication between producers, customers and also the authorities. The majority of them are also available as national standards.

The set of standards was acquired by all stakeholders involved in the production and application of pigments in open and well-organized processes. They cover all aspects of interest mentioned above. Additionally, delivery standards are available for certain pigments (e.g., titanium dioxide, iron oxide, chromium oxide). These standards define the properties of the pigments inquestion and form a basis for communication between producer and customer.

The standards are the result of the continuous dedicated work of several decades. They require a constant effort by the stakeholders to keep them up to date.

58.4.1 Particle size distribution

As already mentioned beside the index of refraction the particular properties are the defining properties for the color and application characteristics of pigments: particle size and morphology. As it is technically next to impossible to produce pigment particles of uniform size and shape the distribution of different particle sizes over a certain range has also to be taken into account.

Both the particle size and shape in itself together with the width of the particle size distribution determine the color and characteristics of a given pigment. Particle sizes of pigments are usually in the range from 0.01 to 10 µm. For the vast majority of pigments, they lie between 0.1 and 1 µm.

There are several ways to present the results of a particle size measurement: It must be differentiated between number, length, surface and volume distributions, depending on the parameter chosen as criterion. Technically the volume distribution is the most important as well as the standard result of next to all measurement instruments as most of the effects used (sedimentation, light scattering etc.) scale by volume resp. mass and not by number. For regulatory purposes the number distribution is often chosen for political reasons.

When discussing the size of pigments as well as every powdery material it has to be recognized that practical all powders consist of primary or constituent particles as well as aggregates and agglomerates, which were defined [1]:

- primary particles: single particle recognizably as individual detectable particle
- aggregates: unit grown together of facially contiguous primary particles, the surface is smaller than the sum of surfaces of the primary particles
- agglomerates: unit not grown together of primary particles and/or aggregates contiguous, e.g., on corners and edges, the total surface does not basically deviate from the sum of the single surfaces

The differentiation was used for the first time by Honigmann und Stabenow to describe the dispersion of pigments in 1962 [18].

The main interest in most cases is the measurement of the size and the size distribution of the primary particles. When delivered pigments and fillers are typically highly aggregated and so there is a need to disperse them rather energetically and for a sufficient time to ascertain that the end of the dispersion process for the given energy is obtained and the results are not dependent on the time any more.

It is of utmost importance to define this process as exact as possible to get results which are reproducible as well as comparable with other laboratories or even methods. The start of this process, this is often forgotten, is the “sample preparation” bitte durch “sampling” ersetzen where well-known separation processes of the coarse and the fine particles have to be conspicuously avoided.

For the measurement of the volume distribution, centrifuges or instruments which use light scattering are established offering quick and reproducible results. Both kind of instruments need the input of optical constants of the material in question to give correct results.

The software of those instruments uses theoretical models that assume all particles being spheres. Sometimes there is a provision in the software of a so-called shape factor, but typically details about this calculation as well as the probable effect remain unclear.

Electron micrographs are the main tool for obtaining information about the morphology of pigments and filler particles as well as information about the number distribution. The physical effects used in centrifuges as well as instruments based on light scattering scale with mass resp. volume of the particles, so the particles size distribution which is directly obtained by the measurement is the volume distribution. While the calculation of the number distribution out of a volume distribution is straightforward some simple mathematics shows that the accuracy of the result is very limited.

In all technical matters this poses not much of a problem as the number distribution is of very limited interest, but in case of regulatory demands this renders these instruments rather useless.

The counting of an electron micrograph is time consuming and rather tedious task, albeit there is an increasing amount of automatization possible.

Nevertheless, due to rather limited numbers of particles selected for enumeration compromises in accuracy and reproducibility must be accepted. As the number distribution is mandatory by the European Chemical Agency (ECHA) for the registration process under the REACH Regulation, the use of electron micrographs is without alternative.

Another and very old method to get information on the particle sizes of pigments and fillers is the sieve analysis. As already mentioned, the particle size of pigments and fillers is well below the smallest sieve size available (5 µm). But of course, this technique is valuable in case of granules.

Sieve analysis can be utilized to determine the particle size distribution by using a set of sieves, often in so-called sieving machines. Sieving is mostly applied to determine the sieve residue at 45 µm as this is an important indicator of quality. Typical components of the residue are aggregates, agglomerates and coarse impurities.

There are two important methods for determining the sieve residue:

- Wet sieving by hand: Drying of the pigment followed by washing with water on a sieve with a specified mesh width. Weighing of the dried residue on the sieve
- Wet sieving using a mechanical rinsing process: The steps are naturally identical but a special instrument is used (Mocker's apparatus).

Additionally, wet sieving down to a mesh size of 5 µm can be realized by applying special sieves of pure nickel membranes. The material is fluidized in an ultrasonic bath.

The specific surface area describes the size of a surface area per given quantity of a solid. Next to always gas adsorption is used and the methods of Brunauer, Emmett and Teller (BET) is used in combination with nitrogen acting as the adsorbent. The limitations of the methods concerning microporous materials should be aware as well as the existence of other calculation methods, like de Boer's t-method.

58.4.2 General methods for characterization

The production of pigments is strictly controlled to achieve the constant quality which is necessary for products of high quality. Quantitative chemical and X-ray analysis is carried out on raw materials, intermediates, and substances used for aftertreatment, but most importantly on the final products, byproducts, and waste products (wastewater and exhaust gas). This serves not only to fulfil quality requirements but also the demands of environmental protection.

The following analytical techniques are used today for the characterization of pigments:
- Atomic absorption spectroscopy (AAS) and inductively coupled plasma (ICP), for the quantitative determination of a variety of metals in the pigments and of impurities.
- Differential thermal analysis (DTA) and thermogravimetric analysis (TGA), for the investigation of reaction sequences and of decomposition temperatures, e.g., thermal decomposition of precursors of inorganic pigments, determination of the temperature for water release, crystallization processes.
- Diffuse reflectance spectroscopy, for the coloristic investigation of pigments and for the monitoring of the pigment production (visible and in some cases also near infrared part of the spectrum).
- Electron diffraction analysis, for the structural identification of microcrystals and crystallites.
- Electrokinetic sonic amplitude (ESA), for the measurement of the charge on the surface of pigments.
- Helium pycnometry, for the determination of the specific gravity or density of pigment powders.
- Nitrogen adsorption, for the determination of the surface area and porosity of pigment powders.
- Nuclear magnetic resonance spectroscopy (NMR), for the identification of pigments and related materials.
- Scanning electron microscopy (SEM) and transmission electron microscopy (TEM) for the determination of particle sizes and shapes of pigments.
- VIS spectroscopy, for the identification of a specific pigment composition.
- IR- and FT-IR spectroscopy, for the identification of a specific pigment composition.
- Raman spectroscopy, for the identification of a specific pigment composition.
- X-ray diffraction measurement (XRD) and X-ray fluorescence spectroscopy (XRF), for the determination of crystal structures and the identification and (semi-) quantification of certain elements.
- Mass spectroscopy, for the identification of organic pigments or of organic components in inorganic pigments.
- Wet chemical analysis, for the qualitative and quantitative determination of certain pigment components

In a given ideal solid ionic compound, the absorption of light is caused by the spectra of the individual ions, comparable to the situation in an ionic solution. For metal ions where s, p, or d orbitals are filled, the first absorption band is in the ultraviolet range as for the heights of corresponding excited energy level. Oxide and fluoride compounds are mainly white for this reason. Transition metal chalcogenides with their partly occupied d and f orbitals are characterized by absorption spectra dominated by the charge-transfer spectrum of the respective chalcogenide ion with its noble gas structure. In transition metals, lanthanides, and actinides, the energy difference between ground state and first excited state corresponds to the energy of visible light, which leads to colored compounds [17].

X-ray investigation is one of the most important tools to determine structure of inorganic pigments. In this way, fine structure and lattice defects can be analyzed as well as the crystallite size can be determined. It is important to note that crystallite size need not be identical with particle size as measured by the electron microscope or other means. But it can be related to the magnetic properties of the pigment.

In the following, some important methods to determine some general physical characteristics of pigments are described.

58.4.2.1 Volatile matter and loss on ignition

The content of volatiles in a pigment is determined by drying a sample in an oven at 105 ± 2 °C. This normally gives a measure of the moisture content.

The determination of the loss on ignition varies between different pigments due to their chemistry and physical characteristics. Essential for all methods is: a sample is weighed, heated to a predetermined temperature, cooled in a desiccator, and reweighed. There are special standard procedures for iron oxide pigments and carbon black.

Quality control instruments using IR-radiation for heating are often employed due to their ease of handling. In these cases, it must be established that no irreversible changes in the pigment are caused as sometimes the IR-radiation heats the sample to significantly higher temperatures than 105 °C which may cause irregular results or even hazards at the workplace.

58.4.2.2 Aqueous extracts, pH values and conductivity

There are defined test procedures for pigments for the determination of the content of matter soluble in water either by hot or by cold extraction.

The pH value of a pigment is naturally not the pH of the powder but that of an aqueous suspension of the pigment which is prepared in a precisely described way.

The electrical conductivity (or resistivity) of an aqueous pigment extract is determined again starting with a suspension of a pigment but here the suspension is centrifuged before measurement.

The procedures to determine the content of water-soluble ions such as sulfates, chlorides and nitrates in pigments and extenders are described in several international standards.

In case of matter which is soluble in hydrochloric acid, the pigment is extracted with 0.1 mol/L hydrochloric acid under prescribed conditions. The content of a large number of metal ions such as barium, cadmium, cobalt, copper, nickel, mercury, or zinc can be determined in this way. The results are important for regulatory purposes.

58.4.2.3 Pigment density

The density of pigment powders is determined by a common pycnometric method. A standard temperature of 25 °C is used. The apparent density of a powdered material (mass in gram of 1 cm^3 of the material) is measured after tamping of the powder in a tamping volumeter under defined conditions. It depends mainly on the true density, but also on the shape, and the size of the particles.

58.4.3 Stability towards light, weather, heat and chemicals

Color loss or structural changes may happen when pigmented systems are exposed to intense radiation and weathering. Yellowing, chalking and loss of gloss are common signs of such reactions. Photochemical degradation of the binder is the main reason behind such phenomena. Pigments can have a catalytic effect in the interaction with the binder or suffer degradation themselves.

Inorganic pigments are chemically inert in almost all cases and can be regarded as the most stable colorants. Particularly pigmentary metal oxides can have a protective effect on the substrate beside the barrier effect of the coating. Sulfides can be oxidized which is one cause of the degradation of some rather famed piece of art dating in the nineteenth century [19–22].

If a binder breaks down there is a significant loss of gloss. Surface gloss gives an attractive appearance and is usually desired. One prerequisite for an aspired gloss is the good dispersion of the pigment particles. A phenomenon called chalking occurs when the degradation of the binder is so severe that the pigment particles come off at the surface of the application system.

The stability towards light and weather depends predominantly on the composition of the binder and of the pigments used. Other factors are defects in the coating, the morphology of the pigment particles as well as the pigment concentration in the application medium. It goes without saying that the medium in which the pigments are used is the most important factor for any eventual degradation. Stability investigations are mostly done by open-air weathering (accelerated weathering) or in weathering test chambers in the laboratory.

Special apparatuses are available for weathering tests, where samples are exposed to a predetermined exposure of light and rain. The light should be simulated to match the direct solar radiation together with the indirect radiation from the sky. The standard illuminant D65 bitte ohne Leerzeichen is mainly used. Color changes during the test duration are measured using standard methods [23, 24].

Unfortunately, these accelerated tests only give rather limited insight into the performance of a given pigment/binder performance under real conditions. It is not possible to simulate all weather conditions much less to accelerate them: Good correlation in one binder system does not guarantee a similar result in another.

Even natural weathering tests can be problematic. As there can be a considerable difference of the weather of consecutive years the results can vary considerably.

Thermal or heat stability of pigments is of great importance in many applications, e.g., in coatings or plastics. Pigments based on metal oxides are characterized by the highest thermal stability. Most of the sulfide pigments are also sufficiently stable in the relevant application systems, but do not achieve the stability of oxide pigments. Some sulfide pigments are even stable enough to be used in enamels, and glass melts (e.g., cadmium sulfide pigments). Pigments based on oxide hydroxides or carbonates are normally less stable than those based on pure oxides. As in the case of weathering the binder has also a significant influence on the heat stability. Another factor is the duration of the heating.

The thermal stability of pigments is usually determined by the measurement of the hue values of pigment-containing coatings or paints at the beginning, during and at the end of a defined heat treatment. The yellowing of white coatings is often used for such measurements.

Inorganic pigments are characterized by high chemical resistance. Metal oxides are also more stable in this respect than other pigment compositions.

Other fastness properties that are often investigated are the overspray fastness, the fastness to blooming, and the fastness to plasticizers These investigations play a more important role when organic pigments are used in the application medium.

58.4.4 Pigment-binder interaction

A pigment-binder dispersion can be regarded as a suspension from the physical-chemical point of view. After drying it can be regarded as solid sol. Three main steps can be defined for the dispersion of pigments and fillers in a binder system:

- Wetting: the air attached on the surface of the pigment particle is removed and a solvate layer is formed.
- Disintegration: agglomerates are destroyed with the help of suitable dispersing technique.
- Stabilization: preservation of the dispersed state taking advantage of repulsive forces between the particles, if necessary, additives must be used.

The dispersing is a time and energy intensive step in the production of paints. The characterization of the dispersing process is therefore an important undertaking. The viscosity of the pigment/binder mixture is typically non-Newtonian, which has to be taken care during manufacturing.

The degree of dispersing can be followed by measuring color, tinting strength, gloss or gloss haze all these properties being sensitive to the dispersing process. There are standards which can be used to describe and compare the dispersing process.

In practice more often than not mixtures of different pigments and extenders are employed and detrimental effects such as flooding, the separation of different pigments, or flocculation must be taken into account. A so-called rub-out test can show the tendency for such an effect.

There are established tests to characterize the behavior of pigments and fillers in binders:

Oil absorption: This old but nevertheless very accurate method determines the volume or the mass of a special linseed oil, which is used to form a coherent putty-like substance. The amount of oil is slowly added via a burette while constantly working the mixture with a spatula with a steel blade. Typically, a single drop changes the mixture to smearing. The result is given per 100 g pigment.

Binder absorption, smear point, and yield point: These parameters are determined to get knowledge about the amount of binder needed to for the formulation of a workable mill base for grinding. Here dissolvers, attritors, roll mills, ball mills, sand mills, and pearl mills can be used. The procedure is similar to the one described above.

Viscosity: The viscosity of a paint is a very important parameter for application and handling of paints. Based on the knowledge about the viscosity, the necessary forces in the initial brushing phase of a coating can be estimated. As all paints show non-Newtonian behavior a rotational viscosimeter is necessary for describing it properly. Here the torque applied to a rotating cylinder or disk in or on the surface of a suspension is measured. In non-Newtonian suspensions, the apparent viscosity is calculated by dividing the shear stress by the shear rate.

Fineness of grind: This parameter is determined using a special grind gauge (grindometer) consisting of a steel block and a groove. Typical values for the depth of the groove are 50, 100 or 250 µm. The pigment sample is placed for the measurement at the deep end of the groove and is drawn to the other end of the groove with a scraper. The fineness of grind is characterized by the depth at which a large number of particles become visible on the surface of the sample as pinholes or scratches.

There are also test methods dedicated to low viscosity and high viscosity media: The measurements determine the time needed to make homogeneous suspensions with different dispersing methods. For low viscosity media an oscillatory shaking machine is often used. The apparatus has the advantage to use several containers so parallel testing is possible.

The development of fineness of grind is a parameter often successfully employed. Samples of the product are investigated in this case with the grind gauge at various stages of the dispersion process (after predefined times).

The determination of the fineness of grind is very suitable for the comparison of diverse pigment qualities and of different dispersing aggregates. This is relevant as coarse particles are often the cause of complaints in the final good.

The dispersion behavior of different organic or inorganic pigments can be compared by additional parameters beside the fineness of grind. These include the change in tinting strength and parameters known from chemical reaction kinetics [23].

Pigment Volume Concentration (PVC) and Critical Pigment Volume Concentration (CPVC) are additional parameters to characterize the behavior of pigments in a binder. The pigment volume concentration (σ) is the fractional volume of pigment in the total solids volume of the dry paint film (V_p pigment volume, V_b binder volume):

$$\sigma = \frac{V_p}{V_p + V_b}$$

The determination of the PVC is carried out by removing the pigment share from a weighed binder sample. Methods such as the extraction of the (liquid) binder with suitable organic solvents or the combustion of the pigmented binder followed by the analysis of the residue are used. The extraction is sometimes carried out with the help of centrifuges, while sulfuric acid is used in the combustion process.

The pigment particles are under CPVC conditions in the state of the maximum packing density. At the same time, the interspaces between the pigment particles are completely filled with binder. When using smaller binder quantities, the interspaces are only incompletely filled. This results in noticeable differences for the pigmented system. Including this consideration, it can be stated that the CPVC represents a pigment concentration boundary at which significant property changes of the pigmented binder are observed.

There are two groups of determination methods for the CPVC:
- Determination by observation of a sudden property change (e.g., color, gloss) in a series of samples with different pigment volume concentrations.
- Determination by experimental generation of the pigment concentration boundary in a selected pigment binder system (dense pigment packing corresponding with the CPVC, measured by means of oil absorption and others).

Standardized regulations describe test methods for the investigation of the dispersion behavior and of further properties of pigments and fillers in other application systems than coatings and paints, in particular in plastics, building materials, cosmetics as well as in printing inks:
- Plastics. Defined mixtures of pigments and polymers are produced for the specific tests followed by processing to pigmented plastics. The test systems are based on polyvinyl chloride, polyolefins, polystyrene and other technically relevant polymers [2]. Bleeding, i.e., the migration of pigment particles from inner parts of the polymer sample to the surface, may cause application problems. The bleeding behavior has therefore to be tested over a specific period of time with a

suitable equipment before the technical use takes place. Other test procedures for pigmented plastic systems are the determination of the heat stability and of the increase in strength caused by cold rolling.

- Building materials: Several methods exist for the characterization of pigmented cement and concrete systems. Of particular importance are tests concerning the influence of pigment type and quantity on setting properties and on the hardness of the building material. Other relevant tests include the relative tinting strength, the lightfastness, the colorfastness in cement, the colorfastness in lime, and the heat stability.
- Cosmetics: Most of the tests in this application field have to be done in the final cosmetic products such as lipsticks, concealers, foundations, face powders, contour powders, eye liners, or nail polishes. In order to obtain reproducible results, the pigments have to be ground and dispersed very carefully as in all other applications. The measurement of rheological properties by means of viscometers is of great importance for cosmetic products. Statements on the stability of products, on the flow properties and on shelf life are possible on the basis of periodically measured viscosities. Dynamic and oscillatory rheological measurements are applied.
- Printing inks: A variety of specified test methods exist in the application segment printing. The measurement of the reflectance of pigmented paper and board (non-fluorescent) and of the opacity or transparency of paper are of high application-technical interest [2]. Other methods of investigation include scanning electron microscopy (topography of the printed film), abrasion testing (abrasion resistance), viscosity measurement of the printing ink, contact angle measurement of the printing ink, determination of the slip and anti-slip properties of the ink (static and dynamic coefficient of friction), and gravimetric measurement (determination of water vapor permeability).

References

1. DIN EN ISO 18451–1. Pigments, dyestuffs and extenders – terminology, part 1: general terms. Sep 2019.
2. DIN EN ISO 18451–2. Pigments, dyestuffs and extenders – terminology, part 2: classification of colouring materials according to colouristic and chemical aspects. Dec 2018.
3. Henshilwood CS, d'Errico F, Van Niekerk KL, Coquinot Y, Jacobs Z, Lauritzen S-E, Menu M, Garcia-Moreno R. A 100,00-year-old ochre-processing workshop at Blombos Cave, South Africa. Science. 2011;334:219.
4. Tilley RJ. Colour and the optical properties of materials. 3rd ed. Wiley & Sons Ltd, 2020.
5. Nassau K. The physics and chemistry of color. 2nd ed. New York: Wiley & Sons Inc., 2001.
6. Elias M, Lafait J, editors. La Couleur - Lumière, vision et matériaux. Paris: Editions Belin, 2006.
7. Winter G. Anorganische Pigmente: Disperse Festkörper mit technisch verwertbaren optischen und magnetischen Eigenschaften. Fortschr Miner. 1979;57:172.

8. Zilhão J, Angelucci DE, Badal-Garcia E, d'Errico F, Daniel F, Dayet L, Douka K, Higham TFG, Martinez-Sánchez MJ, Mondes-Bernádez R, Muzia-Mascarós S, Pérez-Sirvent C, Roldán-Garcia C, Vanhaeren M, Villaverde V, Wood R, Zapata J. Symbolic Use of Marine Shells and Mineral Pigments by Iberian Neandertals. Proc Nat Acad Sci. 2010;107:1023.
9. Pomiès MP, Menu M, Vignaud C. TEM observations of goethite dehydration: application to archaeological samples. J Eur Ceram Soc. 1999;19:1605.
10. Helwig K. A note on burnt yellow earth pigments: documentary sources and scientific analysis. Stud Conserv. 1997;42:181.
11. Noll W. Anorganische Pigmente in Vorgeschichte und Antike. Fortschritte der Mineralogie. 1979;57:203.
12. Ball P. Bright earth: art and the invention of color. Farar Straus Giroux, 2002.
13. Smith HM. High performance pigments. Weinheim: Wiley-VCH, 2002.
14. Pfaff G, Gabel P, Kieser M, Maile FJ, Weitzel J. Special effect pigments. 2nd ed. Hannover: Vincentz, 2008.
15. Smith AE, Mizuguschi H, Delaney K, Spaldin NA, Sleight AW, Subramanian MA. Mn^{3+} in trigonal bipyramidal coordination: a new blue chromophore. J Am Chem Soc. 2009;131:17084.
16. Smith AE, Comstock MC, Subramanian MA. Spectral properties of the UV absorbing and near-IR reflecting blue pigment, $YIn_{1-x}Mn_xO_3$. Dyes Pigm. 2016;133:214.
17. Pfaff G. private communication.
18. Honigmann B, Stabenow J. Über Teilchengrößenbestimmungen an Pigmenten. In: VI. Fatipec congress. Wiesbaden: Verlag Chemie, 1962:89.
19. Berns RS. Color and the visual arts. Los Angeles: Getty Trust Publications, 2016.
20. Kirchner E, Van Der Lans I, Ligterink F, Hendriks E, Delaney J. Digitally reconstructing Van Gogh's'Field with Irises near Arles'. part 1: varnish. Color Res Appl. 2018;43:150.
21. Kirchner E, Van Der Lans I, Ligterink F, Geldof M, Gaibor AN, Hendriks E, et al. Digitally reconstructing Van Gogh's'Field with Irises near Arles'. part 2: pigment concentration maps. Color Res Appl. 2018;43:158.
22. Kirchner E, Van Der Lans I, Ligterink F, Geldof M, Megens L, Meedendorp T, et al. Digitally reconstructing Van Gogh's'Field with Irises near Arles'. part 3: determ.ining the original colors. Color Res Appl. 2018;43:311.
23. Völz HG. Industrial colorimetry. Weinheim: Wiley-VCH, 2001.
24. Kettler W, et al. Colour technology of coatings. Hannover: Vincentz Network, 2016.

Heinz Mustroph

59 Polymethine dyes

Abstract: There are a lot of different dye classes with complete different chemical structures. On the basis of a common classification the term “polymethine dyes” was created in 1926. Common criteria are a conjugated polymethine chain with terminal functional groups and an odd number $2n + 3$ of π-centers and $2n + 4$ π-electrons (n is the number of vinyl groups), the vinylene shift of the 0–0 vibronic transition of about 100 nm per vinyl group, the influence of substituents or heteroatoms according to Dewar´s Rules and a bathochromic shift by twisting of the polymethine chain according to the Brunings–Corwin effect.

Keywords: Brunings–Corwin effect, Dewar’s Rules, polymethine chain, polymethine dyes, electronic spectroscopy, vinylene shift

The term *polymethine dyes* was introduced by Walter G. König to systematize the structural relationship between different dye classes [1]. Nowadays this term is often used in dye chemistry to characterize and define dye classes with special electronic structures and whose electronic spectra are influenced in particular ways by structural changes [2–6].

The eponymous structural element of polymethine dyes is a conjugated chain of methine (–CX=) groups, which connects terminal functional groups. X is usually H, but can represent substitution with alkyl, halogen, aryl or heteroaryl systems, as well as functional groups based on nitrogen, oxygen or sulfur heteroatoms.

Most importantly, polymethines are characterized by an odd number $2n + 3$ of π-centers and $2n + 4$ π-electrons (where n is the number of vinyl groups –CH=CH–). This special feature has marked impact on their electronic structure and thus their equilibrium structure. The π-charge density distribution along the polymethine chain alternates markedly, while the differences in the carbon–carbon equilibrium bond lengths in symmetrical dyes are small, but they are not completely identical. According to the alternating π-charge density distribution along the polymethine chain Dewar’s Rules predict an alternating influence of substituents on the chain or heteroatoms in the chain on the 0–0 vibronic transition depending on their electronic properties [7]. Each additional vinyl group in polymethines gives a bathochromic shift of the 0–0 vibronic transition of about 100 nm (the so-called vinylene shift) [1, 5, 8, 9] and twisting of the polymethine chain results in a bathochromic shift of the 0–0 vibronic transition (Brunings–Corwin effect) [10].

This article has previously been published in the journal Physical Sciences Reviews. Please cite as: H. Mustroph, Polymethine Dyes *Physical Sciences Reviews* [Online] 2019, 4. DOI: 10.1515/psr-2020-0084

https://doi.org/10.1515/9783110587128-059

In the strict meaning of the term polymethines, it would include also *polyenes* **1**, which consist of methine groups only. However, polyenes are characterized with an even number of π-centers (methine groups), the same number of π-electrons, equal π-charge densities on every methine group and alternating single and double bonds with substantial differences in the carbon–carbon equilibrium bond lengths. Each additional vinyl group incorporated into the polyene chain gives a bathochromic shift of the 0–0 vibronic transition of about 30 nm [8, 9]. Twisting of the (nearly) single bonds results in a hypsochromic shift of the 0–0 vibronic transition. The most commercially important group of polyene colorants are *carotenoid dyes*.

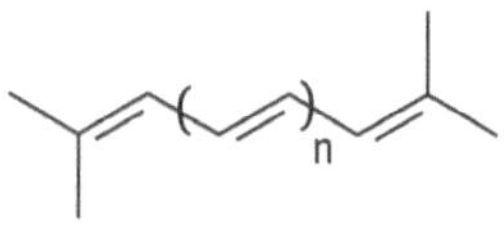

1

Due to the considerable difference in features and independent from the only common structural element, the methine groups, polyenes stand apart from polymethines in the sense of the above definition! Therefore, polymethines cannot be reduced to that of donor/acceptor-substituted polyenes, as it is often done.

For further illustration, in the following classification only nitrogen and oxygen as heteroatoms in the terminal functional groups are considered. The polymethine dyes cover a wide variety of structural types and roughly they can be sub-divided into three categories, cationic, **2**, anionic **3** and neutral (having formal unit electrical charges of opposite sign) **4**.

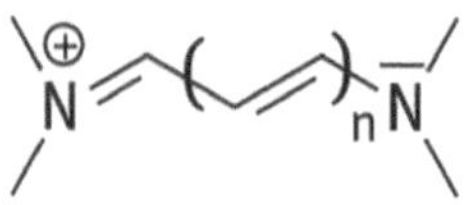

2

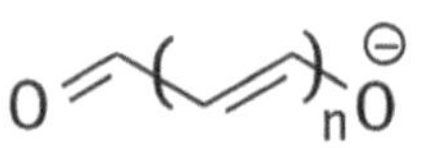

3

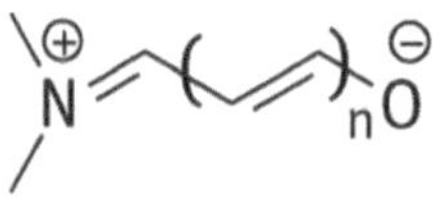

4

59.1 Cationic polymethine dyes

The first synthesized cationic polymethine dye exhibited a cyan color, which gave the whole dye class the name [9]. Today all substances in which both nitrogen atoms in **2** form part of heterocyclic rings that are in conjugation with the polymethine chain are called *cyanine dyes*. There are many examples in which the two end groups are either the same (symmetrical cyanine dyes like **5**) or different (unsymmetrical cyanine dyes such as **6**).

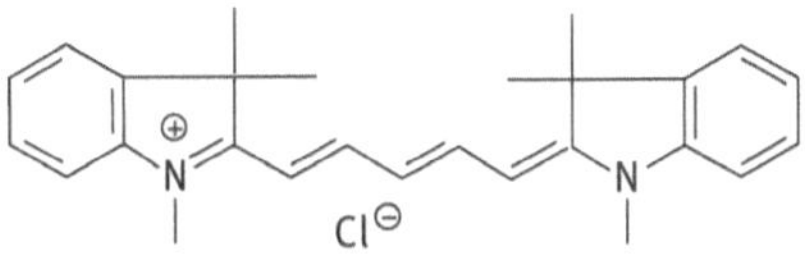

5

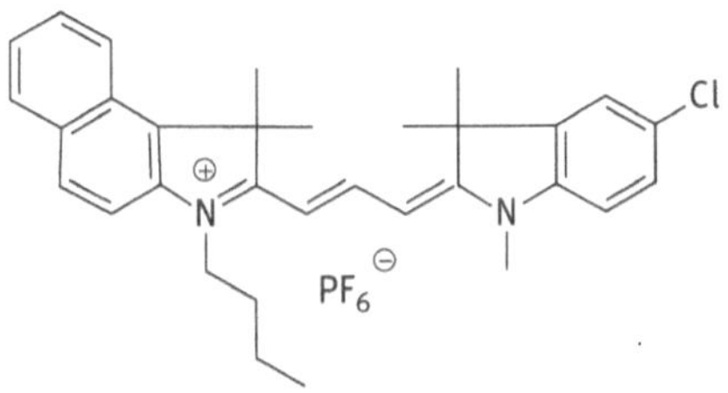

6

Dyes in which a 4-R_2N-aryl group replaces one of the two terminal heterocyclic rings form a subclass known as *hemicyanine dyes* **7**. The name alludes to their *half-cyanine* substitution pattern.

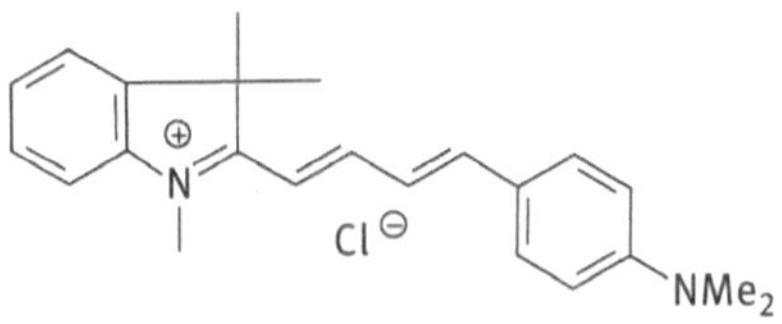

7

Switching the second heterocycle for another 4-R_2N-aryl group gives the *diarylmethine dyes*. The name points out that the two aryl rings are linked with just a single non-aromatic methine fragment. The reader should pay attention that an old and wrong terminology is still widely in use, e. g. diarylmethine and triarylmethine dyes are often still referred to as diarylmethane and triarylmethane dyes, respectively. However, the central C-atom is a sp^2 hybridized methine and not a sp^3 hybridized methane atom.

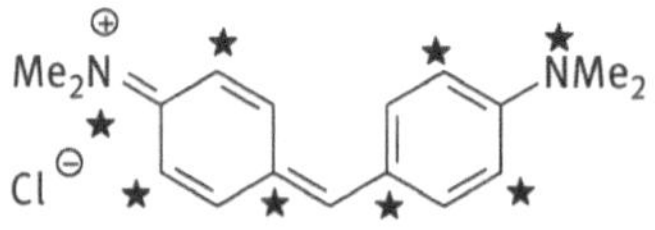

8

λ_{max} = 607 nm (ε_{max} = 148,000 M^{-1} cm^{-1})

At first glance it might be not very obvious that the diarylmethines belong to the polymethines. With the sample of *Michler's Hydrolblue* **8** the three basic features of polymethines shall be illustrated. According to Dewar's Rules substitution in un-starred positions by an electron donating substituent causes a hypsochromic and by an electron accepting substituent a bathochromic shift. Vice versa, substitution in starred positions by an electron donating substituent causes a bathochromic and by an electron accepting substituent a hypsochromic shift.

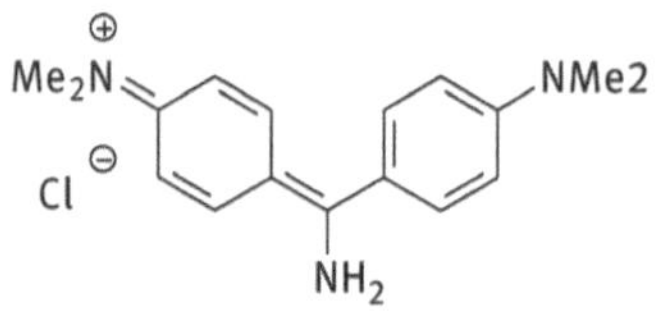

8a

λ_{max} = 431 nm (ε_{max} = 25,300 M^{-1} cm^{-1})

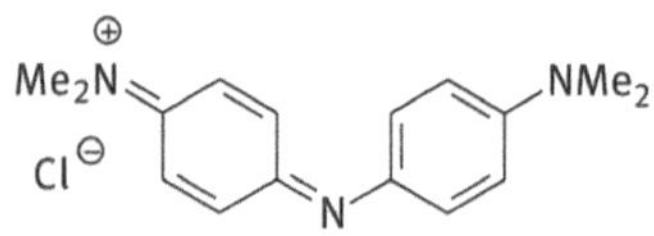

8b

λ_{max} = 728 nm

The central methine C-atom is on an un-starred position. Substitution of the strong electron donor NH_2 in **8a** (*Auramine O*) leads to a strong hypsochromic shift of the absorption maximum. Vice versa, replacement of –CH= by the strong electron acceptor –N= in **8b** (*Binschedler's Green*) results in a substantial bathochromic shift.

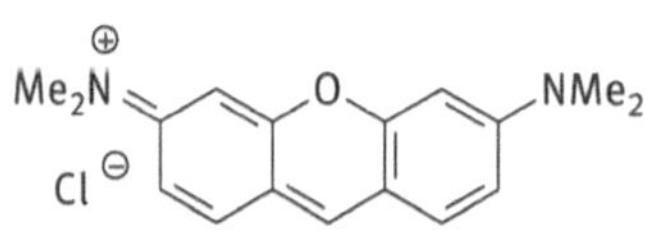

8c

λ_{max} = 546 nm (ε_{max} = 70,000 M^{-1} cm^{-1})

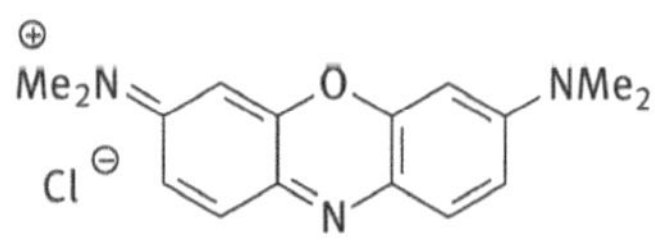

8d

λ_{max} = 642 nm (ε_{max} = 123,000 M^{-1} cm^{-1})

According to these rules substitution in un-starred positions of *Michler's Hydrolblue* and *Bindschedler's Green*, respectively, by the electron donating O to the *pyronine dyes* (*Pyronine G*, **8c**) and *oxazine dyes* (*Oxazine 1*, **8d**), respectively, causes a hypsochromic shift.

However in these cases it is not the only effect. The two *ortho*-methyl groups in **8e** causes a twisting of the C–aryl bond, which results in a bathochromic shift in comparison with **8** (Brunings–Corwin effect) [10]. Vice versa, planarization due to the cyclyzation by the oxygen bridge in **8c** and **8d** leads to an additional hypsochromic shift.

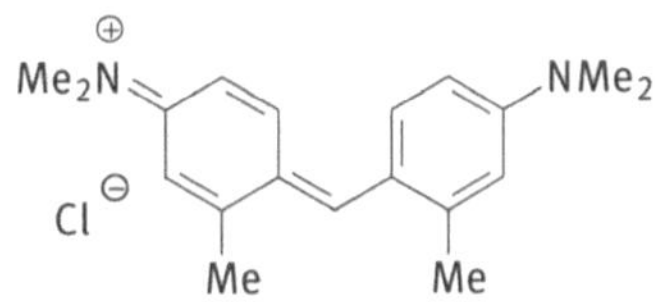

8e

λ_{max} = 623 nm (ε_{max} = 121, 000 M^{-1} cm^{-1})

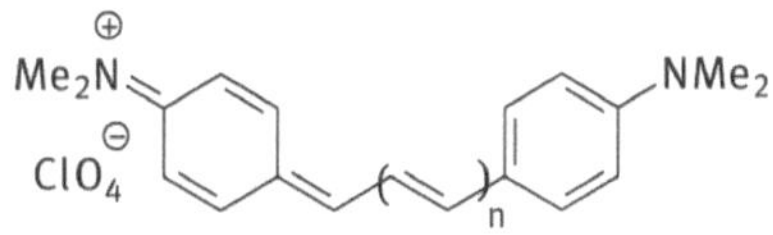

8f

(n = 1) λ_{max} = 693 nm
(n = 2) λ_{max} = 790 nm
(n = 2) λ_{max} = 883 nm

Last but not least, very typical for polymethine dyes is the vinylene shift of extended conjugation as illustrated by **8** and **8f** [11].

Addition of another directly conjugated aryl substituent onto the methine fragment produces the technically important *triarylmethine dyes* like e. g *Crystal Violet* **9** and *Brilliant Green* **10**.

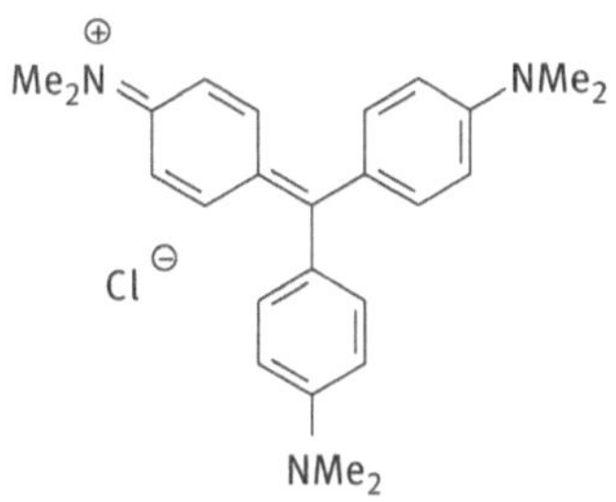

9

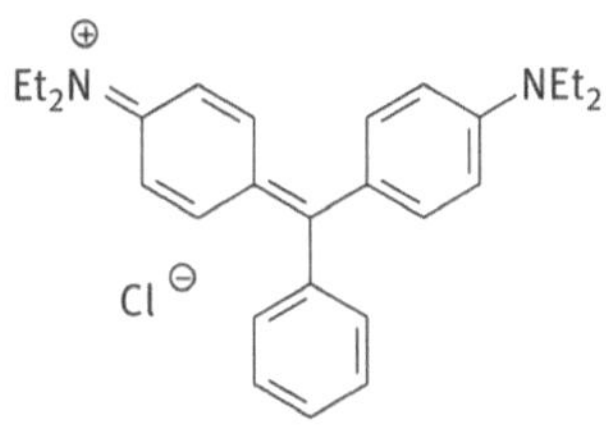

10

Introduction of an oxygen bridge between both rings *ortho* to the methine fragment of triarylmethine dyes produces *rosamine dyes* like e. g. *Rosamine B* **11**. Rosamine derivatives with a carboxylic group in the 2-position of the free rotating phenyl ring are called *rhodamine dyes* like e. g. *Rhodamine B* **12**.

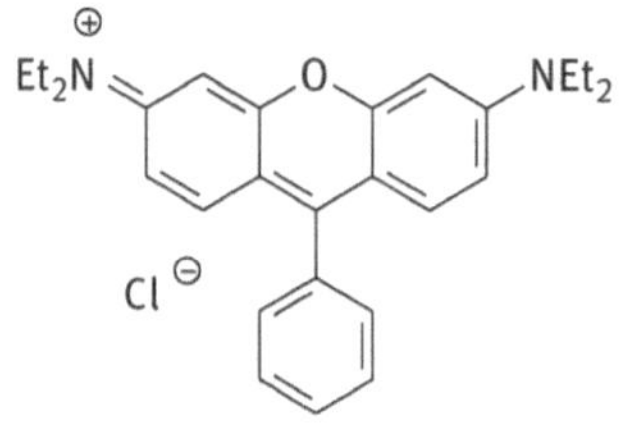

11

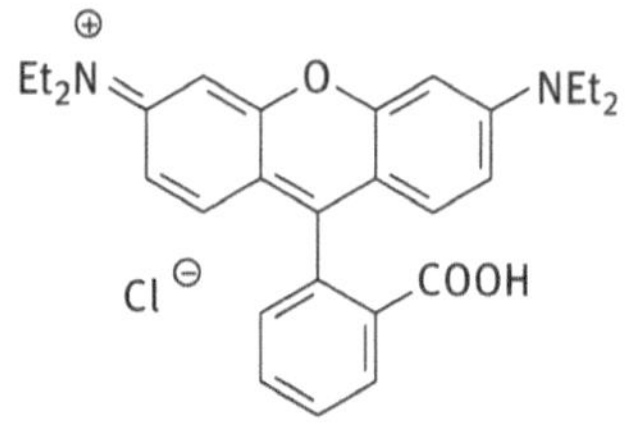

12

Dyes containing the *xanthylium* heterocycle, such as the pyronine, rosamine, and rhodamine dyes, are collectively known as *xanthene dyes*. The outstanding property of the xanthene dyes is their strong fluorescence and, therefore, many of which are prized commercially for their fluorescence. It has been known for a long time that a rigid planar molecular structure favors a strong fluorescence. The background is, the non-radiative deactivation of S_1 to S_0 is mostly responsible for the loss of fluorescence quantum yield, which is caused by intramolecular movements of the conjugated π-system. Most important in this process are the Ph–C torsional vibrations, which deactivate S_1 to S_0 by vibrational relaxation. While **8** is practically non-fluorescent, the introduction of the oxygen bridge in **8c** leads to a highly fluorescent dye due to hindrance of Ph–C torsional vibrations. The carboxylic group in the 2-position of the free rotating phenyl ring in **11** to **12** hinders the rotation of the phenyl ring which decreases vibrational relaxation from S_1 to S_0 and increases the fluorescence quantum yield of **12** in comparison with **11**.

Polymethine dyes terminated only with two open-chain R_2N groups ($R = H$, alkyl, aryl) are called *streptocyanine dyes* **13**.

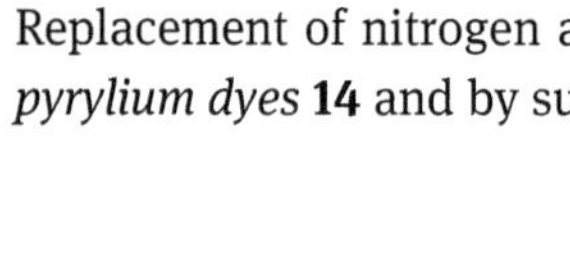

13

Replacement of nitrogen as heteroatom in the heterocycles by oxygen leads to the *pyrylium dyes* **14** and by sulfur to the *thiopyrylium dyes*.

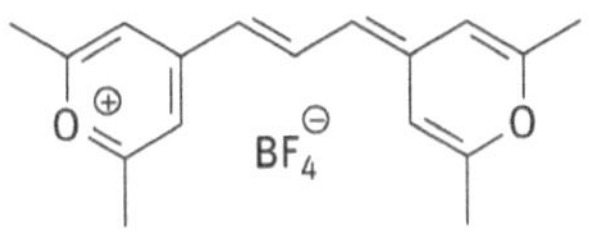

14

59.2 Anionic polymethine dyes

In anionic polymethine dyes there are two different charge-carrying terminal atoms. The most commonly encountered anionic polymethine dyes in the sense of the general structure **3** are the *oxonol dyes*. They are named as oxonol dyes because oxygen is the charge-carrying terminal atom. The simplest oxonols are the anions of malondialdehyde and glutacondialdehyde.

In industrial oxonol dyes the terminal oxygens are often constituents of heterocycles like e. g. in **15**. Another dye class with terminal oxygens are the trioxytriphenyl-methine (*Corallin*) and dioxytriphenylmethine dye **16**, mostly used with a carboxylic group in the 2-position of the phenyl ring. Alkali opens the lactone ring of *Phenolphthalein* resulting in the colored dianion **17**.

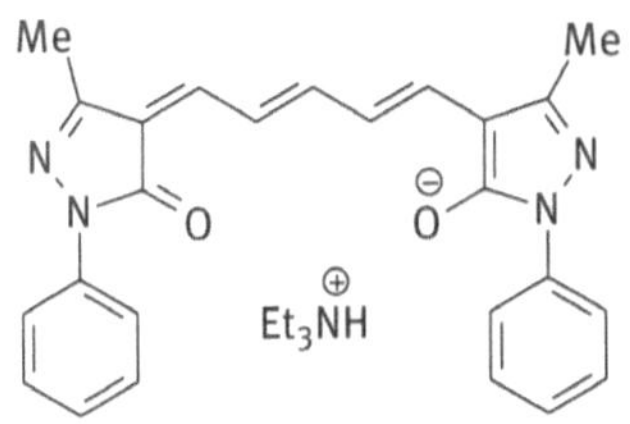

15

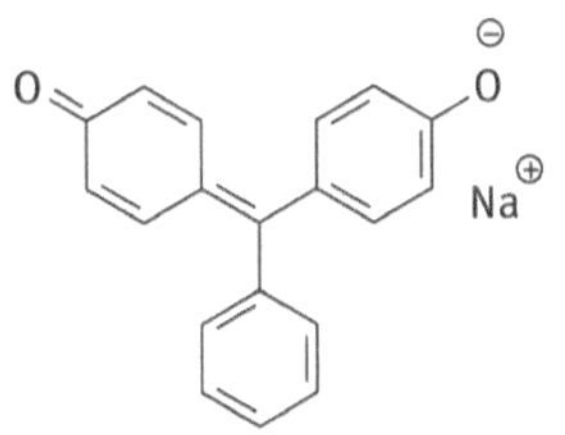

16

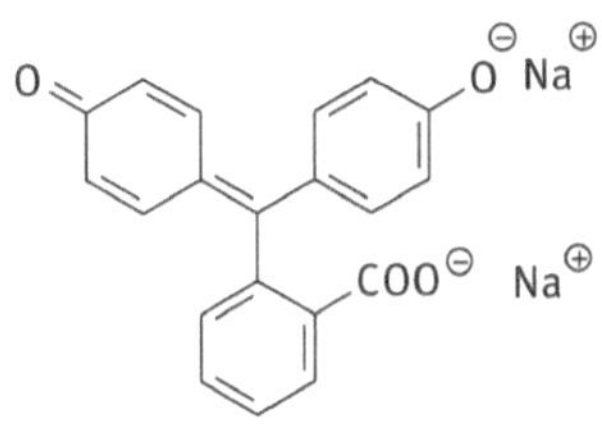

17

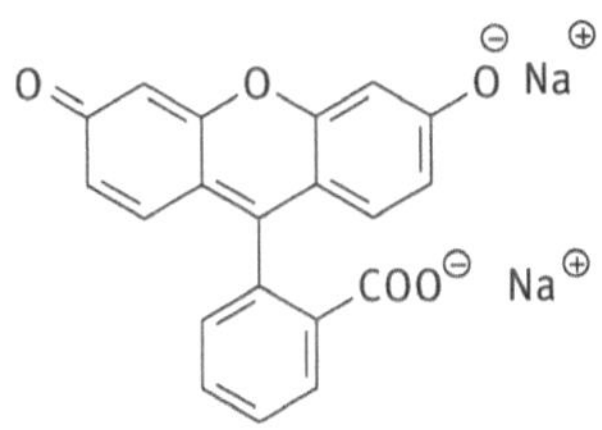

18

In the dianion of Phenolphthalein **17** the 4-oxyphenylrings can free rotate. Therefore it is practically non-fluorescent. As discussed with the xanthene dyes pyronine, rosamine, and rhodamine the vibrational relaxation caused by Ph–C torsional vibrations is reduced in *Fluoresceine* **18** and the fluorescence quantum yield is increased substantially.

The second anionic charge-carrying terminal atom is the carbanion, substituted with two electron acceptors like e. g. in malonodinitrile, cyanoacetates, or cyanoacetamides. In line with the definition of the open-chain cationic *streptocyanine dyes* these dyes are named *anionic streptocyanine dyes* **19**.

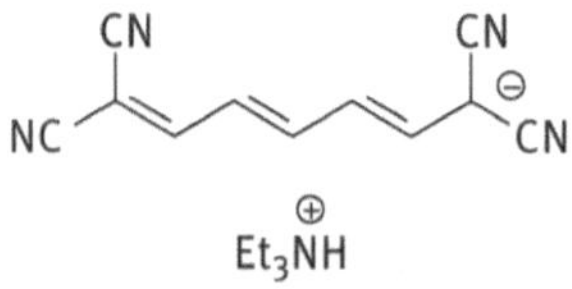

19

59.3 Neutral polymethine dyes

There are some ways to get neutral polymethine dyes. One obvious option is the formal combination of a cationic with an anionic dye.

If one terminal component is a heterocyclic ring and the second an active methylene compound the dyes can be considered as a part of a cyanine dye. Therefore, the corresponding dyes are called *merocyanine dyes* **20** (*meros* = part).

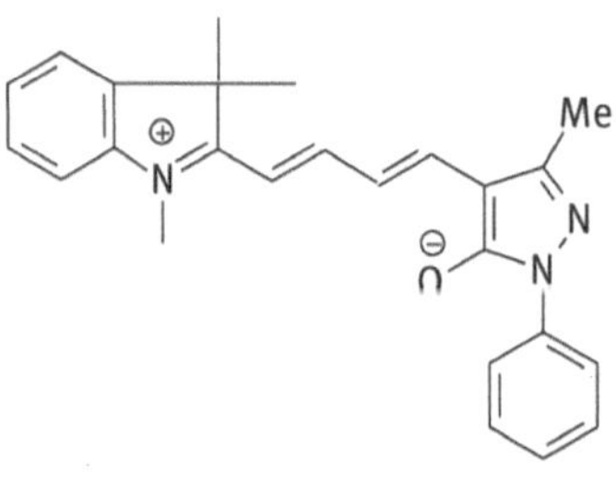

20

Following the classification of *hemicyanine dyes* the replacement of the terminal heterocyclic ring in oxonols by a 4-R_2N-aryl group forms *hemioxonol dyes*. If there is only one formal open-chain methine group like e. g. in **21**, these dyes are often called *methine* dyes. However, the assumption of one methine group as the polymethine chain is wrong. The polymethine chain in **21** range from 4-Me_2N(+) to O(−). In addition, "methine" dye is a less systematic term and is used also in di- and triarylmethine dyes.

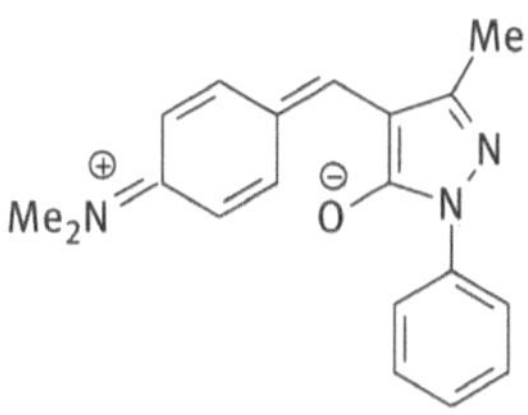

21

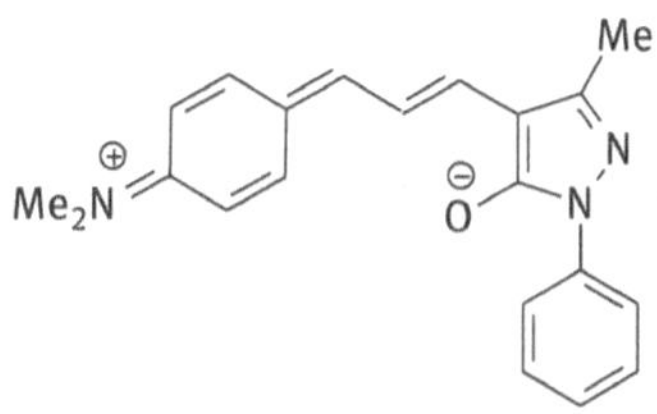

22

Obviously, a dye formed by the reaction of 4-*N*,*N*-dialkylaminocinnamaldehyde with the same pyrazolone to **22** belongs to the same dye class, the hemioxonols, but is not a "methine" dye in the sense of one methine group.

To round off the topic, the open-chain compounds e. g. *N*,*N*-dimethylformamide and 3-dimethylaminoacrylaldehyde are the simplest *streptomerocyanines*.

Also the dyes formed from the reaction of a heterocyclic aldehyde (e. g. Fischer's Aldehyde) with an open-chain active methylene compound (e. g. malonodinitrile) to **23**, belong to the merocyanines [12].

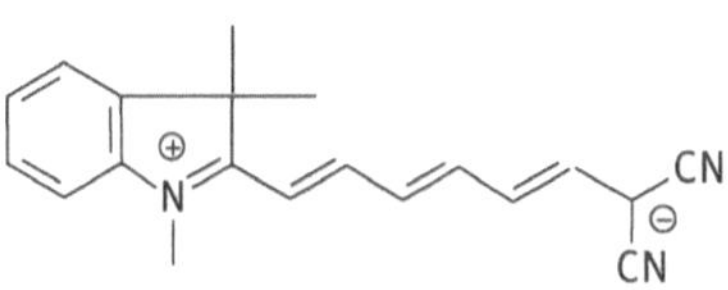

23

There are no doubts that the neutral merocyanines are classified as part of the polymethine dyes.

Neutral polymethine dyes formed from the reaction of e. g. 4-*N*,*N*-dialkylaminobenzaldehyde or 4-*N*,*N*-dialkylaminocinnamaldehyde with an open-chain active methylene compound (e. g. malonodinitrile, cyanoacetates, or cyanoacetamides) are often called *styryl dyes* (owing to the presence of the >C=C< fragment drawing the non-charge-separated contributing structure **24N** only) and sometimes not classified as polymethine dyes: “Styryl dyes have been a separate classification from polymethine dyes because they are not polymethine dyes” [13]. Is this statement correct? In the following it will be shown whether this is true or not:

24S **24N**

Based on a simple Valence Bond (VB) model the electronic structure of such compounds can be described as a *resonance hybrid* between a *charge-separated contributing structure* **S** and a *non-charge-separated contributing structure* **N**. The degree of mixing of both contributing structures determines the electronic structure in the ground electronic state (S_0) as well in the excited electronic state (S_1) [14]. In turn, the electronic structure in S_0 and in S_1 determines the equilibrium nuclear coordinates in both electronic states $R_e(S_0)$ and $R_e(S_1)$. If e. g. **S** contributes mainly in S_0 to the resonance hybrid, in S_1 the opposite is the case, that is, **N** contributes mainly in S_1 and vice versa. The consequence is, the difference ΔR between the equilibrium nuclear coordinates in the excited and ground electronic state $R_e(S_1)$ and $R_e(S_0)$, respectively, increases with increasing nonsymmetric distribution of **S** and **N**. The vibronic 0–0 transition energy is not influenced by such effects, whereas λ_{max} can be substantially influenced by them [9, 15]. A material and widespread error is to discuss electronic spectra in terms of λ_{max} only, without paying attention to the vibronic fine structure of the spectra. The value of λ_{max} is nothing other than the intensity maximum of an absorption band and can correspond to the 0–0, 0–1 or any other vibronic transition. The discussion of λ_{max} often leads to wrong conclusions regarding the electronic structure.

This shall be illustrated by the Franck–Condon principle. If the difference between $R_e(S_0)$ and $R_e(S_1)$ is substantial, then a vibronic transition other than the 0–0 transition represents λ_{max}. It is only in the case that the difference in $R_e(S_0)$ and R_e (S_1) is small, and that λ_{max} corresponds to the 0–0 vibronic transition [9, 15, 16]. In other words, the contribution of both contributing structures **N** and **S** to the electronic ground state should be similar.

The equilibrium **N** ↔ **S** is determined by the electron donating and accepting power of the terminal groups as well the polarity of the solvent [14]. Using a powerful electron donating and electron accepting group like e. g. in **25** and a polar solvent like e. g. DMSO λ_{max} corresponds to the 0–0 vibronic transition and one can clearly demonstrate on the basis of the vinylene shift that the “styryl dyes” belong to the polymethine dyes [17].

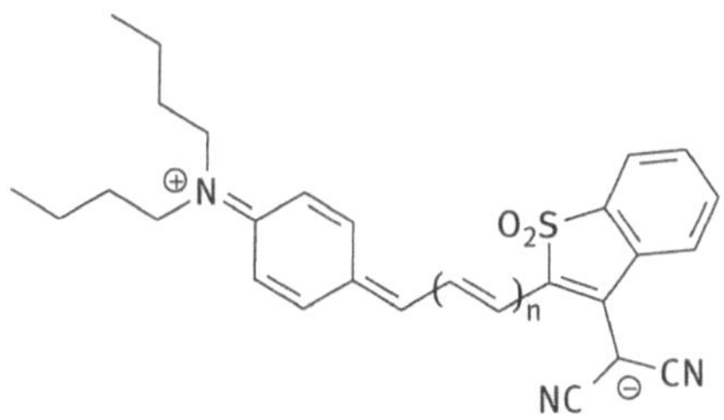

25

(n = 0) λ_{max} = 609 nm
(n = 1) λ_{max} = 709 nm
(n = 2) λ_{max} = 805 nm
(n = 3) λ_{max} = 886 nm

In summary, a dye in which a 4-R_2N-aryl group replaces one terminal open-chain carbanion in an anionic streptocyanine dye the dye remains part of the polymethines. By analogy to hemicyanines and hemioxonols, this subclass should be better named as *hemistreptocyanine dyes* **24**, **25**.

Another important group is the *squaraine* (or *squarylium*) *dye* class as exemplified by **26** and **27**, which are similar to cyanine dyes and *diarylmethine dyes*, respectively.

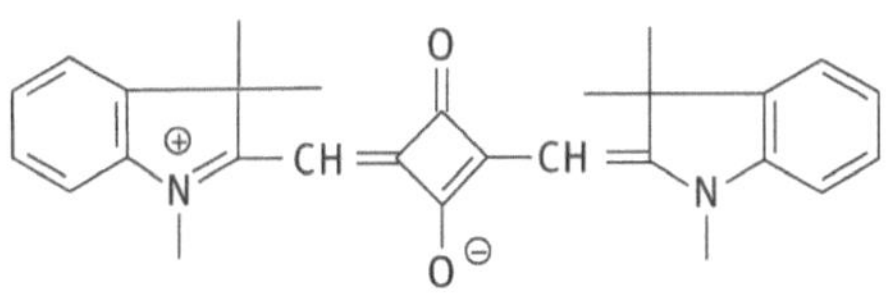

26

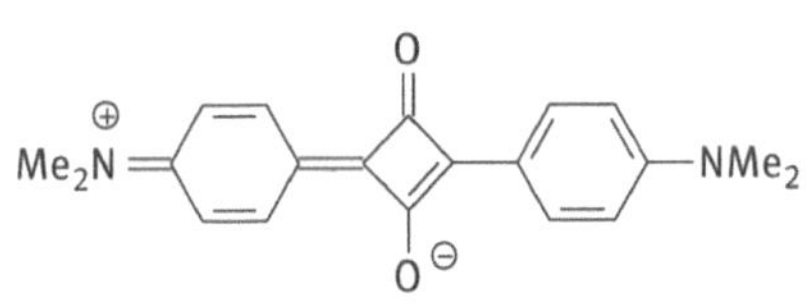

27

Last but not least, also the chelates of dipyrromethens like e. g. the 4,4-difluoro-4-bora-3a,4a-diaza-s-indacene **28** (often abbreviated as BODIPY derived from Borondipyrromethene) are members of the polymethine class. These dyes are highly fluorescent materials [18]. Similar to the xanthene dyes their great structural rigidity reduces the vibrational relaxation of S_1 to S_0 by Ar–C torsional vibrations.

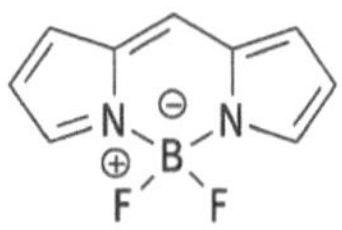

28

In summary, there are a lot of different dye classes [19], and many of them are part of the polymethine dyes. However, this is not obvious at first glance, because the polymethine dyes comprise a wide range of dyes with a diverse array of chemical structures. Common criteria are a conjugated chain with terminal functional groups and an odd number $2n + 3$ of π-centers and $2n + 4$ π-electrons, the vinylene shift of the 0–0 vibronic transition of about 100 nm per vinyl group, the influence of substituents or heteroatoms according to Dewar´s Rules, and a bathochromic shift by twisting of the polymethine chain.

References

1. König W. Über den Begriff der „Polymethinfarbstoffe" und eine davon ableitbare allgemeine Farbstoff-Formel als Grundlage einer neuen Systematik der Farbenchemie. J Prakt Chem [2]. 1926;112:1.
2. Herzfeld KF. On the absorption spectrum of some polymethine dyes. J Chem Phys. 1942;10:508.
3. Sklar AL. Energy levels and color of polymethine dyes. J Chem Phys. 1942;10:521.
4. Dähne S. Systematik und Begriffserweiterung der Polymethinfarbstoffe. Z Chem. 1965;5:441.
5. Tyutyulkov N, Fabian J, Mehlhorn A, Dietz F, Tadjer A. Polymethine dyes – structure and properties. Sofia: St Kliment Ohridski University Press, 1991.
6. Strekowski L, editor. Topics in heterocyclic chemistry, vol. 14, heterocyclic polymethine dyes: synthesis, properties and applications. Berlin: Springer, 2008.
7. Dewar MJ. Colour and constitution. Part I basic dyes. J Chem Soc. 1950;2329.
8. Hallas G. Effects of extended conjugation in some basic dyes. J Soc Dyers Colourists. 1970;86:237.
9. Mustroph H, Ernst S, Senns B, Towns AD. Molecular electronic spectroscopy: from often neglected fundamental principles to limitations of state-of-the-art computational methods. Color Technol. 2015;131:9.
10. Brunings KJ, Corwin AH. Steric influences on the aromaticity of dipyrrylmethenes. The synthesis and study of the properties of a di-N- methyldipyrrylmethene. J Am Chem Soc. 1942;64:593.
11. Schmidt H, Wizinger R. Über die Vinylenhomologen der Di- und Triphenylmethanfarbstoffe. III. Liebigs Ann Chem. 1959;623:204.
12. Mustroph H, Mistol J, Senns B, Keil D, Findeisen M, Hennig L. Relationship between the molecular structure of merocyanine dyes and the vibrational fine structure of their electronic absorption spectra. Angew Chem Int Ed. 2009;48:8773.
13. Gregory P. Classification of dyes by chemical structure. In: Waring DR, Hallas G, editors. The chemistry and application of dyes. New York: Plenum Press, 1990.

14. Mustroph H, Reiner K, Senns B, Mistol J, Ernst S, Keil D, et al. The effects of substituents and solvents on the ground-state π-electronic structure and electronic absorption spectra of a series of model merocyanine dyes and their theoretical interpretation. Chem Eur J. 2012;18:8140.
15. Mustroph H. Dyes: quantum chemical calculation of electronic spectra. Phys Sci Rev. 2019:4. DOI:10.1515/psr-2019-0040.
16. Mustroph H, Towns A. Fine structure in electronic spectra of cyanine dyes: are sub-bands largely determined by a dominant vibration or a collection of singly- excited vibrations? ChemPhysChem. 2018;19:1016.
17. Blanchard-Desce M, Alain V, Bedworth PV, Marder SR, Fort A, Runser C, et al. Large quadratic hyperpolarizabilities with donor-acceptor polyenes exhibiting optimum bond length alternation: correlation between structure and hyperpolarizability. Chem Eur J. 1997;3:1091.
18. Loudet A, Burgess K. BODIPY dyes and their derivatives: syntheses and spectroscopic properties. Chem Rev. 2007;107:4891.
19. Towns A. Colorants: general survey. Phys Sci Rev. 2019:4. DOI:10.1515/psr–2019–0008.

Robert Christie and Adrian Abel

60 Quinacridone pigments

Abstract: This chapter surveys the structural and synthetic chemistry and the industrial applications of quinacridones, a small but extremely important group of high-performance carbonyl (or polycyclic) organic pigments. They are based on one of the most important new chromophoric systems developed specifically for pigment applications after the introduction of the phthalocyanines, and currently occupy a prominent position in the red to violet shade areas. A historical perspective on the discovery and commercial development of the quinacridones is presented initially. There then follows an illustrated discussion of the structural chemistry of the pigments, encompassing both molecular and crystal structures. Throughout the chapter, specific features of their molecular structures and the nature of the intermolecular association within the crystals are related to their influence on the color and technical performance in application, in which they exhibit some of the highest standards of heat stability, solvent resistance, and fastness to light and weather encountered in organic pigments. Finally, a survey of the principal current applications of the specific individual commercial quinacridone pigments is presented.

Keywords: quinacridone pigments, quinacridones, polymorphism, solid solutions, CI Pigment Violet 19, CI Pigment Red 122, CI Pigment Red 202, CI Pigment Red 206, CI Pigment Red 207

60.1 Fundamentals

The quinacridones represent one of the most important chromophoric systems specifically developed for pigment applications after the industrial introduction of pigments based on the phthalocyanines. They form a relatively small group of high-performance carbonyl pigments, occupying the yellowish-red to violet shade areas. In terms of chemical structure, they are often alternatively referred to as polycyclic pigments, as they are based on a linear arrangement of five fused 6-membered rings. They exhibit some of the highest standards of heat stability, solvent resistance, and fastness to light and weather encountered in organic pigments, similar to those provided by copper phthalocyanine blues and greens [1–4].

This article has previously been published in the journal Physical Sciences Reviews. Please cite as: R. Christie, A. Abel, Quinacridone Pigments *Physical Sciences Reviews* [Online] 2021, 6. DOI: 10.1515/psr-2020-0195

https://doi.org/10.1515/9783110587128-060

60.2 History

The term quinacridone dates back to 1896, first used by Niemantowski, a Polish academic. However, the products that he reported adopted an angular structure, which ultimately were found to lack the excellent color and technical performance properties of the linear trans structure, on which the high-performance pigments that we recognize today are based. Linear trans quinacridone was first described by Sharvin in 1915 [5]. The first documented synthesis of the compound was reported by Liebermann in 1935 [6]. When originally investigated in the laboratories of DuPont, New Jersey in the US, their potential was not initially recognized, at least in part because there was little demand for pigments of this quality at the time, especially at the costs associated with their production. The potential was realized in the late 1950s, when they were introduced commercially by Du Pont. In 1955 the chemists at DuPont, led by Struve, discovered two pure shade pigments, polymorphic forms of the parent compound, with excellent fastness properties and developed methods for their synthesis [7].

60.3 Structure and properties

The molecular structures of quinacridone pigments are based on a system of five fused alternate benzene and 4-pyridone rings. The intense colors and outstanding technical properties are given by compounds which possess the linear trans arrangement (**1**) illustrated in Figure 60.1. Compounds (**1a-c**) (Table 60.1) are the industrially important products. Different geometrical arrangements of the system, structures (**2**), (**3**) and (**4**) as illustrated in Figure 60.2 are known, but none of these offer the color and technical performance provided by arrangement (**1**). The systematic chemical name of this structure is 5,12-dihydroquino[2,3-b]acridine-7,14-dione. Over 120 substituted quinacridones have been identified, many in more than one crystal modification, but only a few substituted quinacridones have yielded commercial products [8].

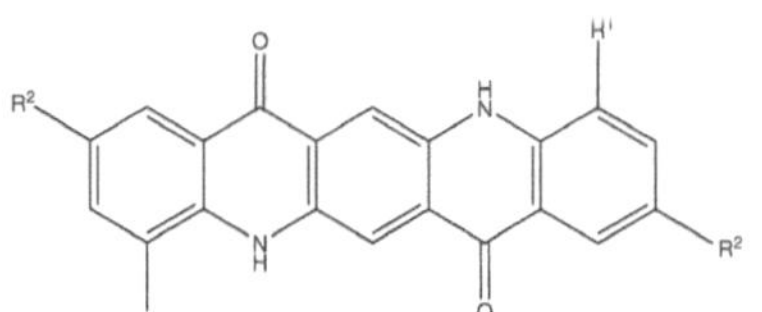

1

Figure 60.1: Structure of linear trans quinacridones 1.

Table 60.1: Substituent patterns in linear trans quinacridones.

Compound	C. I. Pigment	R^1	R^2
1a	Violet 19	H	H
1b	Red 122	H	CH_3
1c	Red 202	H	Cl
1d	–	Cl	H

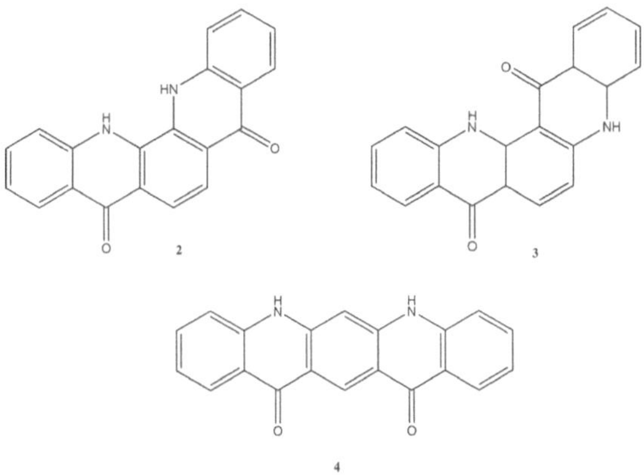

Figure 60.2: Alternative geometrical arrangements of quinacridone.

It may appear rather surprising that these relatively small molecules should provide such high levels of thermal and chemical stability, insolubility, and fastness to light. However, these properties are explained mainly by strong intermolecular association in the crystal structure, due to a combination of hydrogen bonding between N–H and C = O groups and dense van der Waals packing [9–12]. The quinacridones show polymorphism and this has a profound effect on the color and technical properties of the pigments. The parent pigment (**1a**) exists in four polymorphic modifications each with its own characteristic X-ray powder diffraction pattern [9]. The α-form is not of industrial interest as it exhibits weak coloristic and technical performance. It has been shown that there are two versions of this form, referred to as α^{I} and α^{II} [9]. In contrast, the reddish-violet β-form and the bluish-red γ -form are both highly important commercial organic pigments. Other polymorphs have been proposed, including δ- and ε-forms, although it has subsequently been demonstrated that these products were either the γ -form or mixtures of β- and γ-forms. In all quinacridone polymorphs, the molecules are connected to neighboring molecules by four hydrogen bonds of the type N–H … O = C. In the β-form, as illustrated in Figure 60.3, each planar molecule is bonded to two neighboring molecules via two hydrogen bonds of almost 180°, a strong hydrogen bonding configuration. In terms of crystal packing, half of the chains run in one direction and the other half in a different direction. The stacked chains form layers held together by van der Waals interactions only. In the γ-form, there is a quite different intermolecular hydrogen bonding arrangement whereby each quinacridone molecule forms individual hydrogen bonds to four adjacent molecules. The molecules adopt a criss-cross packing arrangement (sometimes referred to as a "hunter's fence") with individual stacks of quinacridone molecules rotated with respect to each other. The origin of the color in quinacridone pigments provides a striking example of the influence of crystal structural effects. It is interesting that in solution the quinacridones exhibit only weak yellow to orange colors. The intense red to violet colors of the pigments in the solid state are therefore due to interactions between molecules in the crystal structure. It has been postulated that this may involve a π-σ correlation, specifically a correlation between π-π interactions and the hydrogen bonding, in which the hydrogen bonds act as channels for intermolecular donor/acceptor charge transfer [11].

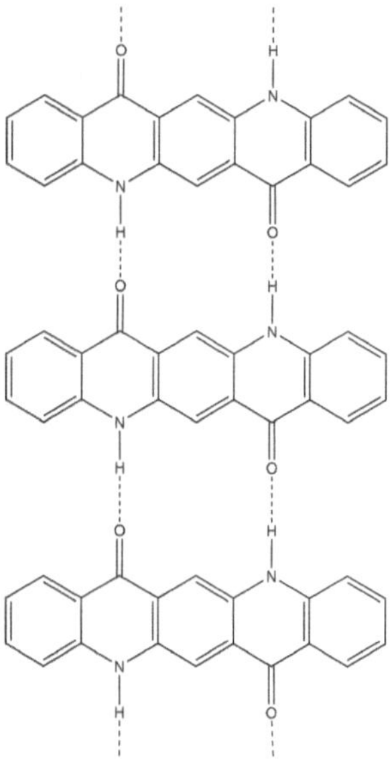

Figure 60.3: Intermolecular association in the crystal structure of β-quinacridone.

Most notable among the substituted quinacridones is the dimethyl substituted version, known as Quinacridone Magenta (**1b**). There is also a series of mixed crystal products, existing as solid solutions, that extend the color range of quinacridone pigments from yellowish-red through to violet. The concept of mixed crystal pigments is utilized in other pigment classes, including phthalocyanines and some azo pigments, but the concept has arguably played a more important role with quinacridones. Two species that have been used to form solid solutions with quinacridone (**1a**) are quinacridonequinone (**5**), illustrated in Figure 60.4, and the dichloroquinacridone (**1d**).

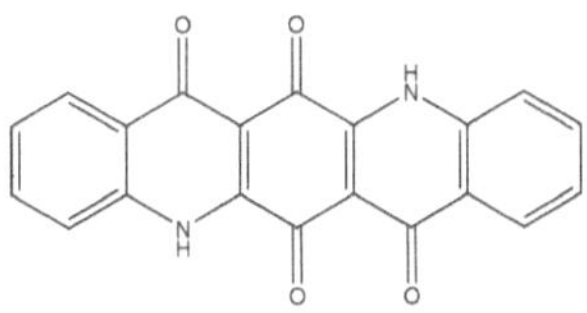

5

Figure 60.4: Structure of quinacridone quinone (5).

60.4 Synthesis and manufacture

Several methods of synthesis of the quinacridones have been reported [4, 8, 13, 14], all involving multiple stages, a complicating feature that accounts, at least in part, for their higher cost. The two industrially important methods are outlined in Figure 60.5. The starting material for both routes is diethyl succinylosuccinate (**6**) which is reacted with the appropriate primary aromatic amine (2 mol) to form intermediate 2,5-diarylamino-3,6-dihydroterephthalic acid diester (**7**). Compound (**7**) then undergoes a ring closure reaction, carried out at elevated temperatures in a high boiling solvent, leading to the dihydroquinacridone (**8**). This species is then oxidized to quinacridone (**1**), for example using sodium 3-nitrobenzene-1-sulphonate. In an alternative route, intermediate (**7**) is oxidized to the 2,5-diarylaminoterephthalate diester (**9**). Subsequently alkaline hydrolysis of diester (**8**) leads to the corresponding carboxylic acid which undergoes acid catalyzed ring closure by treatment at an elevated temperature with, for example, polyphosphoric acid to form quinacridone (**1**). In both routes, the cyclization reaction is an example of an intramolecular Friedel Crafts acylation reaction. The mechanism of these reactions has been discussed in detail in a review article [15]. In the case of the parent quinacridone (**1a**), the synthesis leads to a crude pigment in its α-form, which may be converted to a pigmentary form by appropriate conditioning processes. The crude product may be subjected to milling with an inorganic salt in the presence of organic solvents, not only to reduce the particle size but also to form the desired polymorphic form. The use of non-polar solvents (e.g., xylene) leads to the β-form while, the γ-form is produced using more polar solvents (e.g., DMF). The β-form may also be formed by acid pasting with concentrated sulfuric acid. These techniques lead to fine particles which are prone to flocculation when incorporated into certain application media. This undesirable feature may be alleviated by treating the pigments with derivatives of quinacridones, or other substances, which attach to the particle surfaces, stabilize the dispersions and enhance compatibility with the application media. Unfortunately, there is no single treatment that works ideally for every binder system, and so application chemists must evaluate which product works best in their system.

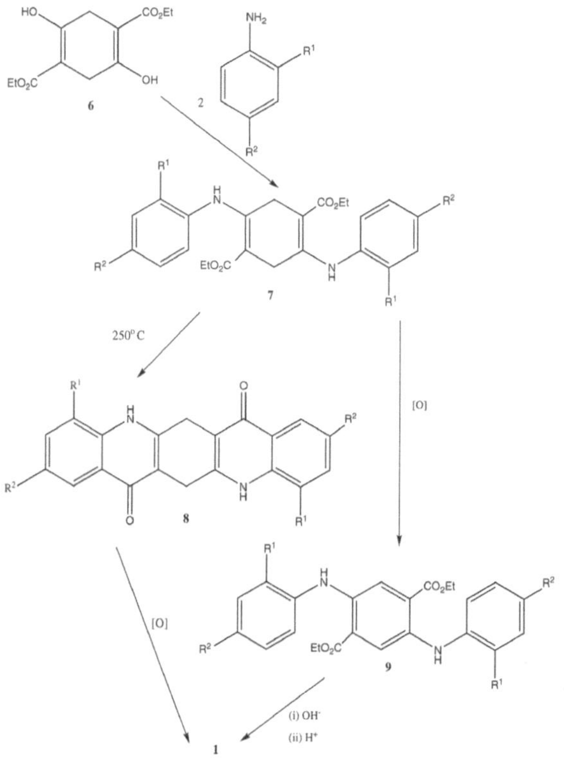

Figure 60.5: Synthetic routes to linear trans quinacridones.

60.5 Applications

CI Pigment Violet 19 (1a) β-modification. This pigment has a reddish-violet shade with high tinctorial strength. It is used mainly in paints but also experiences significant use in plastics. Its lightfastness is very good to excellent in full shade and in reductions, although it darkens with prolonged exposure to light [15]. In paints it is used for all industrial finishes. Its main traditional use was to produce mid-red shades in combination with molybdate red, although currently other opaque yellowish red or orange organic pigments are used to completely replace molybdate red for automotive original equipment finishes. The pigment has good heat stability but bleeds slightly when over-coated at 160 °C. The β-form has slightly inferior resistance to polar solvents than the γ-modification. Products are manufactured with different particle size distributions. The finer grades are more transparent and are used in metallic paints. These grades maintain the excellent lightfastness of the coarser grades. In plastics, the pigment shows heat stability up to 300 °C, so that it is suitable for most common polymers, with the exception of polyamides and PET. In polyolefins, it causes significant dimensional instability in injection moldings. It is used for the spin dyeing of fibers, especially polypropylene. Its use in inks is limited by its high price, but it is used for applications where high levels of fastness properties are required, including metal decorative inks, in which it is stable to 180 °C and fast to sterilization and calendering.

CI Pigment Violet 19 (1a) γ-modification: When the Colour Index originally allocated this pigment its generic name, the rules determined that identical chemical compounds within a particular class of colorants were to be given the same name. Although the rules have changed to allow compounds outside the original defined color area to be allocated a different CI Generic Name, in the case of this pigment its original name had become entrenched and there was a reluctance to change it! This pigment is a bluish-red, but quite definitely red and not violet. It has a remarkably clean shade, although it can vary in color considerably depending on the particle size distribution. The larger particle size version is opaque, has a yellower shade and the best fastness properties of all grades in this crystal modification. The average size of the primary particles is around 220 nm with a specific surface area of around 30 m^2g^{-1}. In coatings, its excellent outstanding fastness properties means that it can be used in full shades, reductions and as a shading component. It gives brilliant yellowish reds when used in combination with opaque organic and inorganic orange pigments. It is suitable for coatings for automotive original equipment. When used in paints applied to an alkaline substrate, for example plaster, the lightfastness can be significantly reduced. The smaller particle size version is transparent, has a bluer shade and a higher tinctorial strength. Its fastness to light and weather is a little lower than the opaque grade. The mean size of the primary particles is around 70 nm with a specific surface area of around 90 m^2g^{-1}. Both grades are stable to 200 °C and show excellent fastness to solvents. In plastics, it is stable up to 300 °C, so that it is suitable for most polymers, including PVC, in which it does not bleed, polystyrene, ABS, and polyolefins. The larger particle size version has excellent lightfastness, whereas the smaller size is rated as very good to excellent. It does not cause warping in injection moldings. Some grades are suitable for cable sheathing. Both grades have limited application in polyamides. In inks, it is used when very high fastness performance, including resistance to powerful solvents, is required, for example in floor coverings, laminates, wall coverings, and vinyl or nitrocellulose inks. It is fast to sterilization and calendering. The finer particle size version is usually preferred for inks due to its higher color strength and transparency.

CI Pigment Red 122 (1b). This brilliant pink shade pigment, often known as Quinacridone Magenta or Quinacridone Pink, has grown in popularity as the demand for a pure shade magenta with high levels of fastness performance has developed. It finds widespread use in coatings, inks, and plastics. It has higher tinctorial strength than the γ-modification of CI Pigment Violet 19 in all but ink applications. In coatings, it is widely used for both architectural and industrial finishes including automotive original equipment. Special transparent grades have been developed for metallic finishes. Lightfastness ranges from very good to excellent, in full shades and pale shades, with slight darkening on prolonged exposure. Its fastness to solvents is excellent. It is a very important component of tinting systems, where its purity of shade means that it is useful to produce clean shades from scarlet to violet as blends with other pigments. In inks, it is quite close to the standard magenta color, and is recommended for both paste and liquid inks. It is used for packaging gravure, solvent-based, water-based and UV cured inks, and for flexographic inks. It is stable to 180 °C and fast to sterilization and calendering. Special grades have been developed for inkjet and toner applications. It is suitable for most plastic applications, as it is stable to 300 °C. It does not bleed in plasticized PVC and is suitable for most polyolefin injection moldings, showing only a slight tendency to cause warping. It can be used for cable sheathings. It is particularly recommended and useful for spin dyeing of fiber, coloring thin-walled articles and films. The pigment is a favorite of artists for its brilliant shades, and can also be used for the pulp coloration of paper when intended for high-quality laminate papers. An interesting potential use for this pigment is in photoelectric cells, where its high absorption, good stability and good photovoltaic properties are reported to improve the performance of Schottky type solar cells [15]. One of the main components as a starting point for the synthesis of this pigment is succinic acid. Some CI Pigment Red 122 grades are now being made from succinic acid obtained from renewable natural resources, with assurances provided that the succinic acid is not being obtained agriculturally at the expense of food crops [16].

CI Pigment Red 202 (1c). The Color Index designation and REACH registration of this pigment both identify it as 2,9-dichloroquinacridone (**1c**). However, there is some confusion over its precise composition as it has been considered as a mixed crystal product [14]. The pigment is sometimes referred to as Quinacridone Magenta, offering properties very similar to CI Pigment Red 122 (**1b**) with a slightly bluer and somewhat duller shade. It is used in paint, inks and plastics and is offered in a variety of forms mainly distinguished by differences in particle size. It is specifically recommended for metallic finishes in the automotive original finish market, where it is considered to give superior weathering properties compared with CI Pigment Red 122. It satisfies all the usual requirements for plastics, as it is heat stable to 300 °C, and provides very good to excellent lightfastness. In inks, the pigment provides high tinctorial strength and has created some interest for non-impact printing processes.

CI Pigment Red 206. This pigment is of historical interest as an early example of a quinacridone pigment formed as a solid solution of the host/guest type. The first discovery of a solid solution came about by accident when it was found that oxidation of 6,13-dihydroquinacridone (**8**), an intermediate in the synthesis of quinacridone (**1a**) (Figure 60.5) led to a mixture of quinacridone (65%) and quinacridonequinone (**5**) (35%). The product after conditioning by milling was brown but in metallic paints gave an interesting maroon color that was suitable for automotive finishes with outstanding weather fastness. These properties may originally have been considered as surprising because quinacridonequinone itself is yellow with poor fastness to light and weather. Although predominantly quinacridone, X-ray diffraction showed that the crystal structure adopted is that of quinacridonequinone. Currently, it has no registered manufacturer.

CI Pigment Red 207: This pigment, identified as a mixed crystal phase pigment derived from the parent quinacridone (**1a**) and the dichloroquinacridone (**1d**) remains a popular scarlet shade commercial product that can be manufactured in different grades with varying properties. However, current commercial grades are semi-opaque and have moderate tinctorial strength. It is used almost entirely for high-quality coatings such as those used on automotive original equipment. As with many quinacridone pigments, it is a popular choice for artists, who like its pure yellow shade red and excellent lightfastness.

There are a few quinacridone pigments which appear to be discontinued, including CI Pigment Orange 48 (originally known as Quinacridone Gold), CI Pigment Orange 49, CI Pigment Red 209, and CI Pigment Violet 42.

References

1. Christie RM. Colour chemistry. 2nd ed. London: RSC, Ch 9 2015.
2. Hunger K, Schmidt MU. Industrial organic pigments. 4th ed. Weinheim: Wiley-VCH Verlag GmbH, Ch 3 2019.
3. Chamberlain TR, In Faulkner EB, Schwartz RJ. High performance pigments. 2nd ed. Weinheim: Wiley-VCH Verlag GmbH, 2009. Ch 18.
4. Jaffe EE. Quinacridone and some of its derivatives. J Oil Colour Chem Assoc. 1992;75:24.
5. Sharvin VV. J Rus Phys Chem Soc. 1915;47:1260.
6. Liebermann H. Über die Bildung von Chinakridonen aus *p*-Di-arylamino-terephtalsäuren. 6. Mitteilung über Umwandlungsprodukte des Succinylobernsteinsäureesters. Annalen Der Chemie. 1935;518:245.
7. Struve WS. US 2,844,485. 22 Jul 1955.
8. Labana SS, Labana LL. Quinacridones. Chem Rev. 1967;67:1–18.
9. Paulus EF, Leusen FJJ, Schmidt MU. Crystal structures of quinacridones. Cryst Eng Com. 2007;9:131–43.
10. Lincke G. Polycyclische pigmente. Farbe und Lack. 1980;86:966–72.
11. Lincke G. On quinacridones and their supramolecular mesomerism within the crystal lattice. Dyes Pigm. 2002;52:169–81.
12. Lincke G, Finzel LUF. Studies on the structure of alpha-quinacridone. Cryst Res Tech. 1996;31:441–52.
13. Altiparmakian RH, Bohler H, Kaul BL, Kehrer F. Quinacridones: structure and mechanism of formation. Helv Chim Acta. 1972;55:85–100.
14. Jaffe EE. US4,541,872. 17 Sep 1985.
15. Tomida M, Kubayashi S, Yokoyama M. Organic solar cell fabrication using quinacridone pigments. Chem Lett Chem Soc Jpn. 1984;13:1305–8.
16. https://www.clariant.com/en/Corporate/News/2015/11/Clariant-highlights-sustainable-solutions-at-ChinaCoat-2015-exhibition.

Gerhard Pfaff

61 Special effect pigments

Abstract: Special effect pigments are in the very most cases synthetic pigments, characterized by high luster, brilliance and iridescent colors known from optically thin films. The visual appearance has its origin in reflection and refraction of light at thin single and multiple layers. The pigments are either transparent, semitransparent or light-absorbing platelet-shaped crystals or layer systems. They can consist of single crystals, but also of monolayer or a multilayer structures in which the layers have different refractive indices and light absorption properties. Pigment based on the layer-substrate principle represent the technically most important class of special effect pigments. There are many variation and combination possibilities for the layer and substrate materials used. Special effect pigments based on platelets of natural or synthetic mica, on alumina, silica, or borosilicate flakes are the main representatives of layer-substrate pigments. They are manufactured in most cases by wet chemical processes combined with high-temperature processes. The production of substrate particles includes mechanical processes, but also crystal growth, glass formation, and web coating.

Keywords: effect pigments, special effect pigments, natural fish silver, basic lead carbonate, bismuth oxychloride, micaceous iron oxide, layer-substrate principle, mica, metal oxide mica pigments, functional pigments, silica flakes, alumina flakes, borosilicate flakes

61.1 Fundamentals and properties

Effect pigments are divided into metal effect pigments (metallic effect pigments) and special effect pigments including pearl luster pigments (pearlescent pigments, nacreous pigments) and interference pigments (see chapters Effect pigments and Metal effect pigments). They consist of µm-sized thin platelets that show strong lustrous effects when oriented in parallel alignment in an application system (paint, coating, plastic, cosmetic formulation, printing ink and others).

Special effect pigments are almost exclusively inorganic pigments. They distinguish themselves by high luster, brilliance and iridescent color phenomena based on optically thin films. Their visual appearance has its origin in reflection and deflection of the light at thin single and multiple layers. Effects of this kind are also found in nature, for example in pearls, clamshells, feathers, fishes scales, gemstones, minerals,

This article has previously been published in the journal Physical Sciences Reviews. Please cite as: G. Pfaff Special Effect Pigments *Physical Sciences Reviews* [Online] 2021, 6. DOI: 10.1515/psr-2020-0196

https://doi.org/10.1515/9783110587128-061

and insects. Studies on the understanding of the optical principles of the natural pearl luster show that the brilliant colors can be attributed to structured biopolymers and layered structures, which are formed by biomineralization [1–7].

A comparison of the optical principles for the interaction of light with absorption (colored and black) and white pigments as well as with metal effect and special effect pigments (pearl luster pigments) can be found in the chapter Effect Pigments.

Pearl luster pigments as typical representatives of special effect pigments simulate the luster of natural pearls or shells of mollusks. These consist of alternating transparent layers with differing refractive indices. The layers consist of $CaCO_3$ in the crystal modification aragonite (high refractive index) and the complex protein conchiolin (low refractive index) [5, 6].

The difference in the refractive indices, arising equally on the interface between air and an oil film or an oil film and water, is a precondition for the known iridescent color phenomena of these systems. Thin highly refractive platelets of pearl luster pigments align themselves parallel in their application media such as coatings, paints, printing inks, or plastics. Interference effects are generated when the distances of the various layers or the thicknesses of the platelets have appropriate values.

Pearl luster pigments are either transparent, semitransparent or light-absorbing platelet-shaped crystals or layer systems. They can consist of single crystals as in the case of $Pb(OH)_2 \cdot 2\,PbCO_3$ and BiOCl or possess a monolayer or a multilayer structure in which the layers have different refractive indices and light absorption properties (Figure 61.1).

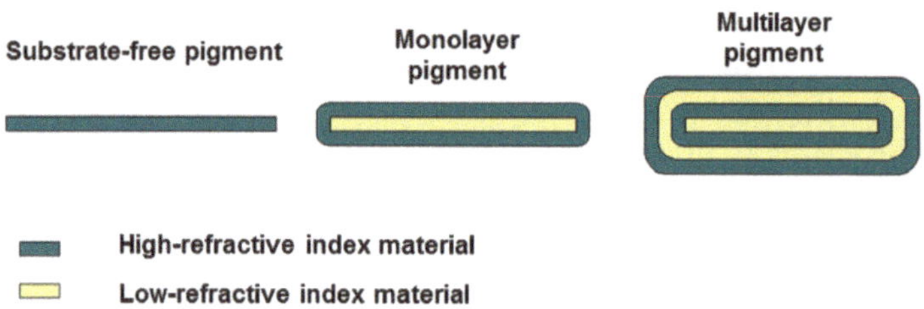

Figure 61.1: Basic structural principles of transparent effect pigments (pearl luster pigments, interference pigments), left: platelet-like particles consisting of an optically homogeneous material (substrate-free); middle and right: platelet-shaped particles, consisting of layer-substrate structures (monolayer, multilayer) [7].

The use of pearls and nacreous shells for decorative purposes started several hundred years ago. First encouraging trials were made around 1650 in France where the rosary maker Jaquin isolated a silky lustrous suspension from fish scales (pearl essence) and applied this to small beads to create artificial pearls [8]. 250 years later, pearl essence material could be isolated from fish scales in form of thin platelets consisting of the two purines guanine and hypoxanthine. In the subsequent time, experiments were carried out to develop synthetic pearl colors as organic or inorganic, transparent, highly refractive formulations. Beginning with 1920, hydroxides, halides, phosphates, carbonates, and arsenates of zinc, calcium, barium, mercury, bismuth, lead, and other metals were synthesized in form of thin platelet-shaped crystals. From all the substances investigated in his period, only natural fish silver, basic lead carbonate and bismuth oxychloride have achieved substantial importance as pear luster pigments. The breakthrough for pearl luster pigments was the development of pearl luster pigments consisting of metal oxide coated mica platelets in the 1960s [1].

Pearl luster pigments, and even more the entire class of special effect pigments, are used to obtain pearl, iridescent (rainbow), or metallic effects. In transparent color formulations, they are applied to obtain brilliance, two-tone color appearance, luster flops, and color travel effects. The most important applications of these pigments are automotive and industrial coatings, printing inks, plastics, and cosmetic formulations.

61.2 Optical principles of special effect pigments

The basic optical principles of pearl luster (interference) pigments are illustrated in Figure 61.2 for a simplified case of nearly normal incidence without multiple reflection and absorption. A part of the original light beam L is reflected (L_1) and a part is transmitted (i.e., refracted) (L_2) at the interface P_1 between two materials with the refractive indices n_1 and n_2. The intensity ratios of the partial beams depend on n_1 and n_2. In a multilayer arrangement as found in natural pearl or pearl luster and iridescent materials, each interference produces partial reflection. After penetration through several layers, virtually complete reflection is obtained, in case that the materials are sufficiently transparent.

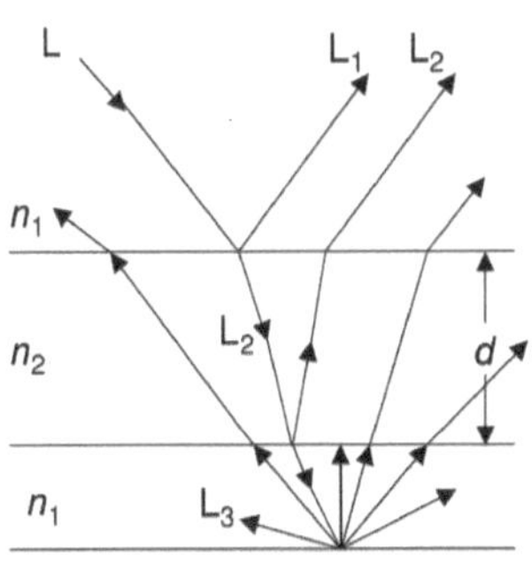

Figure 61.2: Simplified scheme showing nearly normal incidence of a light beam (L) from an optical medium with refractive index n_1 through a thin solid film of thickness d with refractive index n_2. L_1 and L_2 are regular reflections from the phase boundaries P_1 and P_2. L_3 represents diffuse scattered reflections of the light from the ground [7].

Pearl effect pigments are able to simulate natural pearl effects. The simplest case is a platelet-shaped particle with two phase boundaries P_1 and P_2 at the upper and lower surfaces of the particles. Such particles are thin and transparent and are characterized by a refractive index, which is higher than that of its surroundings. The properties of thin platelets with a thickness in the range of up to several hundred nanometers can be understood based on the physical laws of thin optical films.

Multiple reflection of light on a thin solid film with a high refractive index causes interference effects in the reflected light and in the complementary transmitted light. For the simple case of nearly perpendicular incidence, the intensity of the reflectance (I) depends on the refractive indices (n_1, n_2), the layer thickness (d), and the wavelength (λ):

$$I = \frac{A^2 + B^2 + 2AB\cos\Theta}{1 + A^2B^2 + 2AB\cos\Theta} \tag{61.1}$$

$$A = \frac{n_1 - n_2}{n_2 + n_1}, \quad B = \frac{n_2 - n_1}{n_2 + n_1}, \quad \Theta = 4\,\pi\frac{n_2 d}{\lambda} \tag{61.2}$$

With given values for n_1 and n_2, the maximum and minimum intensities of the reflected light, which are visible as interference colors, can be calculated. The calculated data agree well with experimental results. Particularly important values for the refractive index of materials used for pearl luster pigments are the following:

mica (muscovite):	1.55-1.62
mica (phlogopite):	1.53-1.62
natural fish silver:	1.85
basic lead carbonate:	2.0
bismuth oxychloride:	2.15
TiO_2 (anatase):	2.5
TiO_2 (rutile):	2.7
a-Fe_2O_3 (hematite):	2.9
SiO_2 (amorphous, thin layer):	1.46
Al_2O_2 (corundum):	1.77
borosilicate glass:	1.47

Pearl luster, respectively, interference pigments are synthesized with a layer thickness d calculated to produce the desired interference colors (iridescence). Most of the pigments consist of at least three layers of two materials with different refractive indices. Thin flakes (thickness approx. 500 nm) of a material with a low refractive index (mica, silica, alumina, borosilicate glass) are coated with a highly refractive metal oxide (TiO_2, Fe_2O_3, layer thicknesses of 50 to 250 nm). Each pigment particle is therefore a small thin film system with four interfaces. The spectral characteristics of more complex multilayer pigments containing additional thin layers of low or highly refractive materials can also be calculated on the basis of appropriate optical parameters.

The color effects of special effect pigments depend on the viewing angle. The platelets of the pigments split white light into two complementary colors depending on their thickness. The reflected (interference) color dominates under regular (maximum) reflection. This applies when the object is observed at the angle of regular reflection. The transmitted color dominates at other viewing angles under diffuse viewing conditions when there is a nonabsorbing (white) or reflecting background. Variation of the viewing angle leads at one point to a sharp peak reflectance (luster) and the color changes between two complementary colors. The complex interaction of luster and color is measured goniophotometrically in reflection and at different angles. A special effect pigment is characterized by a minimum of three L*a*b* data sets (CIELAB-system) measured under different conditions (e.g., 0 °/45 ° black background, 22.5 °/22.5 ° black background, 0 °/45 ° white background). These data specify such an effect pigment with respect to its hiding power, luster, and hue [1, 9–12].

The transmitted light is absorbed from a black background or in a blend of the effect pigment with carbon black and the reflected interference color is seen as the mass tone (i.e., overall color) of the material. In blends of special effect pigments with absorbing colorants, the particle size of the latter should be very small to reduce scattering to a minimum. The pearlescent effect or the iridescent reflection is otherwise quenched by the hiding pigments. This has also to be taken into consideration for blends with strongly reflecting metal effect pigments. Blends of transparent effect pigments with different interference colors obey an additive color mixing law. For example, appropriate blends of blue and yellow interference pigments generate a green interference color.

61.3 Substrate-free pigments

Only three substrate-free pearl luster pigments have gained historical and commercial importance: natural fish silver (natural pearl essence), basic lead carbonate, and bismuth oxychloride. Further substrate-free effect pigments are micaceous iron oxide (industrially relevant), titanium dioxide (industrially irrelevant due to the absence of an adequate production process), and colored organic flakes (small industrial importance due to limited optical and stability properties) [1].

61.3.1 Natural fish silver

Natural fish silver is an organic effect pigment, but is mentioned here because it was the first pearl luster pigment. It is isolated as a silky lustrous suspension from fish scales (herring, carp, sardine). The pigment particles in the suspension are platelet-shaped with a very high aspect ratio. They consist of 75 to 97% guanine and 3 to 25% hypoxanthine. The manufacture starts from an aqueous suspension of white fish scales, which is treated with organic solvents in a complicated washing and phase-transfer process to remove proteins and irregular guanine crystals. More than four tons of fish are necessary to produce one kilogram of fish silver. The pigment particles tend to agglomeration, which is the reason that they are only handled as dispersions. They have a refractive index of about 1.85, a relative low density of 1.6 g/cm^3, and exhibit high but soft luster. Natural fish silver is almost exclusively used in expensive cosmetic formulations [1–4].

61.3.2 Basic lead carbonate

Basic lead carbonate was the first commercially successful synthetic pearl luster pigment. It consists of hexagonal lead salt crystals, especially basic lead carbonate $Pb(OH)_2 \cdot 2\,PbCO_3$. Basic lead carbonate is precipitated from aqueous lead acetate solutions with carbon dioxide under carefully controlled conditions:

$$3\,Pb(CH_3COO)_2 + 2\,CO_2 + 4\,H_2O \rightarrow Pb(OH)_2 \cdot 2\,PbCO_3 + 6\,CH_3COOH \tag{61.3}$$

The resulting monocrystalline platelets have an aspect ratio of >200. They are less than 0.05 µm thick and show hexagonal dimensions of about 20 µm. With their refractive index of 2.0 and an even surface, they exhibit a very strong luster. An increase of the platelet thickness leads to interference pigments [1]. The crystals are fragile and are therefore handled in dispersions. Their high density of 6.14 g/cm^3 is responsible for an unfavorable fast settling behavior. Further disadvantages are the low chemical stability and the toxicological properties of basic lead carbonate. Especially the risks when working with lead compounds have led to a significant reduction in the use of basic lead carbonate for pigment applications [1].

61.3.3 Bismuth oxychloride

Bismuth oxychloride pigments are synthesized by the hydrolysis of very acidic bismuth salt solutions in the presence of chloride (pH value <1.0):

$$Bi(NO_3)_3 + HCl + H_2O \rightarrow BiOCl + 3\,HNO_3 \tag{61.4}$$

The quality of the BiOCl crystals formed can be varied by careful adjustment of bismuth concentration, temperature, pH value, pressure, reactor geometry, and addition of surfactants. Crystals with a tetragonal dipyramidal structure are usually formed. These can be flattened to platelets with a high aspect ratio by modified reaction conditions. Products with an aspect ratio of 10 to 15 exhibit low luster and very good skin feeling. They are used as fillers in a variety of cosmetic products. Crystals with higher aspect ratios show strong luster. They are mainly used in cosmetics for nail polishes.

Besides the use in cosmetics, the pigments are also applied in interior coatings, printed surfaces, buttons, and jewelry [1, 13, 14].

Disadvantages of BIOCl pigments are low light stability and fast settling caused by the high density of 7.73 g/cm^3. The light stability can be improved to some extent by doping with transition metals and UV absorbers in the application system. The low light stability of BiOCl is characterized by graying of the pigmented system during light exposure. High-lustrous bismuth oxychloride pigments are almost sold in the form of preparations. Special pastes are used with pigment contents of 50 to 80%, adjusted to the specific paint and printing formulations. These pastes can be equipped with suitable UV stabilizers [1].

61.3.4 Micaceous iron oxide

Micaceous iron oxide (MIO) consists of pure or doped hematite (α-Fe_2O_3). There are large natural deposits of MIO. Micaceous iron oxide is also formed by hydrothermal synthesis in alkaline media. The dull dark color of the so obtained products is as unappealing as that of the natural product. The incorporation of substantial amounts of dopants leads to increased aspect ratios of up to 100, resulting in an increased luster. The color can be influenced by the composition and more attractive reddish brown tones are achieved [15–17].

The most important dopants for synthetic MIO are Al_2O_3, SiO_2, and Mn_2O_3. These dopants can enforce the formation of thinner platelets, often crystallizing in the spinel structure. Iron(II) sulfate is the most suitable starting material for the synthesis of micaceous iron oxide pigments. Iron(II) is oxidized with air to iron(III) in an agitated vessel. The so formed iron(III) is precipitated as α-FeOOH by alkali hydroxide addition:

$$2\,FeSO_4 + 4\ NaOH + O_2 \rightarrow 2\,\alpha\text{-}FeOOH + 2\,Na_2SO_4 + 2\,H_2O \tag{61.5}$$

The α-FeOOH in the suspension is converted in an autoclave in the presence of aluminum oxide and potassium permanganate to platelet-shaped Al and Mn containing hematite. The autoclave is equipped with stirrer and heating. The size of the platelets is controlled by the reaction parameters, such as temperature, pressure, reaction time, concentration of alkali, and α-FeOOH content. Type and quantity of the doping agents are very important for the quality of the growing crystalline platelets [49, 50]:

$$2\ FeOOH + (Al_2O_3, NaOH, KMnO_4) \rightarrow Al_xMn_yFe_{2-x-y}O_3 + H_2O \tag{61.6}$$

The α-FeOOH suspension is heated up for several hours together with the doping agents to temperatures above 170 °C, typically to 250 to 300 °C. In a second reaction step, the pH-value is further increased in order to achieve platelets in the required final size and shape. Platelet-shaped doped and undoped iron oxide, which has a refractive index of about 2.9 and a density of 4.6 to 4.8 g/cm^3, is preferably used as effect pigment in paints and coatings [18, 19].

61.3.5 Titanium dioxide flakes

Titanium dioxide flakes can be produced by breaking down continuous films of TiO_2. Such films can be obtained using a web-coating process where $TiOCl_2$ is thermally hydrolyzed on the surface of the web [20]. Substrate-free TiO_2 flakes can also be achieved from TiO_2-mica pigments by dissolving the substrate in strong acids or hydroxides [1, 16]. The so obtained titanium dioxide flakes do not consist of single crystals. They are polycrystalline and quite porous. Their main disadvantage is the insufficient mechanical stability for most applications, especially for technical uses where stress is exerted. TiO_2 flakes show typical interference colors, which are dependent on the respective thickness of the platelets [4, 20]. Further variations of color and luster of these effect pigments can be achieved by coating the platelets with metal oxide layers of different composition [21, 22].

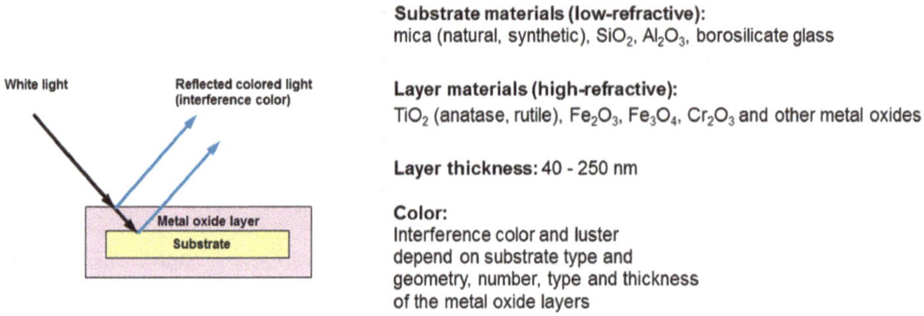

Figure 61.3: Structure of layer-substrate pigments [7].

61.4 Pigments based on the layer-substrate principle

Pigments based on the layer-substrate principle are the dominant representatives of special effect pigments. The variation and combination possibilities of layer and substrate materials are shown in Figure 61.3. Special effect pigments based on natural mica as substrate material have achieved the highest importance amongst the layer-substrate pigments. Other substrates, such as alumina, silica, borosilicate glass, and synthetic mica are used in form of flakes as alternatives to natural mica.

61.4.1 Metal oxide mica pigments

61.4.1.1 Fundamentals

Metal oxide mica pigments consist of platelets of natural mica coated with thin films of transparent metal oxides. The mica platelets act as templates for the synthesis and as a mechanical support for the deposited thin optical metal oxide layers. Most of the pearl luster and interference pigments are based on natural muscovite mica, some of them on synthetic fluorophlogopite mica. Micas belong to the class of layered aluminosilicates. Pure mica samples are optically transparent. Muscovite occurs worldwide in large deposits, but only a few of these are suitable for pigment production. Natural mica is classified as biologically inert and approved for the use as filler and colorant [1–4, 7].

The selection and pre-processing of the mica is decisive for the quality and appearance of the resulting special effect pigments. Rough mica blocks are ground and classified into different particle size ranges at the beginning of the process. Particle size distribution and aspect ratio of the thus obtained mica platelets have a large influence on the properties of the final pigments. The thickness of the mica platelets is typically in the range from 300 to 600 nm. The platelet diameters can be adjusted by the choice of the grinding and classification parameters. Typical ranges are <15, 5 to 25, 10 to 50, or 30 to 110 µm. Light incident on the final pigment particles is regularly reflected from the metal oxide and the mica planes and scattered from the edges. Pigments based on larger mica platelets have therefore high brilliance and low hiding power. Pigments on smaller platelets, on the other hand, have accordingly lower brilliance and higher hiding power.

There are several deposits for natural muscovite mica (basic composition $KAl_2[AlSi_3O_{10}](OH)_2$) with appropriate quality and only slight impurities at various locations around the world. Synthetic fluorophlogopite mica (basic composition $KMg_2[AlSi_3O_{10}]F_2$) is normally produced by a high-temperature crystallization process starting from Al_2O_3, MgO, SiO_2, and K_2SiF_6. The necessary synthesis temperatures are considerably above 1200 °C. The formed mica blocks are ground and classified comparable with the natural mica [25].

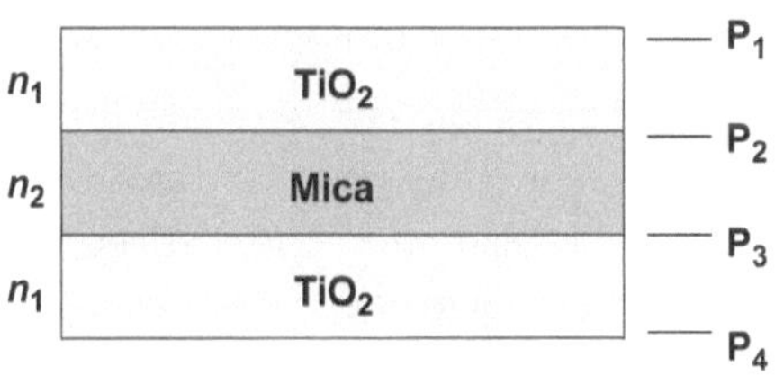

Figure 61.4: Schematic structure of a titanium dioxide mica pigment with the four existing interfaces (P_1-P_4).

A mica platelet coated from all sides with a metal oxide has three layers with two different refractive indices (n_1 and n_2) and four phase boundaries (P_1 – P_4): P_1 / TiO_2 / P_2 / mica / P_3 / TiO_2 / P_4 (Figure 61.4). Interference of light is generated by combined reflections at pairs of phase boundaries, some of which are equal: $P_1P_2 = P_3P_4$, $P_1P_3 = P_2P_4$, P_1P_4, and P_2P_3. The thicknesses of the mica platelets correspond to a statistical distribution. In consequence, interference effects involving the phase boundaries between the mica substrate and the oxide coating add together to generate a white background reflectance. The interference color of a large number of particles depends for this reason only on the thickness of the metal oxide layer on both sides of the mica.

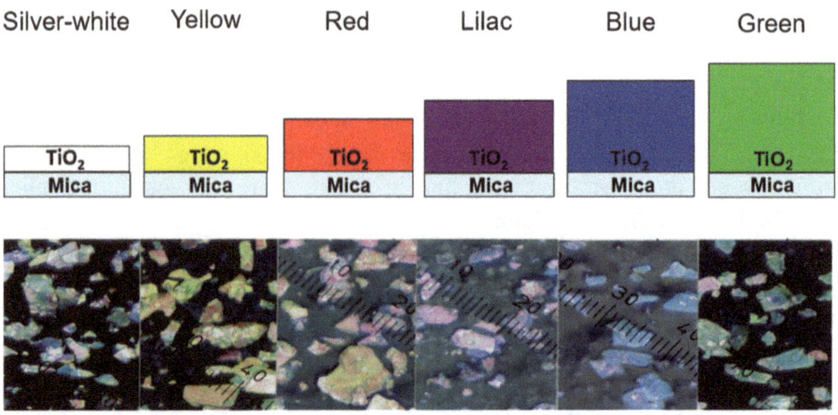

Figure 61.5: Schematic illustration of the structures (only the upper TiO_2 layer is shown) and of the related interference colors of different titanium dioxide mica pigments. The microscopic images illustrate the corresponding interference colors of TiO_2 mica pigments (scale in the images in μm).

The most important effect pigments based on mica consist of TiO_2-mica (TiO_2 in form of anatase and rutile) and Fe_2O_3-mica (Fe_2O_3 in form of α-Fe_2O_3). TiO_2-mica pigments represent the transparent pearlescent and interference types. Fe_2O_3-mica pigments combine interference color with the absorption color of α-iron(III) oxide and cover the range of bronze, copper, red, red-violet, and red-green color shades. So-called combination pigments are brilliant, mass-tone-colored types consisting mostly of TiO_2 and another metal oxide with one color (interference color same as mass tone) or two colors (interference and mass tone different) on the mica. Multilayer structures with more than three layers of different refractive indices extend the series of mica-based effect pigments by very intensive shades and show in part strong angle-dependent color effects. The color of all single pigments depends on the specific composition and on the viewing angle. A schematic illustration of the structures and of the corresponding interference colors of different titanium dioxide mica pigments is shown in Figure 61.5 together with related microscopic images.

61.4.1.2 Production of metal oxide mica pigments

Two different processes are used for the production of TiO_2-mica pigments. They are referred to as homogeneous hydrolysis ("sulfate process") and titration ("chloride process"). Both processes take place in aqueous suspension.

Homogeneous hydrolysis at 100 °C:

$$TiOSO_4 + mica + n\,H_2O \rightarrow TiO_2 \cdot (n-1)\,H_2O/mica + H_2SO_4 \quad (61.7)$$

Titration:

$$TiOCl_2 + mica + H_2O + 2\,NaOH \rightarrow TiO_2(hydrate)/mica + 2\,NaCl \quad (61.8)$$

In both processes, the resulting freshly coated mica particles are separated from the suspension by filtration, washed with water, dried, and calcined in air at temperatures of about 800 °C:

$$TiO_2(hydrate)/mica \rightarrow TiO_2/mica + H_2O \quad (61.9)$$

The titration process is preferred for interference pigments with thick TiO_2 layers because it is easier to control.

When TiO_2 is precipitated onto mica platelets under reaction conditions unfavorable for side precipitation, e.g., pH value >1.5 and temperatures in the range of 60 to 90 °C, only the anatase modification is formed. Even after annealing at 1000 °C, no rutile phase is detected in the TiO_2 layers. On the other hand, titanium dioxide hydrate precipitated in an aqueous solution without the presence of substrate particles, can be completely converted into rutile at about 700 °C [24].

However, the formation of rutile on the mica platelets is also desirable because of the higher refractive index and the better light stability compared with anatase. The higher refractive index leads to stronger reflectivity and pearlescent effects. Consequently, processes have been developed to create rutile layers on mica platelets. Tin dioxide hydrate is initially precipitated as a thin layer onto the substrate, and then the TiO_2 hydrate layer is deposited using the common process. $SnCl_4$ is used in most cases as the precursor for the SnO_2 hydrate pre-coating:

$$SnCl_4 + mica + 4\,NaOH + H_2O \rightarrow SnO_2(hydrate)/mica + 4\,NaCl \quad (61.10)$$

$$SnO_2(hydrate)/mica + TiOCl_2 + 2\,NaOH \rightarrow TiO_2(hydrate)/SnO_2(hydrate)/mica + 2\,NaCl \quad (61.11)$$

SnO_2 formed during the calcination acts as a template for the TiO_2. It crystallizes only in the rutile structure and forces from there the formation of the rutile structure for the titanium dioxide:

$$TiO_2(hydrate)/SnO_2(hydrate)/mica \rightarrow TiO_2(rutile)/SnO_2/mica \quad (61.12)$$

Figure 61.6 (left) shows the SEM picture of a TiO_2-mica pigment. Shape and size of the pigment particles are clearly visible. Figure 61.6 (right) shows the magnified view on the TiO_2 layer of an individual pigment particle. The layer of TiO_2 formed during coating and calcination shows a sintered granular structure. It covers the mica platelets completely. Its thickness can be adjusted with highest precision in the nanometer range during the process of manufacture. The exact control of the

Figure 61.6: Scanning electron micrographs of a titanium dioxide mica pigment; left: overview image; right: TiO_2 surface of an individual pigment particle (source: Merck KGaA).

layer thickness is decisive for the reproducible production of special effect pigments based on the layer-substrate principle.

The production of multilayer pigments needs the precipitation of an additional material with low refractive index. Typically, SiO_2 layers are deposited between TiO_2 layers. An example for a multilayer composition based on mica is the layer structure $TiO_2/SiO_2/TiO_2$/mica/$TiO_2/SiO_2/TiO_2$ [25].

The desired interference color determinates the thickness of the TiO_2 layer on the mica. For the manufacture of a silver-white pigment, 50 nm of anatase is necessary and for a blue interference color about 120 nm. The sequence of interference colors obtained in practice with increasing TiO_2 layer thickness is in accordance with physical color calculations in the color space. The color values for single interference pigments can be calculated using the equations for the optics of thin films and are expressed by the L*a*b*-values according to Hunter [1, 2, 7]. The calculation of the TiO_2 layer thickness necessary to obtain a specific color is done by superpositioning of the a- and b-curves. As an example, to obtain a pure interference blue, a layer thickness should be selected which lies beneath the maximum value for – b, and where the a-value goes through the zero point. The purest blue can therefore be obtained with a TiO_2 layer thickness of about 125 nm. At 110 nm, the combination of the high value for + a and the high value for – b generates a violet shade.

Interference pigments show a distinct color flop (color changing effect, flip-flop effect). The color varies with the angle of observation. The interference color is visible from a certain angle, the glancing angle. The color, which is seen in a specific application of TiO_2 interference pigments depends strongly on the background. Due to the transparency of the pigments, the background color, e.g., white, black, or colored, plays an important role for the entire color appearance. TiO_2 mica pigments as well as other special effect pigments are often used in combination with other pigments, e. g. with metal effect pigments, white, black and colored inorganic pigments, and also organic pigments. Color and effects achieved by the use of special effect pigments in the different application systems have therefore a very complex origin.

Iron(III) oxide is suitable for the coating of mica platelets similar to titanium dioxide. It combines a high refractive index (metallic luster) with sufficient hiding power and excellent weather resistance. Fe_2O_3-mica pigments are manufactured by precipitation of goethite from $FeCl_3$ or $FeSO_4$ solutions onto mica platelets in aqueous suspension followed by filtration, washing, drying, and calcination of the resulting coated particles at 700–900 °C:

$$FeCl_3 + mica + 2\,H_2O \rightarrow \alpha\text{-}FeOOH/mica + 3\,HCl \tag{61.13}$$

$$2\,\alpha - FeOOH/mica \rightarrow \alpha\text{-}Fe_2O_3/mica + H_2O \tag{61.14}$$

Hematite layers are formed on the mica platelets during calcination and water is removed. The thickness of the layers depends on the amount of α-FeOOH deposited during the coating process.

The α-Fe_2O_3 nano-crystallites formed at the beginning of the annealing procedure have diameters of 20 to 40 nm and show a preferred orientation, which has its origin in a template effect arising from the crystallographic structure of the muscovite mica. The initially generated granular layer structure is sintered during the calcination to form a less porous, crystalline, and coherent phase [26].

Fe_2O_3-mica pigments have typical layer thicknesses of 50 to 150 nm and show brilliant, intense colors. Absorption and interference colors are produced simultaneously and vary with the layer thickness. The red shades are as intense because interference and absorption phenomena enhance each other. A green-red flop visible from different viewing angles is possible at a thickness of the α-Fe_2O_3 layer, which corresponds to green interference.

Combination pigments containing TiO_2 and other colorants together on mica platelets have extended the class of special effect pigments. Fe_2O_3 is the most important metal oxide for combination with titanium dioxide for such pigments. Brilliant golden pigments with combined layers can be synthesized using two different routes. The pigment structures formed differ considerably from each other. In the first route, a thin layer of Fe_2O_3 is deposited onto the surface of a TiO_2-mica pigment. The generated interference color is the result of both metal oxide layers. The mass tone of such a pigment is determined by the properties of the Fe_2O_3 layer. Interesting gold pigments, e.g., with reddish-golden color, are possible. In the second route, co-precipitation of hydrated titanium dioxide and iron oxide on mica particles followed by calcination leads to greenish-golden pigments. The mass tone is in this case further modified by the highly refractive yellowish iron titanate phase Fe_2TiO_5 (pseudobrookite), which is formed under these conditions [1–4].

Other inorganic colorants used instead of iron oxide for combination pigments are Cr_2O_3 (green), iron blue (blue), Fe_3O_4 (black), and $FeTiO_3$ (silver-gray). In the case of black materials, the interference color is seen as the mass tone. The optical situation is similar to blends with black pigments in a color formulation where the transmitted part of the light is absorbed. Additional coating of a TiO_2-mica pigment with an organic colorant for a mass-tone or two-tone effect is executed by precipitation and deposition on the surface of the mica pigment in aqueous suspension, assisted by complexing agents or surfactants. Another method is the fixation of the colorant as a mechanically stable layer by using proprietary additives.

Mica platelets can be coated with a variety of other compounds to achieve attractive color effects. Solid-state reactions extend the possibilities for the synthesis of mica-based pigments. Further options are offered by calcination of the materials in the presence of inert (N_2, Ar) or reactive gases (NH_3, H_2, hydrocarbons). The formation of some colored phases, which are not producible by working in air, is possible under these conditions [1, 27–31].

Simple blending of transparent absorption pigments with pearl luster pigments is one way to generate new color effects. Another way is the production of pearl luster pigments coated with a layer of a transparent absorption colorant to realize more pronounced brilliant colors with a sharper color flop. An additional advantage of such pigments is the avoidance of dispersion problems associated with transparent absorption pigments due to their high surface area and small particle size [1, 2, 7].

Mica-based and other special effect pigments appropriate for outdoor applications must meet the highest standards for color fastness and weather resistance. These pigments are coated additionally with thin layers of transparent and colorless inorganic and/or organic compounds (surface treatment). These layers increase the light resistance by reducing the photoactivity of the titanium dioxide surface. In addition, the interaction between pigment and binder is optimized. This also applies to iron oxide or chromium oxide pigment types. Pigments developed with a surface treatment are preferably used in automotive coatings and in industrial outdoor applications. Criterion for the use in these applications is in most cases the "Florida test", which has to be passed.

61.4.1.3 Functional metal oxide mica pigments

Metal oxide mica pigments were developed in the early phase only for their special coloristic properties. They have, however, also the potential for functional pigments. These consist typically of metal oxide layers on platelets of natural mica. The used metal oxides possess electrically conductive, magnetic, infrared reflecting or other physical properties, and give rise to the functional behavior of the pigment.

Light-colored conductive pigments are applied as conductive component in coatings with permanent antistatic or electrostatic-dissipative (ESD) properties. They are also used as an additive for conductive fillers for the electrostatic spraying of paints. The conductive metal oxide layer on the mica consists in most cases of tin dioxide containing antimony with the general formula $(Sn,Sb)O_2$. Such compositions show semi-conductive to conductive properties. Antimony is incorporated into the lattice of tin dioxide on lattice sites of tin. The production of the pigments starts from mica suspensions and soluble tin and antimony salts. The manufacturing process is comparable with that used for the coating of mica platelets with TiO_2 or Fe_2O_3 layers. The conductivity of the pigmented system depends strongly on the pigment volume concentration [1, 32–35].

Magnetic pigments based on mica coated with magnetite can be manufactured either by the reduction of Fe_2O_3 containing mica pigments or by precipitation and oxidation starting from iron(II) salt solutions. The magnetite layer formed on the mica platelets gives these pigments their black absorption color, however, at certain thicknesses they appear colored at the glancing angle. The magnetic properties are less favorable compared with the characteristics of conventional magnetic pigments, such as γ-Fe_2O_3, Fe_3O_4, Co-containing iron oxides, metallic iron, or CrO_2 [1].

Solar heat-reflecting pigments based on mica consist of almost completely transparent platelet-shaped particles, which reflect a certain part of the infrared radiation. They are used for architectural and horticultural purposes, where it is important to have plenty light, but little heat. The pigments absorb only a small amount of light and reduce substantially the heating caused by intense sunlight. Metal oxide layers on mica used for solar heat-reflecting pigments consist mostly of TiO_2 and SiO_2 with precisely tuned thicknesses [1, 36].

Pigments for the laser marking of polymers may also be based on the layer-substrate principle that is used in special effect pigments. In this case, mica platelets are coated with specific layers of either TiO_2 or $(Sn,Sb)O_2$. The laser technology offers several advantages compared with the sharp lettering achieved by printing. The writing speed is very high and the marking is immediately permanent and indelible. Furthermore, it is an ink free marking process. The platelet-shaped pigment particles absorb the incidental laser light (suitable lasers are CO_2-laser with 10.6 µm, Nd:YAG-laser with 1064 or 532 nm, and excimer-laser with 193 to 351 nm) and become themselves a temperature sink. The temperature of the pigment particles increases rapidly and the polymer material surrounding the pigment is carbonized. The polymer turns black specifically at the laser-targeted point. Carbonizing can take place together with the formation of fine gas bubbles, thus producing light gray to whitish markings. The quality of the laser marking depends strongly on the pigment volume concentration, the polymer material, and the laser parameters used [37, 38].

61.4.2 Effect pigments based on silica flakes

Silica flakes (SiO_2 flakes) can be used instead of natural or synthetic mica as a substrate for special effect pigments. The flakes are produced by a web-coating process. SiO_2 flakes offer several advantages over the use of mica. Their thickness can be controlled during preparation so that at the end a pigment with a true optical three-layer system is obtained, the interference color of those systems being stronger than that for the conventional mica pigments for which the effect of the mica is levelled by a broad thickness distribution. As a synthetic substrate material, silica does not have iron impurities that cause the slightly yellow mass tone of some natural mica qualities. Finally, SiO_2 has a lower refractive index (1.46) than mica (about 1.58) and, therefore, leads to stronger interference effects [1, 39–42].

Characteristics of silica flake pigments are uniform and controllable substrate thickness, smooth and uniform substrate surface, absolutely transparent substrate with no mass tone color, excellent reflection and chroma, and intensive color travel.

The silica flakes are manufactured by a web coating process. A polymer web, moving at a controlled speed, is wet with an aqueous solution of a silica compound. The film on the web is dried and a stable coating layer is formed. The thickness of this layer can be selected from 50 to 1000 nm by adjustment of the process parameters. The wet layer is removed from the web and processed further to produce flakes that are fractioned and coated. The transparent flakes can be coated with various metal oxides, such as titanium dioxide and α-iron(III) oxide, using the procedure known from the mica pigments [43].

The color of pigments based on silica flakes can exhibit extreme angle-dependent effects. Objects painted with them will change appearance with the direction of lighting and the location of the observer. Color changes from gold-silver to green, to green-blue, to dark blue are possible. Strong color travel is seen even under subdued lighting conditions. Decisive for the effect are the uniform thicknesses of the silica substrate and the metal oxide layer. Both layer thicknesses have to be adjusted and controlled precisely in a range of less than 5 nm.

Special effect pigments with improved color strength and very high luster are produced on silica flakes with a specific interference color by deposition of titanium dioxide or iron oxide layers of a thickness required to generate the same interference color. Such pigments show stronger chroma than comparable types based on mica.

The scanning electron micrographs in Figure 61.7 provide information about particle size and shape as well as the layer thicknesses of a α-Fe_2O_3 silica flake pigment. It shows clearly that both the SiO_2 flake and the Fe_2O_3 layer thickness are precisely controlled.

Figure 61.7: Scanning electron micrographs of a silica flake pigment coated with iron(III) oxide; left: overview image; right: cross-section through one particle (source: Merck KGaA).

61.4.3 Effect pigments based on alumina flakes

Thin hexagonal monocrystalline alumina flakes (Al_2O_3 flakes) suitable as substrate material for special effect pigments, which show an extreme sparkle in coating applications, can be produced by a specific crystallization process. The flakes are produced starting from an aqueous aluminum salt solution. Small amounts of dopants are added to control the following crystal growth. An aluminum-containing precipitate is formed, which is filtered off, dried and heated up to temperatures above 1000 °C. After washing of the cooled material, very thin flakes are obtained, which consist of corundum and show a high aspect ratio, a very narrow thickness distribution, and very smooth surfaces. The thickness can be controlled by the reaction parameters. The resulting flakes can act as substrate particles for metal oxide layers in the process, which is also used for mica and silica flakes. Special effect pigments of very high luster are obtained [1, 39–42, 44].

Pigments based on thus produced alumina flakes possess a narrow particle size distribution of about 5 to 30 µm as well as a high aspect ratio. The striking sparkle effect has its origin in the optimized thickness of all layers, including that of the Al_2O_3 substrate. The effect is additionally intensified by directed illumination, e. g. by sunlight. The sparkle appearance in coating applications is controlled by the quantity of the added pigment. The effect, which is visible already at concentrations of 0.1% in the paint system can be intensified by an increase of the pigment concentration of up to 2%. The single light spots arising from the pigment structure and orientation seem to spring back and forth when a pigmented surface is tilted.

The scanning electron micrographs in Figure 61.8 show alumina flakes coated with titanium dioxide. The special shape of the particles resulting from the crystal growth process for the alumina flakes and the fine structure of the TiO_2 layer are clearly visible.

Figure 61.8: Scanning electron micrographs of an alumina flake pigment coated with titanium dioxide; left: overview image; right: image of the fine-grained structure of the TiO_2 layer (source: Merck KGaA).

61.4.4 Effect pigments based on borosilicate flakes

Special effect pigments with neutral body color, high luster and attractive sparkling properties can be achieved from borosilicate flake substrates (glass flakes) [1, 17, 45, 46]. Such pigments consist mostly of calcium-aluminum borosilicate platelets coated with metal oxide layers, e.g., SiO_2, TiO_2 or Fe_2O_3. Synthetically manufactured high-quality flakes are absolutely planar and show a very smooth surface. They are characterized by a relatively uniform thickness of each single particle and show therefore typically a homogeneous color of each platelet.

Manufacturing of borosilicate pigments consists of the process steps raw flake production, grinding and classification of the flakes, and coating of the flakes with highly-refractive metal oxides, preferably titanium dioxide and iron(III) oxide. The synthesis of the borosilicate flakes is carried out starting from a glass melt. Raw materials of high purity are used to achieve the desired colorlessness and transparency of the resulting flakes, which are produced directly from the melt [47].

They can be used after cooling, grinding and classification as substrate platelets for effect pigments. Coating of the borosilicate flakes and the following working steps are done analogous to the processes used for mica and other platelet-like substrates.

Special effect pigments based on borosilicate flakes possess in most cases particle diameters below 200 µm and a thickness below 2 µm. Dependent on the flake geometry and the coating, the pigments provide a unique and striking multicolored sparkle effect in application systems. This appearance, which is in particular visible in systems pigmented with silver-white types, is called the multicolor effect.

Figure 61.9 shows scanning electron micrographs of borosilicate flakes coated with titanium dioxide. The flakes are perfectly covered with the TiO_2 layer leading to high light reflection and good skin feeling, which is important for the application of the pigments in cosmetic formulations [45].

Figure 61.9: Scanning electron micrographs of a borosilicate flake pigment coated with titanium dioxide; left: overview image; right: TiO_2 surface of an individual pigment particle (source: Merck KGaA).

61.4.5 Comparison of different pigments based on the layer-substrate principle

Today, a broad number of special effect pigments based on different substrates are available for various applications. Besides natural mica, silica, alumina and borosilicate glass, also synthetic mica (fluorophlogopite) is used as carrier material for special effect pigments [7]. Flakes consisting of synthetic mica are produced using a high-temperature flux crystallization process. Typical starting materials for this process are SiO_2, Al_2O_3, K_2CO_3 and MgF_2. Main advantages of synthetic mica in comparison with natural mica are the very little residual absorption and the extremely low content of heavy metals in the platelets.

Figure 61.10 shows a comparison of the flakes of the five relevant substrate materials for special effect pigments together with bismuth oxychloride platelets. Each of the substrates has specific properties, that are transferred to the resulting pigments when coated with high-refractive metal oxides. Effects like classic silver-white and interference (natural mica, synthetic mica) color travel (silica flakes), sparkle (alumina flakes), transparency (borosilicate flakes) and special skin feel combined with hiding power (BIOCl) can be obtained in dependence on the substrate material chosen for the production of a specific pigment.

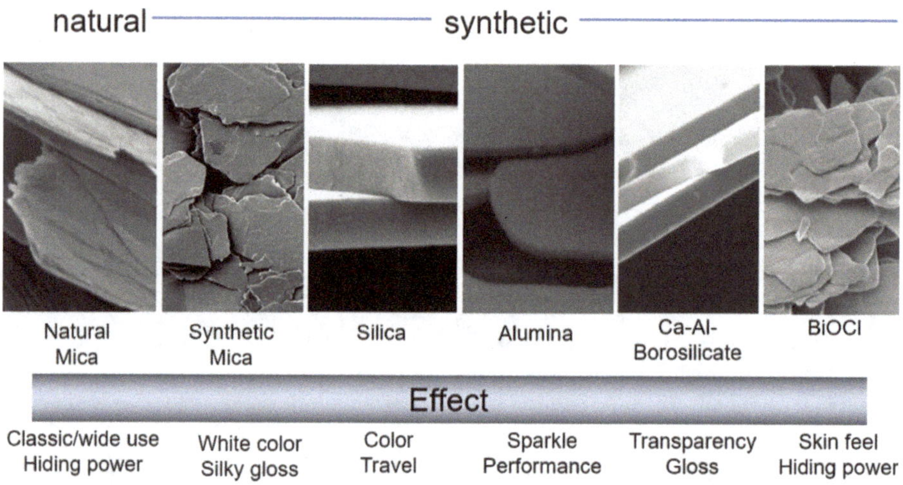

Figure 61.10: Scanning electron micrographs of different substrate flakes for special effect pigments together with an image of BiOCl platelets (source: Merck KGaA).

61.5 Pigment properties and uses

Special effect pigments are used alone or as part of color formulations where color depth, brilliance, iridescence, color travel, and other spectacular effects are required [1–4, 7]. Pearl and interference effects, brilliance, stability and user-friendly behavior in different application systems are specific characteristics of these pigments. Requirements for the application of the pigments are transparent or at least translucent binders. Formulations with absorption pigments are problem-free possible, but have to take the transparency of the effect pigments and the color mixing rules into the consideration.

A basic prerequisite for the achievement of the desired luster and color effects is the parallel alignment of the platelet-shaped pigment particles during their application in coatings, paints, plastics, printing inks, or cosmetic formulations. A homogeneous dispersion of the pigments is possible in most cases by careful stirring in the liquid application system.

Special effect pigments are broadly used as standard components of solvent- and water-based automotive and industrial coatings. The decorative character and the technical quality make them attractive for these application areas. Automotive designers know the special features of the pigments very well and have a wide range of color and composition possibilities at their disposal. Attractive and individual coatings are achieved not only by using the various color possibilities of the effect pigments, but also by the different particle sizes available and by combination with suitable absorption pigments. Typical examples for industrial paint applications of special effect pigments are bicycles, airplanes, rail wagons, cans and containers, building exteriors, interior decoration, furniture, artificial pearls, and jewelry [16, 48].

Special effect pigments are used in nearly all kinds of thermoplastic and thermosetting polymers. All common techniques for polymer processing are applied for adding color and special effects to plastic parts. The orientation of the pigment platelets parallel to the object surface happens in almost all applications through flow movements in the fluid polymer matrix. Typical effect pigment concentrations in polymer applications are in the range from 0.5 to 2% [38].

The pigments are also used in a broad variety of printing applications where they fulfill the highest esthetic requirements, for example, when articles should reflect a touch of luxury and artistic style. The pigments are processed with all common printing techniques such as offset, screen, gravure and flexographic printing. Coating techniques such as overprint-varnishing (OPV) and paper coating are also used. Special effect pigments can be applied in solvent- and water-based printing formulations as well as in UV inks [49–51].

Cosmetic applications necessitate specific effect pigment qualities approved for the use according to cosmetic regulations. The applications include color cosmetics (e.g., lipsticks, eye shadows, blushers, eye pencils, make-up, mascaras, nail lacquers), personal care products (e.g., shampoos, body washes, lotions, creams, oral care products, hair gels) and special effect items (e.g., theatrical and costume make-up, temporary hair color sprays) [13, 14].

Special effect pigments for ceramic and glaze applications are often coated with additional layers to achieve stabilization against the aggressiveness of the frits at high temperatures [52].

Some of the special effect pigments, mainly those consisting of mica coated with TiO_2 or Fe_2O_3, are approved for food and pharmaceuticals. Various color shades ranging from silver and gold up to red and bronze are available for these specific fields of application. Such pigments are used in confectionery (e.g., liquorice, fruit gums, jelly beans, chewing gums), chocolate, biscuits, ice cream and beverages. They meet all international standards for food and drug safety [53].

No toxic effects were found in scientific examinations for special effect pigments based on mica. The pigments do not show acute toxicity (LD_{50} value rat oral: >15,000 mg/kg). Metal oxide mica pigments demonstrate no irritation or sensitivity to skin or mucous membranes. Investigations on comparable products based on other substrate platelets such as silica, alumina, or borosilicate flakes did also not find any irritation to the skin or mucous membranes [54].

There is no indication of hazards to humans through contact with substrate-based effect pigments in the working place. Studies on the genetic toxicity of mica-based and other comparable pigments have not shown mutagenic effects *in vitro* and *in vivo*. Ames and micro-core tests have not shown a mutation of the inheritance. Chronic toxicity tests with effect pigments based on mica ended with the result that there are no ill effects on rats over a period of 2 years [7, 54, 55].

Special effect pigments based on mica, silica, alumina, and borosilicate are highly stable and inert. Virtually no metal ions can be leached from their surface in aqueous suspensions. These pigments are therefore regarded as nontoxic and harmless with regard to ecological hazards. As in water insoluble inorganic materials, the pigments are nonbiodegradable [54].

All titanium dioxide and iron oxide mica pigments without an additional surface treatment are approved for the application in food packaging, e.g., as a colorant in related printing inks and plastics. Some of the pigments are even approved as food and pharmaceutical colorants.

Special effect pigments consisting of bismuth oxychloride are considered to be nontoxic and nonhazardous. Extensive and direct contact of the pure BiOCl powder with the skin should be avoided due to possible skin irritation.

Basic lead carbonate is classified like other lead compounds as dangerous for the environment. Lead compounds are treated as very toxic to aquatic organisms. They may cause long-term adverse effects in the aquatic environment.

References

1. Pfaff G. In special effect pigments. Pfaff G, editor. 2nd ed. Hannover:Vincentz Verlag, 2008:16.
2. Franz H-D, Emmert R, Nitta K. Interference pigments. Kontakte (Darmstadt). 1992;(2):3.
3. Pfaff G. Perlglanzpigmente. Chem Unserer Zeit. 1997;31:6.
4. Pfaff G, Reynders P. Angle-dependent optical effects deriving from submicron structures of films and pigments. Chem Rev. 1999;99:1963.
5. Snow MR, Pring A, Self P, Losic D, Shapter J. The origin of the color of pearls in iridescence from nano-composite structures of the nacre. Amer Miner. 2004;89:1353.
6. Katti KS, Katti DR. Why is nacre so tough and strong? Mater Sci Eng C. 2006;26:1317.
7. Pfaff G. Inorganic pigments. Berlin/Boston: Walter de Gruyter GmbH, 2017: 205.
8. Wiegleb JC. Handbuch der allgemeinen Chemie. vol. Band 2. Berlin/Stettin: Friedrich Nicolai, 1781:636.

9. Hofmeister F, Pieper H. Remissionsmessungen von Interferenz-, Aluminium- und Körperfarbenpigmenten. Farbe + Lack. 1989;95:557.
10. Pfaff G, Gabel P. Colour measurement from many angles. Eur Coat J. 2005:(6)30.
11. Gabel P. In special effect pigments. Pfaff G, editor. 2nd ed. Hannover:Vincentz Verlag, 2008:197.
12. Gabel P. In colour technology of coatings. Hannover: Vincentz Verlag, 2016:61.
13. Thurn-Mueller A, Hollenberg J, Listen J. Special effect pigments and fillers for cosmetics. Kontakte (Darmstadt). 1992:(2)35.
14. Pfaff G, Becker M. Household personal care today. Special effect pigments in cosmetic applications. Monogr Suppl Ser "Colour Cosmet". 2012:(4)12.
15. Ostertag W, Mronga N. Properties and applications of luster oigments. Macromol Chem Macromol Symp. 1995;100:163.
16. Maile FJ, Pfaff G, Reynders P. Effect pigments – past, present and future. Progr Org Coat. 2005;54:150.
17. Pfaff G. In industrial inorganic pigments. Buxbaum G, Pfaff G, editor. 3rd ed. Weinheim: Wiley-VCH Verlag, 2005:230.
18. Patent EP. 0 068 311 (BASF). 1982.
19. Ostertag W. Effektpigmente. Nachr Chem Tech Lab. 1994;42:849.
20. Patent EP. 0 837 911 (Merck). 1997.
21. Patent EP. 0 931 112 (Merck). 1997.
22. Patent EP. 0 944 677 (Merck). 1997.
23. van Valkenburg A, Pike RG. Synthesis of mica. J Res Natl Bur Stand. 1952;48:360.
24. Kumar KN, Keizer K, Burggraaf A, Okubo T, Nagamoto H. Textural evolution and phase transformation in titania membranes: part 2. supported membranes. J Mater Chem. 1993;3:1151.
25. Patent EP. 1 025 168 (Merck). 1997.
26. Hildenbrand VD, Doyle S, Fuess H, Pfaff G, Reynders P. Crystallization of thin anatase coatings on mica. Thin Solid Films. 1997;304:204.
27. Ostertag W, Mronga N, Hauser P. Eisenoxidbeschichtete Aluminiumpigmente. Farbe + Lack. 1987;93:973.
28. Patent EP. 0 632 821 (Merck). 1992.
29. Hauf C, Kniep R, Pfaff G. Preparation of various titanium suboxide powders by reduction of TiO2 by with silicon. J Mater Sci. 1999;34:1287.
30. Patent EP. 0 354 374 (Merck). 1988.
31. Patent EP. 0 246 523 (Merck). 1986.
32. Brueckner H-D, Glausch R, Maisch R. Neuartige helle, leitfähige Pigmente auf Glimmer/Metalloxid-Basis. Farbe + Lack. 1990;96:411.
33. Hocken J, Griebler W-D, Winkler J. Elektrisch leitfähige Pigmente zur antistatischen Ausrüstung von polymeren Beschichtungsstoffen. Farbe + Lack. 1992;98:19.
34. Pfaff G, Reynders P. Neue Entwicklungen bei Metalloxid-Glimmer-Pigmenten. Chem Rundschau (Schweiz, Jahrbuch). 1993:42:31.
35. Vogt R, Neugebauer E, Pfaff G, Stahlecker O. Eur Coat J. 1997:(7–8):706.
36. Patent EP. 0 659 198 (Hyplast, Merck). 1992.
37. Pfaff G. Funktionelle anorganische Pigmente. Phänomen Farbe. 2003;23:21.
38. Kieser M. In special effect pigments. Pfaff G, editor. 2nd ed. Hannover:Vincentz Verlag, 2008:137.
39. Teaney S, Pfaff G, Nitta K. New effect pigments using innovative substrates. Eur Coat J. 1999;(4):90.
40. Pfaff G. Innovationen bei Effektpigmenten. Welt Der Farben. 2000;1:16.

41. Sharrock S, Schül N. New effect pigments based on SiO_2 and Al_2O_3 flakes. Eur Coat J. 2000; (1–2):20.
42. Pfaff G. Special effect pigments based on silica flakes. Inorg Mater. 2003;39:123.
43. Patent EP. 0 608 388 (Merck). 1991.
44. Patent EP. 0 763 573 (Merck). 1995.
45. Rüger R, Oldenburg N, Schulz E, Thurn-Schneller A. Star effect from borosilicates. Cosmet Toiletries Manufacture Worldwide. 2004;(5):133.
46. Huber A, Pfaff G. Spezielle Effektpigmente. Phänomen Farbe. 2005;25:34.
47. Weitzel J. In special effect pigments. Pfaff G, editor. 2nd ed. Hannover:Vincentz Verlag, 2008:171.
48. Böhm K, Pfaff G, Weitzel Ein effektvolles Farbenspiel: Wie Perlglanzpigmente in Druckfarben für verschiedene Druckverfahren eingesetzt werden können. J. Farbe + Lack. 1999;105:30.
49. Pfaff G, Rathschlag T. New design possibilities with special effect pigments in printing inks. Paintindia. 2002;(4):65.
50. Pfaff G, Hechler W, Brabänder C. Keram Z. Anwendungen von Perlglanzpigmenten im Dekorbereich. 2001;53:1002.
51. Patent EP. 1 469 745 (Merck). 2002.
52. Pfaff G. In special effect pigments. Pfaff G, editor. 2nd ed. Hannover:Vincentz Verlag, 2008:209.
53. Bernard BK, Osheroff MR, Hofmann A, Mennear JH. Toxicology and carcinogenesis studies of dietary titanium dioxide-coated mica in male and female Fischer 344 rats. J Toxicology Environm Health. 1989;28:415.

Andrew Towns

62 Spirooxazine dyes

Abstract: This article provides an introductory overview of a class of colorant that underpinned the establishment of the most industrially important application of photochromic dyes. It outlines the properties that led to the rise (and subsequent decline) in commercial significance of spirooxazine colorants, and which attract researchers seeking to put them to work in new technologies. Within this survey, the chief means of synthesising such dyes and adjusting their photochromism are briefly discussed. The reader is directed to useful sources of information concerning not only these aspects, but also potential applications for spirooxazine-based photochromism.

Keywords: photochromism, photochromic, colorant, dye, merocyanine

The first commercially significant exploitation of photochromic dyes – sun-responsive spectacles – depended upon the photochromism of the spirooxazine (SO) class of colorant. While the industrial importance of SOs declined since the mid-1990s owing to advances made with naphthopyran photochromism [1], they continue to be marketed for a range of applications. This article will focus on the SO subtypes of greatest historical and current use, outlining their key molecular features and the principal means by which they are manufactured. It will also briefly look at the most industrially important ways of manipulating the photochromism of SOs through modification of dye structure, before surveying existing and potential uses.

62.1 Spirooxazine dye characteristics and molecular structure

As one might guess from the name of the class, the essential component of a commercially useful SO dye is an oxazine ring that bears a spiro linkage. The latter feature is located adjacent to the oxygen atom of a 2*H*-1,4-oxazine ring, which is itself fused with an aromatic system. A general SO structure, **1a**, is shown in Figure 62.1.

The choice of cyclic structure on the opposite side of the spiro link to the oxazine moiety, i. e. which makes up the remaining half of **1a**, is also very important in order to weaken the C_{spiro}–O bond: these components are depicted in yellow and green, respectively, in Figure 62.1. It has long been known that enough strain must be introduced so that this part of the oxazine ring breaks heterolytically when the molecule is photoexcited, but not too much such that the rupture is permanent [2].

This article has previously been published in the journal Physical Sciences Reviews. Please cite as: A. Towns, Spirooxazine Dyes *Physical Sciences Reviews* [Online] 2020, 5. DOI: 10.1515/psr-2020-0013

https://doi.org/10.1515/9783110587128-062

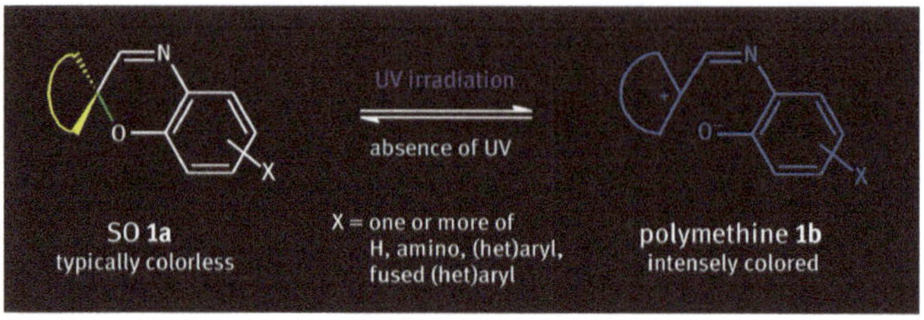

Figure 62.1: Generalized representation of SO structure and photochromism.

Consequently, industrial SO dyes exhibit positive T-type photochromism [3]: absorption of ultraviolet (UV) radiation, and in certain instances short-wavelength visible light, by the colorless (or weakly colored) ring-closed form **1a** ultimately generates ring-opened polymethine [4] chromophores **1b.** These species are intensely colored but thermally unstable. Their concentration will drop unless there is constant irradiation with UV to drive the photochemical ring-opening and continually replenish the colored form of the dye **1b** that thermally bleaches back to the colorless ring-closed molecule **1a** (see Figure 62.1). SO dyes are thus heliochromic [3]: they become intensely colored in strong sunlight, but fade back to colorless at low ambient light levels.

In the case of industrial SO dyes, the spirocyclic motif is almost always a substituted indoline system. Most typically, the oxazine moiety consists of the heterocycle fused to naphthalene. Colorants containing this combination are often referred to as spiroindolinonaphthoxazine (SINO) dyes, although the abbreviations NISO and NOSI are sometimes encountered. Two sub-classes of SINO command special interest: naphth[2,1-*b*] [1,4]oxazines**2** ("[2,1-*b*]SINOs") and naphth[1,2-*b*] [1,4]oxazines**3** ("[1,2-*b*]SINOs") (see Figure 62.2). Industry prizes their bluish-red to greenish-blue photochromic effects. As we shall see in the next section, the potential of SOs took years to be recognised and then realised, but such colorants played a crucial role in the first substantial commercial success of photochromic dyes.

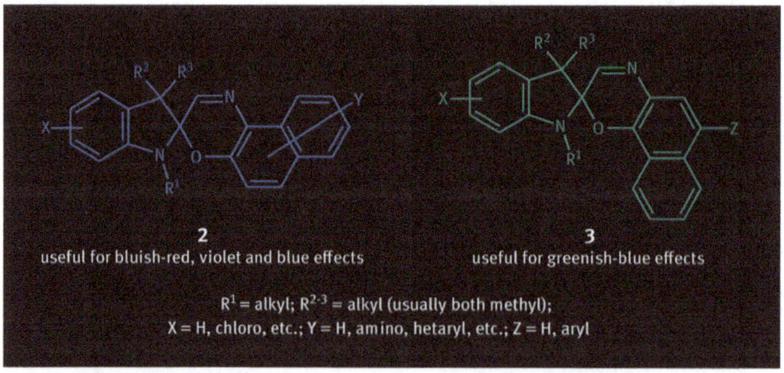

Figure 62.2: Generalized structures of industrially important SINO dyes.

62.2 The rise and (partial) fall of spirooxazines

The ascent of SOs to prominence resulted directly from the discovery of their relatively good fatigue resistance in comparison to the more extensively investigated analogous spiropyran (SP) class [5]. The lack of robustness of SPs was lamented half a century ago [6, 7]. It prevented their exploitation in applications such as sun-responsive ophthalmic lenses. SOs differ from SPs only by the presence of the nitrogen atom in their oxygen heterocycle. Consequently, the first examples of SOs were made not long after the pioneering studies into SP photochromism of the 1950s. Although they were initially investigated for the US military around 60 years ago [8] and [2,1-*b*]SINOs débuted in the patent literature during 1967–1971 [9, 10], it was not until the late 1970s that the much superior fatigue resistance of certain SOs relative to that of SPs was recognised [8]. The photodecomposition rates of SINOs in solution by UV was found in some cases to be orders of magnitude lower than those of corresponding SPs. This discovery prompted a flurry of industrial activity, culminating in the launch of the first photochromic plastic ophthalmic lenses onto the US market in the early 1980s by American Optical [8, 11]. However, they did not live up to customer expectations owing to their blue photocoloration and inadequate service life [12]. A subsequent attempt by PPG Industries based upon the analogous spiroindolinopyridobenzoxazine dye type **4a** (see Figure 62.3), upon which work started in 1983, did not meet with success either [13].

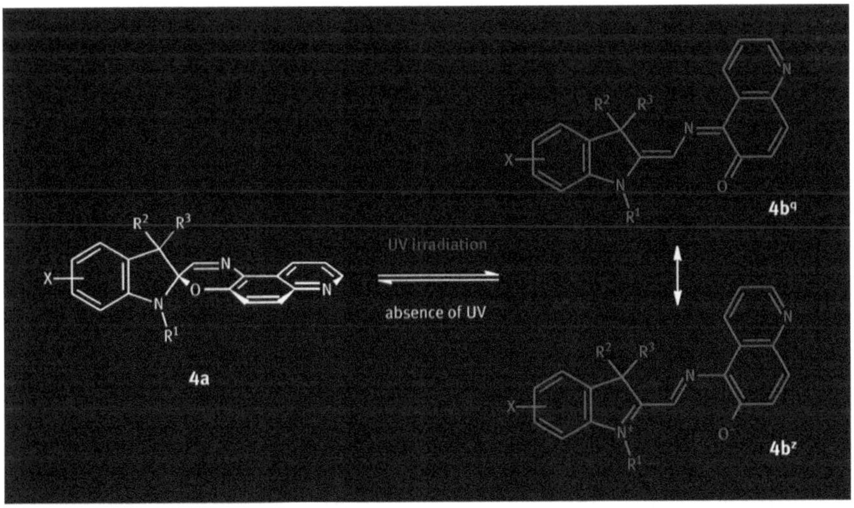

Figure 62.3: Generalized structure for commercialized spiroindolinopyridobenzoxazine dyes (**4a**) and their photomerocyanine forms (**4b**) of which just one isomer is shown.

Nevertheless, development of SO-based lenses continued in the USA, Europe and Japan with the aim of producing marketable, robust, and responsive photocoloration in shades (gray and brown) preferred by consumers. These efforts were rewarded commercially in the 1990s by the establishment of PPG's Transitions® and Rodenstock's Colormatic® and Photocolor® lenses in the American market [11], while in Japan Tokuyama manufactured lenses relying on SO technology [14]. The race to create better quality lenses continued. The TransitionsPlus® lens, launched onto the US market in 1992, depended upon a mixture of two blue SO and two more hypsochromic naphthopyran dyes [15, 16]. Using blends of dyes drawn from two different families of photochromic colorant proved problematic. The difference in the response of their photochromism to variations in ambient temperature caused unwanted changes in hue during photoactivation and/or thermal reversion [17]. (The photocoloration of SINOs is generally more temperature sensitive than that of naphthopyran dyes. With both families of colorants, the rate of thermal ring-closure is accelerated as ambient temperature increases, leading to weaker photochromism. The effect tends to be more pronounced with SOs.) The development of naphthopyrans that produced dull bathochromic photocoloration of bluish and greenish shades [1] with acceptable kinetics marked the beginning of the end of the SO class's dominance. The Transitions III® lens system, brought to market in 1996, employed an all-naphthopyran formulation whose components were better matched in terms of temperature sensitivity [16]. All commercial photochromic polymer lenses are now based primarily upon naphthopyran dyes. Today SOs are firmly of secondary importance for lens production. Nonetheless, as will be discussed later, they still find use in novelty and specialised applications. SO dyes also continue to attract the interest of researchers seeking to exploit photochromic molecules for new technologies.

SOs are available off-the-shelf in research quantities [18] as well as in larger amounts from specialist chemical manufacturers [19]. Pricing is typically of the order of "dollars per gram" rather than the "dollars per kg" of commodity colorants, owing to a combination of the small production volumes, the relative expense of the precursors, and in many cases the multiple number of synthetic steps. The last two aspects arise from the need to manipulate SO dye structure to tune color, kinetics, solubility, and other properties. The next section will examine the chief approaches of industrial interest to molecular design.

62.3 Influence of structure on photochromism of spirooxazine dyes

The photochromism of a SO dye centers on the ability of its colorless spiro form to undergo a reversible molecular rearrangement. The colorant's oxazine system ring opens to generate metastable colored "photomerocyanine" species whose structures resemble that of merocyanine dyes [20]. Figure 62.3 illustrates the transformation.

The reaction is triggered by the absorption of electromagnetic radiation within the near-UV region (typically of wavelengths 320–380 nm), although in the case of numerous commercial dyes, violet light of wavelengths as long as 410–420 nm can stimulate photocoloration. One of the potential photochemical pathways that follows the absorption of this light energy involves cleavage of the bond between the spiro carbon atom and the oxygen heteroatom of the oxazine ring. The two separate π-systems, held out of each other's plane by the spiro link (as depicted in the representation of **4a** in Figure 62.3), become part of a more planar extended conjugated system that stretches the length of the molecule. Ultimately, the breaking of the C_{spiro}–O bond leads to the formation of a mixture of photomerocyanine isomers [21]. Ring-opening proceeds initially to metastable cisoid species, which rearrange to more stable transoid isomers. Figure 62.3 shows what is believed to be the most stable conformation (*trans-trans-cis*, 'TTC') [22, 23]. Substituents on the dye influence the extent to which the electronic structures of the photomerocyanines have quinoidal (**4b**^{q}) and zwitterionic-like (**4b**^{z}) character, which in turn affects their solvatochromism and thus color [24]. The fatigue-resistance of commercial SOs is ascribed to the photoisomerization occurring via singlet photo-excited states rather than those of a triplet nature which are more prone to undergo undesired non-reversible photochemical reactions [25].

The λ_{max} values of the metastable ring-opened forms of SO dyes span ~ 480–670 nm (yellowish-red to bluish-green), although the range of readily available commercial dyes is narrower (*ca.* 560–620 nm in toluene, i. e. bluish-red to turquoise). The photochromism of SINOs is predominantly T-type, although many such dyes exhibit a minor element of P-type photochromic behaviour. The latter phenomenon involves photomerocyanine species absorbing visible light whose energy drives ring-closure, supplementing the effect of thermal bleaching. However, the quantum yields for cyclisation of commercial colorants are at least an order of magnitude lower than those for ring-opening. Irradiation by strong sunshine thus results in intense photocoloration despite simultaneous absorption of UV and visible light by ring-opened and ring-closed isomers, respectively. More detail concerning structural and environmental influences in respect of this balance can be found in [25].

Industrial SINOs feature patterns of substitution in either or both the indoline and naphthalene moieties that alter color and enhance their photochromism, i. e. increase intensity of photocoloration as well as rates of photoactivation and thermal fading. We shall now look at the most important of these design strategies. (Numerical data will be used illustratively, but please bear in mind that the environment in which a SO is placed will greatly affect its photochromism. Solvent, temperature and other factors greatly influence properties such as color, rate of thermal fade, and fatigue. Caution must be exercised when comparing spectroscopic and kinetics data for SOs from different sources. Even small variations in experimental conditions can make a marked impact on photochromism and potentially mislead.)

The influence of the indoline *N*-substituent on kinetics was first recognised during the late 1970s [26]. Increasing its size constitutes an effective means of retarding thermal fade and increasing photocoloration intensity, but without substantially affecting color. The parent [2,1-*b*]SINO **5a** exhibits weak blue photocoloration owing to rapid thermal reversion of its colored merocyanine forms to its colorless ring-closed isomer. Replacement of the *N*-methyl group with substantially bulkier functions tends to slow the thermal fade rate, giving colorants with more pronounced photochromism and greater industrial utility. This effect is illustrated by the "residual photocoloration" figures in Table 62.1: a larger number signifies a slower rate of ring closure, which tends to be accompanied by an increase in photocoloration strength (see "relative intensity").

Table 62.1: Effects of varying substituents at the 1-position of [2,1-*b*]SINOs on intensity and thermal fade of photocoloration [27].

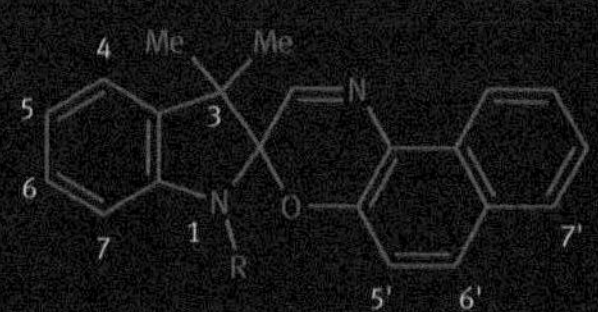

5

Dye	R		Relative intensity*	Residual photo-coloration**
5a	CH_3	Me		
5b	CH_2CH_3	Et	0.9	0.9
5c	$CH_2CH_2CH_3$	*n*-Pr	0.9	0.9
5d	$CH_2CH(CH_3)_2$	*i*-Bu	1.4	1.2
5e	$CH_2C(CH_3)_3$	Np	1.7	1.3

* absorption intensity at photostationary state relative to that for **5a**
** residual photocoloration 10s into thermal fade relative to that for **5a**

The *N*-ethyl and *N*-propyl chains of **5b** and **5c** make little difference to kinetics. In contrast, isobutyl (*i*-Bu) and neopentyl (Np) groups as *N*-substituents of dyes **5d** and **5e** markedly boost photocoloration intensity. Their bulk sterically hinders the ring-closure reaction. Consequently, the rate of thermal fading is lowered: for a given intensity of incident UV, the equilibrium merocyanine concentration increases. Rate of fatigue is not adversely impacted, while the absorption maximum shifts bathochromically by just a few nm – for comparative examples involving derivatives of **5a** and **5e**, see Table 62.2. Relative to their *N*-methyl analogues, SINOs with bulky 1-substituents produce stronger photocoloration of similar hue and robustness.

Table 62.2: Effects of varying substituents at the 5- and 5′-positions of [2,1-*b*]SINOs **5a** and **5e** on absorption maximum and half-life of thermal fade of photocoloration in toluene solution at 20 °C [24, 25, 28].

Substitution		5a		5e	
5	5′	λ_{max} (nm)	$t_{1/2}$ (s)a	λ_{max} (nm)	$t_{1/2}$ (s)
NO_2	H	566	9		
H	H	596	4	600	31
Me	H			608	21
OMe	H	612	4	615	10
H	CO_2Me	612	2	620	12
H	$CH=C(CN)_2$	668	17		

calculated from rate constant assuming ideal first order rate of absorbance.

The presence of *i*-Bu or Np as *N*-substituents confers another advantage. Chains that branch at the carbon atom *beta* to the indoline nitrogen atom tends to reduce the temperature sensitivity of photochromism. Dyes bearing such groups thus develop a similar hue when photoactivated, but of greater intensity that is less prone to weakening at high ambient temperatures. SOs **5d** and **5e** are commercially successful colorants because of these advantages. They furnish blue hues when photoactivated. The thermal fade rates of their photocoloration, often quoted in terms of half-life ($t_{1/2}$, time taken for absorbance to halve [3]), are industrially useful. In addition, the strength of coloration developed under irradiation with light of a given intensity decreases when ambient temperature rises from 10 °C to 40 °C at a lower rate than that of **5a.** While the tabulated illustrative data relate to [2,1-*b*]SINOs, the above approach also applies to [1,2-*b*]SINOs.

Modifying the 3,3-dialkyl fragment of the indoline system often constitutes an effective (albeit lesser known) strategy to manipulate the kinetics of photochromism whilst making only a modest impact on photomerocyanine λ_{max}. The most common pattern of substitution in industrial SINOs is *gem*-dimethyl for reasons of cost and synthetic accessibility. However, alternative patterns have been explored by industry [29] and successfully exploited on a commercial basis. When coupled with crowding at the 1-position, a minor change to one or both 3-alkyl groups can make a big impact on rate of thermal fade and photocoloration intensity without appreciably altering hue. For example, lengthening a single 3-methyl group of **5e** to ethyl makes no difference to the λ_{max} of photocoloration, but produces a more intense colorant by increasing $t_{1/2}$ substantially from 31s to 52s at 20 °C [28]. Variations in 3-substituents were initially explored in the early 80s [30] and exploited commercially soon afterwards [14, 27].

Industry makes use of several strategies to manipulate photocoloration hue through molecular design, although these inevitably influence kinetics too. In addition to considering the synthetic difficulty (= cost!) of introducing structural changes, one must also bear in mind their impact on rate of photocoloration, thermal fade, as well as fatigue – and try to mitigate unwanted effects with other alterations to structure. For example, placing electron donors onto the 4- to 7-positions of the indoline ring of SINOs leads to bathochromic shifts of ring-opened photoisomers, but at the expense of reducing $t_{1/2}$ and photocoloration intensity [22]. Examples are shown in Table 62.2. Since a methoxy function is a stronger electron donor than a methyl group, it produces a bigger red-shift in absorption maximum and shortens half-life. The converse occurs when electron-acceptors are placed in these positions. For example, introduction of a nitro group into the 5-position of **5a** shifts photomerocyanine λ_{max} hypsochromically in toluene solution by 30 nm and slows down the rate of thermal reversion, i. e. $t_{1/2}$ increases (see Table 62.2).

Making changes to the naphthoxazine moiety of SINOs also produces substantial variations in photochromic properties. The orientation of the aromatic system fused to the oxazine ring makes a profound impact. Whereas simple [2,1-*b*]SINOs **5** photoactivate to blue, industrial [1,2-*b*]SINOs generally ring open to furnish turquoise photomerocyanines with slower rates of thermal fade. For example, **6a** and **6b** (see Figure 62.4) are commercial dyes in their own right (λ_{max} 617 nm and 618 nm, respectively with half-lives of 38s and 120s in toluene at 20 °C). The bathochromic photocoloration of a [1,2-*b*]SINO **6** relies on the extended π-structures of its photomerocyanines. The relatively large size of the conjugated system located over its naphthoxazine moiety also leads to the dye's spiro form being intrinsically pale yellow. Such [1,2-*b*]SINOs thus tend to suffer from slight residual color even when not activated by UV.

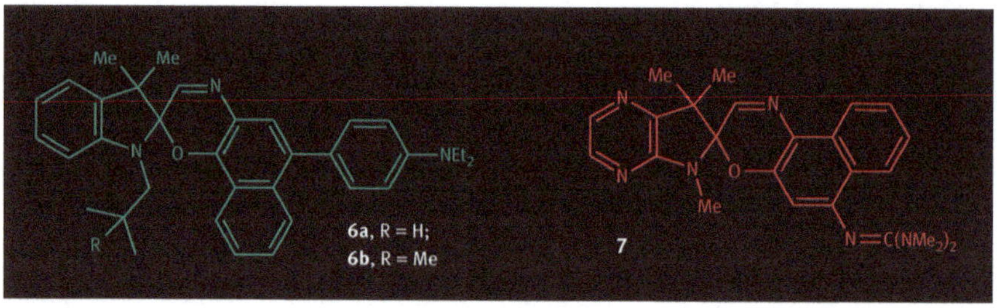

Figure 62.4: Examples of commercial [1,2-*b*][1,4]oxazine photochromic dyes **6** and a colorant **7** that produces extremely hypsochromic photocoloration for the SO class.

The effect on photomerocyanine λ_{max} of locating electron acceptors and donors on the naphthalene system is the converse of that noted above for indoline substituents. Table 62.2 shows two examples which feature electron-withdrawing functions in the 5′-position of naphthoxazine moiety ($-CO_2Me$; $-CH{=}C(CN)_2$): the ester function brings about a bathochromic shift and shortens half-life. The more strongly accepting *gem*-dicyanoethylene group produces an even greater red-shift (although $t_{1/2}$ is increased – an example of how substituent effect trends in SOs are not always straightforward). Conversely, placing the very powerful tetramethylguanidino donor onto the

6′-position, whilst making the indoline benzo ring electron deficient by introducing nitrogen heteroatoms at positions 4 and 7, produces one of the most hypsochromic SOs known, dye **7** (see Figure 62.4). It becomes yellowish-red when photoactivated, absorbing maximally at 478 nm in toluene solution [31].

While hypsochromic dyes like **7** were not commercialised, two related sub-classes of SO took on industrial importance. Introduction of a nitrogen atom into the 7′-position of the naphthalene system of [2,1-*b*]SINO colorants furnishes the first kind: spiroindolinopyridobenzoxazines**4a.** This type of colorant was commercialised in the 1980s and early 1990s. The extra heteroatom makes only a modest difference to hue (e. g. λ_{max} of photocoloration of **5a** and **4a** (R^1–R^3 = Me; X = H) are 596 nm and 594 nm in toluene respectively) [32]. Its benefit was said to lie in better rates of activation and enhanced intensity [22].

The second type of interest is a [2,1-*b*]SINO featuring an amino group in the 6′-position. Industry has been making use of this substitution pattern for over a quarter of a century to enhance photochromic properties and expand the commercially viable range of photocoloration hues towards the red. Increasing the basicity of the 6′-amino group induces greater hypsochromism and a lengthening of half-life in comparison to the corresponding unsubstituted dye **8a.** This trend is exemplified by data in Table 62.3. The increasingly shorter wavelengths of λ_{max} in the order **8b**, **8c** and **8d** correspond with the relative basicity of the 6′-groups in these derivatives, i. e. indolin-1-yl < morpholino < dimethylamino, respectively.

Table 62.3: Effect of introducing various amino functions to 6′-position of a [2,1-*b*]SINO on absorption maximum and half-life of thermal fade of photocoloration in toluene solution at 20 °C [27].

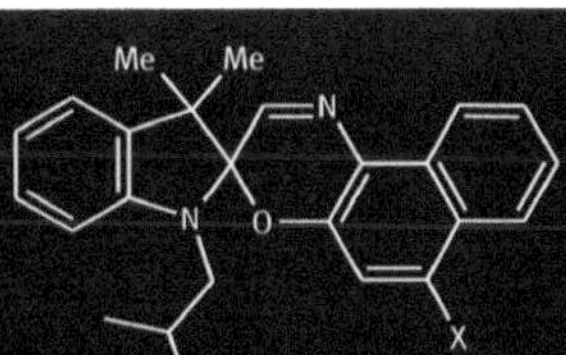

8

Dye	X	λ_{max} (nm)	$t_{1/2}$ (s)
8a	–H	599	9
8b	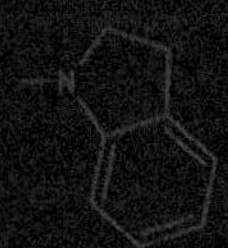	590	25
8c		576	33
8d	–NMe$_2$	564	32

Further benefits claimed for this substitution pattern are: (i) bigger quantum yields for ring-opening; (ii) enlarged photomerocyanine extinction coefficients; (iii) less competition between photomerocyanines and their spiro counterparts for absorption of those wavelengths of UV light that trigger ring-opening [33, 34]. Industry has made use of all of the 6′-amino functions listed in Table 62.3 to create SINOs that are very responsive to both near UV and violet light, developing strong purple to bluish-red photocoloration. The main disadvantage of this kind of SO is that they tend to have a yellowish “base color”, i. e. the UV absorption of their spiro forms tends to bleed into the visible region, so rather than being colorless when unactivated, they are intrinsically pale yellow.

For a more detailed discussion of structural influences on the photochromic properties of SOs (as well as a fuller picture of the diversity of SO structures reported in the academic and patent literature), the excellent review [25] is recommended.

All of the aforementioned colorants exhibit positive photochromism, i. e. photoisomerization brings about absorption of longer wavelength light [3]. While negative photochromism is not unknown in SINOs and related SOs, the phenomenon is rare. There are only a few examples of dyes whose spiro forms are metastable minor components in a thermal equilibrium mixture skewed towards merocyanines which are reversibly photobleached with visible light [35] Certain conventional SOs are known to become negatively photochromic under special circumstances [36]. Complexation and/or protonation can lower the ground state energy of merocyanine isomers sufficiently for them to dominate the equilibrium. Light of visible wavelengths causes photoisomerization to their colorless ring-closed forms which are thermally unstable.

This section described some of the most common strategies employed by industry to devise SO dyes that possess commercially useful photochromic properties. It demonstrated that the modifications to colorant structure often do not impact upon color and kinetics in a mutually exclusive manner. Application of these approaches to fine tune SO photochromism is complicated by the profound influence exerted by the environment into which dyes are placed. Consequently, while dye chemists can fall back on useful rules of thumb to guide their molecular design decisions, the inherent complexity of illuminant–SO–substrate systems means that there is no substitute for synthesising and physically testing candidate dyes to evaluate their color, kinetics and longevity. We shall briefly look next at the principal ways in which industrially important SO colorants are prepared.

62.4 Synthesis of spirooxazines

Of the preparative pathways to SOs, one dominates both academic research and industrial practice: reaction of *o*-hydroxynitrosoarenes with alkylidene heterocycles. Given the commercial importance of SINOs, the most employed procedures involve condensation of Fischer’s base (2-methylidene-1,3,3-trimethylindoline; **9**,

$R^1 = R^2 = R^3 = X = Me$) or its derivatives with nitrosonaphthol-derived compounds (see Figure 62.5). 1-Nitroso-2-naphthol (**10**; Y = H) furnishes [2,1-*b*]SINOs **2** that produce blue photocoloration. 5-Nitroso-6-hydroxyquinoline **11** affords spiroindolinopyridobenzoxazines **4a**, which also photoisomerise to blue species – while λ_{max} values of their long wavelength absorption bands differ by only a few nanometers from those of the corresponding SINO derivatives **10**, this family is claimed generally to become colored more rapidly. 2-Nitroso-1-naphthols **12** give [1,2-*b*] SINOs **3** [28, 37]. There is evidence that all these *o*-hydroxynitrosoarenes react as their quinone oxime tautomers [25, 38].

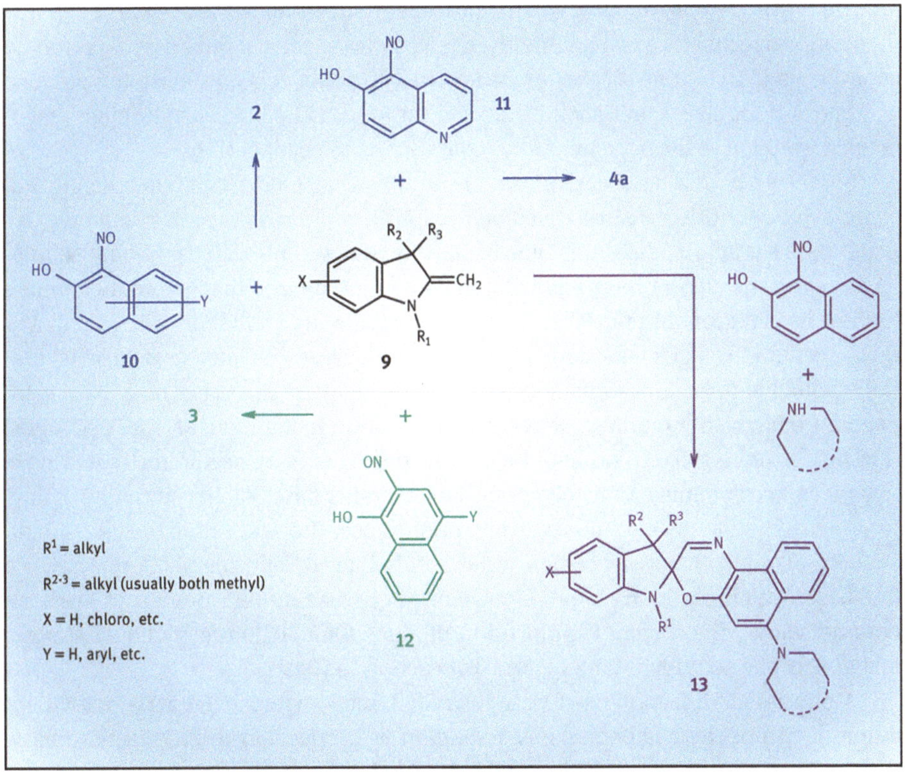

Figure 62.5: Generalized commercial synthetic pathway to SINOs.

In its simplest form, synthesis involves refluxing a stirred mixture of the Fischer's base derivative and nitroso compound in an alcohol. SO either precipitates during reaction or when the mass is cooled. The colorant is filtered off from the resultant suspension, washed and then purified. (Drying may not be required before purification if the residual solvent does not interfere with product clean-up.) Methanol is often a good choice of reaction solvent, since not only do SOs possess low solubility in alcohols, but it also encourages impurities to remain in solution. Selection of other solvents in which these materials have greater solubility may be advantageous if the product and/or impurities precipitate as tars during alcohol-based reactions, hindering isolation. For example, toluene is an effective reaction solvent from which product can be isolated by controlled gradual addition of methanol which acts as a "non-solvent".

Despite process optimisation, the "crude" SO isolated from reaction mass usually must be purified to remove discoloration caused by contamination with impurities and/or residual starting material. Column chromatography is occasionally used for this purpose on a commercial basis, but only as a last resort owing to the technique's high cost. Recrystallisation is the preferred option. An effective procedure for cleaning up SOs resembles that outlined above for precipitating crude dye. Dissolution in an aromatic hydrocarbon, an ester or an ether (such as toluene, ethyl acetate or tetrahydrofuran, respectively) in which the dyes tend to have good solubility is followed by controlled addition of an alcohol as non-solvent to selectively crystallise the product. Treatment of the dye solution with activated carbon may be beneficial through preferential adsorption of the trace impurities contributing to residual color

Simple Fischer's base derivatives are liquids that tend to generate small, but significant, quantities of strongly colored impurities upon storage. A common practice is thus to vacuum distil this type of starting material prior to use. Doing so minimises carry-through of intensely colored contaminants into the SO product, which hinders its isolation and purification as well as reducing yield. Following their synthesis, certain Fischer's base derivatives may be efficiently isolated as acid salts, e. g. in hydroiodide form. Salt formation leads to precipitation of the Fischer's base compound, from which impurities are readily washed out. In addition, as pure dry solids, they are far more stable than liquid bases. Acid salts may be used directly in a reaction when accompanied by a solvent-miscible organic base, such as a trialkylamine, to scavenge liberated acid. Alternatively, prior to use, the acid salt is treated with an aqueous alkali in the presence of a water-immiscible solvent: the alkali releases the free base, which partitions into the solvent. After washing the solution of Fischer's base derivative, the solvent is stripped to produce the reaction-ready liquid starting material, or the solution is simply used for reaction "as is".

A process modification used by industry to improve yield is the slow gradual addition of one of the components as a solution or suspension to the reaction mass, thereby minimising decomposition and side-reactions. Adjusting stoichiometry is another approach, e. g. employing an excess of nitroso compound when it is suspected of having some instability to the conditions of reaction and monitoring for complete consumption of the Fischer's base derivative by thin layer chromatography.

One surprisingly straightforward variation is effective in producing the highly photoresponsive [2,1-*b*]SINOs **13**, which bear 6′-amino functionalities and activate to intense bluish-red and purple hues (see Section 62.3). These dyes are built up from three components in a solvent-based one-pot reaction: (i) Fischer's base derivative **9**; (ii) nitrosonaphthol **10** (Y = H); (iii) an amine which may be cyclic (e. g. indoline, piperidine or morpholine) or acyclic (e. g. diethylamine). The yield and quality of product is highly dependent on reaction conditions and stoichiometry as well as factors such as the order of addition of the components. In some instances, **10** is reacted with the amine before addition of **9**. In others, a slurry of **10** in solvent is added in a steady controlled manner to mixtures of **9** and amine.

Using the above strategies, 50–70% yields of purified SINO derivatives **10–13** with respect to Fischer's base derivatives are possible. Reports of work that target optimised conditions of SO synthesis rarely appear in the open literature: one such example is [39]. A useful broader overview of SO synthesis is given in [25]. While Fischer's base is readily available, its analogues are much less accessible commercially and must usually be synthesised. For an excellent review on the synthesis of Fischer's base compounds from the perspective of photochromic dye preparation, see [40]. An overview of SO synthetic chemistry with emphasis on routes to Fischer's base analogues and *o*-hydroxynitrosoarene precursors is available in Japanese [41].

Commercial SO dyes are almost invariably supplied as powders, which have shelf-lives of years when protected from light and stored in cool dry conditions. SO dyes do not tolerate prolonged exposure to acidity and moisture very well. Typically, they are off-white or pale-coloured non-toxic solids that melt above >100 °C, although all [1,2-*b*]SINOs and certain [2,1-*b*]SINOs with extended conjugated π-systems (e. g. dyes **13**) are intrinsically yellow solids owing to the absorption of their ring-closed forms bleeding into the visible region from the near UV. Industrial SOs do not dissolve in water and possess poor solubility in alcohols (≤1%w/v at 20 °C). Solvents of low to medium polarity are far more accommodating. Aromatic hydrocarbons like toluene act as good solvents for both [2,1-*b*] and [1,2-*b*]SINOs (typically 6–30%w/v at 20 °C) as do ethers such as THF (typically 10–50%w/v at 20 °C). These classes of dye exhibit moderate solubility in common aliphatic esters and ketones, e. g. around 2–8%w/v and 1–3%w/v in ethyl acetate and acetone, respectively, at 20 °C. Concentrated SO solutions tend to exhibit thermochromism: for example, the colors of toluene-containing reaction masses or solutions of crude dye prepared during purification tend to deepen as they are heated towards reflux. Neat commercial SO colorants also become intensely colored on melting.

Industrial SOs do not exhibit visible color changes when irradiated in pure solid form – they must be in solution, either dissolved in a solvent or a suitable polymeric matrix, like poly(ethylene), for their photochromic properties to become manifest. The environment of the dye must provide sufficient free volume not to restrict the geometry change that occurs upon ring-opening. As is typical for SOs, the ring-closed form **4a** can be thought of consisting of two halves roughly perpendicular to each other: the π-systems of each moiety are isolated, hence they absorb predominantly in the UV. Photoisomerisation ultimately generates a merocyanine mixture, which is believed to comprise predominantly the TTC isomer – a relatively planar structure that permits conjugation along the whole length of the molecule's π-system, resulting in color. Too little free volume in the matrix in which the dye resides will thus impede or even prevent photochromism. The crystal lattice of a SO colorant in neat solid form is usually just such an environment, hence the lack of visible photoresponse displayed by commercial SO dyes as powders. However, instances of SOs which exhibit photochromism in a crystalline state are rare but not unknown, e. g. *N*-methylation at the quinolinic nitrogen atom of **4a** (R^1–R^3 = Me; X = H) gives a yellow solid that reversibly turns green when irradiated with UV [42]. There is even one case of a SO whose crystals exhibit P-type photochromism: instead of an indoline structure, it sports an azahomoadamantane moiety [43]. This bulky tricyclic system is thought to create sufficient volume within the crystal lattice for ring-opening and -closing to occur upon exposure to UV (~320 nm) and visible light (~540 nm) respectively. Such behaviour is very unusual within the SO class. Consequently, the "conventional" P-type classes (e. g. diarylethenes [44]) will maintain a stranglehold on usage in the development of applications that demand thermally irreversible switching. SO colorants are likely to remain restricted to technologies which call for T-type photochromism. The next section will outline current and potential uses of the SO class.

62.5 Applications of spirooxazine dyes

SO dyes got their major commercial break as sun-responsive colorants in ophthalmic lenses. A detailed insight into how SOs have been used in this application can be gleaned from a good, but now dated, review [11] dedicated to lens use of this dye class.

Industry continues to make use of the responsive and intense heliochromism of SO colorants in inks and mass-colored thermoplastics. A small amount of SO goes a long way: in correctly formulated products, dye concentrations of only around 500 ppm generate intense photocoloration. Products manufactured with SO colorants include promotional novelty items and toys. They have even been used in artwork and jewellery. However, their lack of photostability and high cost in relation to conventional colorants remain a major obstacle to more widespread aesthetic use outside of niche applications. Academic studies into the application of commercial SOs as textile dyes merely demonstrate their unsuitability: as disperse dyes for nylon [45], acrylic [46] and polyester [47], SOs produce dyeings of significant residual color and/or weak photocoloration. These difficulties stem from SOs tending to degrade upon exposure to aqueous environments at elevated temperatures in addition to the bulky nature of the dye molecules hindering their diffusion into fibers. Also, the polymer matrices of these textiles are not conducive to good photochromic performance. Work-arounds to some of these problems include application of SOs from solvents [48–50] or supercritical carbon dioxide [51, 52]. In addition, analogues of commercial SO dyes have been reported that bear water-solubilising sulfonic acid groups to confer substantivity towards wool [53] and silk [54]. Reactive photochromic dyes which bind covalently to cotton have also been briefly studied [55, 56]. Nevertheless, these approaches all suffer from significant shortcomings that prevent commercialisation. Radical breakthroughs in SO technology will be needed to usurp the current chief means of introducing photochromic effects to textiles. These techniques comprise: (i) screen-printing of dye dissolved in a coating which is then thermally cured; (ii) "pigment coloration" using a binder into which microencapsulate containing solutions of SO has been dispersed; (iii) mass-colored polyalkylene thread. While these methods have been used commercially with SO colorants for many years, they are only suited to niche markets where poor fastness, particularly a lack of photostability, is not of importance.

SOs continue to be explored as functional dyes. For example, they are of interest for optical devices owing to the change in refractive index that accompanies photoisomerisation [57]. The ability of SINOs **13** to be activated by violet light attracted researchers seeking to develop novel molecular switches for use in super-resolution microscopy [58], a Nobel Prize-winning technique employed in creating images of objects whose dimensions are smaller than the wavelengths of visible light. The ability to utilise visible light rather than UV for photoactivation is advantageous in several ways, especially when scrutinising living tissue. Researchers are also seeking to exploit SOs as light-actuated functional materials for far-distant potential uses, such as photoresponsive bioreceptors of metal ions to regulate metabolic pathways and storage of data in biological systems [59]. An SO colorant has even been investigated as a possible component of neurons for the construction of physical neural networks in artificial intelligence systems [60].

Despite these pioneering efforts into developing uses for SOs as functional colorants, no application which would surpass the existing industrial outlets of SOs looms on the commercial horizon.

62.6 Summary

SOs remain a commercially important photochromic dye class, although they sit in a distant second position to naphthopyrans in terms of market volume and value. Industrially useful SINO dyes owe their utility to structural features in one or both of the indoline or naphthalene moieties of their parent compounds. These motifs (i) adjust hue; (ii) retard thermal fading and thus intensify photoactivated color; (iii) increase quantum yield of photomerocyanine formation and thereby improve the efficiency of their interaction with UV light; (iv) reduce fatigue and so lengthen service life. The choices of substitution at the 1- and 3-positions of SINO dyes are powerful tools with which to manipulate dye kinetics. Unfortunately, these design features are often overlooked by academia. Some studies that sought to exploit SINO-based photochromic units achieved only limited success owing to their rapid thermal fade and weak photocoloration. The more pronounced photochromism brought about by appropriate functionalisation of the five-membered ring of the indoline moiety, or substitution at the 6′-position of the naphthoxazine system, would likely afford better colorants with which to perform investigations and demonstrate proof-of-concept. The design rules outlined earlier are thus worth bearing in mind when attempting to employ SOs as functional dyes in novel applications. Commercialisation of a new use of SO dyes that would eclipse the value of photochromic dye currently consumed in ophthalmic lens manufacture remains distant, especially given the attention devoted by researchers to light-responsive dye types other than the SO class. Wider adoption of the approaches exemplified in this chapter which enhance SO photochromism through molecular design could increase this class's chance of regaining the figurative place in the sun that it once had.

References

1. Towns A. Naphthopyran dyes. Phys Sci Rev. 2020;5, in press:20190085.
2. Nakatani K, Piard J, Yu P, Métivier R. Introduction: organic photochromic molecules, Chapter 1. In: Tian H, Zhang J, editors. Photochromic materials: preparation, properties and applications. Weinheim: Wiley-VCH, 2016:1–45.
3. Towns A. Photochromic Dyes, to be submitted.
4. Mustroph H. Polymethine dyes. Phys Sci Rev. 2020;5:20190084. DOI: 10.1515/psr-2019-0084.
5. Towns A. Spiropyran Dyes, to be submitted.
6. Bertelson RC. Reminiscences about organic photochromics. Mol Cryst Liq Cryst. 1994;246:1–8.
7. Bertelson RC. Chapter 3 ("Photochromic process involving heterolytic cleavage"). In: Brown GH, editor. Photochromism – Techniques of chemistry, vol. III. London: Wiley, 1971:45–431.
8. Chu NY. Chapter 10 ("4n+2 Systems: Spirooxazines"). In: Dürr H, Bouas-Laurent H, editors. Photochromism molecules and systems. Amsterdam: Elsevier, 1990:493–509.
9. Ono H, Osada T, Kosuge K (Fuji Shashin Film KK). Photochromic compound, Japanese Patent Application 42/55662 (1967); US Patent 3578602, 1971.
10. Arnold G, Vollmer HP, Gabor P (Licentia GmbH). Spiro-1:4-oxazines, German Patent Application DE 1927849A1, 1970.
11. Crano JC, Kwak WS, Welch CN. Chapter 2 ("Spiroxazines and their use in photochromic lenses"). In: McArdle CB, editor. Applied photochromic polymer systems. Glasgow: Blackie, 1992:31–79.
12. Bruneni JL. More than meets the eye: the stories behind the development of plastic lenses. Pittsburgh: PPG Industries, 1997.
13. van Gemert B. The commercialization of plastic photochromic lenses: a tribute to John Crano. Mol Cryst Liq Cryst Sci Technol Sect A Mol Cryst Liq Cryst. 2000;344:57–62.
14. Kobayakawa T. Photochromism of spiro-oxazines. In: Third International Symposium on Photochromism, Fukuoka, 1999.
15. Crano JC, Flood T, Knowles D, Kumar A, Van Gemert B. Photochromic compounds: chemistry and application in ophthalmic lenses. Pure Appl Chem. 1996;68:1395–8.
16. Van Gemert B. Commercialization of plastic photochromic lenses. In: Third International Symposium on Photochromism, Fukuoka, 1999.
17. Van Gemert B, Kish DG. The intricacies of color matching organic photochromic dyes. In: ChemiChromics '99 conference, New Orleans, 1999.

18. For example: Tokyo Chemical Industry. Photochromic Dyes. www.tcichemicals.com/eshop/en/gb/category_index/12995/. Accessed: 06 Mar 2020; Merck, Photochromic and Thermochromic Dyes, www.sigmaaldrich.com/materials-science/material-science-products.html?TablePage=9541081. Accessed: 06 Mar 2020.
19. For example: Vivimed Labs Europe. www.vivimedspecchem.com/vivimed-products/reversacol-photochromic-dyes. Accessed: 06 Mar 2020.
20. Mustroph H. Cyanine dyes. Phys Sci Rev. 2020;5:20190145. DOI: 10.1515/psr-2019-0145.
21. Wilkinson F, Worrall DR, Hobley J, Jansen L, Williams SL, Langley AJ, et al. Picosecond time-resolved spectroscopy of the photocolouration reaction of photochromic naphthoxazine-spiro-indolines. J Chem Soc Faraday Trans. 1996;92:1331–6.
22. Maeda S. Chapter 2 ("Spirooxazines"). In: Crano JC, Guglielmetti RJ, editors. Photochromic and thermochromic compounds, vol. 1. New York: Plenum, 1999:85–109.
23. Minkin VI. Chapter 2 ("Photoswitchable Molecular Systems Based on Spiropyrans and Spirooxazines"). In: Feringa BL, Browne WR, editors. Molecular switches, 2nd ed. Weinheim: Wiley-VCH, 2011:37–80.
24. Metelitsa AV, Lokshin V, Micheau JC, Samat A, Guglielmetti R, Minkin VI. Photochromism and solvatochromism of push–pull and pull–push spiroindolinenaphthoxazines. Phys Chem Chem Phys. 2002;4:4340–5.
25. Lokshin V, Samat A, Metelitsa AV. Spirooxazines: synthesis, structure and photochromic properties. Russ Chem Rev. 2002;71:893–916.
26. Hovey RJ, Chu NY, Piusz PG, Fuchsman CH (American Optical Corp.). Photochromic compounds, US Patent Application US4342668A, 1982.
27. Towns AD. Chapter 5 ("Industrial Photochromism"). In: Bergamini G, Silvi S, editors. Applied Photochromism. Switzerland: Springer, 2016:227–79.
28. Partington SM, Towns AD. Photochromism in spiroindolinonaphthoxazine dyes: effects of alkyl and ester substituents on photochromic properties. Dyes Pigm. 2014;104:123–30.
29. For example: Momota J, Kobayakawa T (Tokuyama Corp.). Spirooxazine Compound, Japanese Patent 3227062, 2001; Krongauz V, Chif A, Aizikovich A, Tchernoivanov V. (Yeda Research and Development Co. Ltd.). Photochromic Spiro(Indoline)Naphthoxazines, World Patent Application WO 01/77112A2, 2001.
30. Melzig M, Martinuzzi G (Optische Werke G Rodenstock). Photochromic substances, World Patent Application WO 85/02619A1, 1985.
31. Rickwood M, Marsden SD, Ormsby ME, Staunton AL, Wood DW, Hepworth JD, et al. Red colouring photochromic 6′-Spiroindolinonaphth[2,1-b][1,4]oxazines. Mol Cryst Liq Cryst Sci Technol Sect A Mol Cryst Liq Cryst. 1994;246:17–24.
32. Kellmann A, Tfibel F, Dubest R, Levoir P, Aubard J, Pottier E, et al. Photophysics and kinetics of two photochromic indolinospirooxazines and one indolinospironaphthopyran. J Photochem Photobiol A: Chem. 1989;49:63–73.
33. Wilkinson F, Hobley J, Naftaly M. Photochromism of spiro-naphthoxazines: molar absorption coefficients and quantum efficiencies. J Chem Soc Faraday Trans. 1992;88:1511–17.
34. Hobley J, Wilkinson F. Photochromism of naphthoxazine-spiro-indolines by direct excitation and following sensitisation by triplet-energy donors. J Chem Soc Faraday Trans. 1996;92:1323–30.
35. Aiken S, Edgar RJL, Gabbutt CD, Heron BM, Hobson PA. Negatively photochromic organic compounds: exploring the dark side. Dyes Pigm. 2018;149:92–121.
36. Barachevsky VA. Negative photochromism in organic systems. Rev J Chem. 2017;7:334–71.
37. York M, Evans RA. Synthesis of green colored photochromic 6′-arylamino spiro [2H]naphth[1,2-b]oxazines. Synth Commun. 2010;40:3618–28.

38. Christie RM, Agyako C, Mitchell K, Lyčka A. The formation of dihydrohydroxyspiro[1,2]oxazine from the reaction of Fischer's base with some isonitroso compounds. A multinuclear NMR study. Dyes Pigm. 1996;31:155–70.
39. Pottier E, Sergent M, Phan Tan Luu R, Guglielmetti R. Synthese De Quelques Spiro[Indoline-Naphtoxazines] Et Spiro(Indoline-Pyridobenzoxazines] Photochromiques. Application De La Methodologie De La Research Experimentale. Bull Soc Chim Belg. 1992;101:719–39.
40. Bertelson RC. Chapter 1 ("Spiropyrans"). In: Crano JC, Guglielmetti RJ, editors. Photochromic and thermochromic compounds, vol. 1. New York: Plenum, 1999:11–83.
41. Nakamura M, Taniguchi T. Synthesis and characteristic of spiro-oxazine compounds. J Synth Org Chem Jpn. 1991;49:392–402.
42. Bénard S, Yu P. A spirooxazine showing crystalline state photochromism. Chem Commun. 2000:65–6.
43. Patel D, Benedict JB, Kopelman RA, Frank NL. Photochromism of a spirooxazine in the single crystalline phase. Chem Commun. 2005:2208–10.
44. Towns A. Diarylethene dyes. Phys Sci Rev. 2020;5:20190146, (in press).
45. Lee SJ, Son YA, Suh HJ, Lee DN, Kim SH. Preliminary exhaustion studies of spiroxazine dyes on polyamide fibers and their photochromic properties. Dyes Pigm. 2006;69:18–21.
46. Billah SM, Christie RM, Shamey R. Direct coloration of textiles with photochromic dyes. Part 1: application of spiroindolinonaphthoxazines as disperse dyes to polyester, nylon and acrylic fabrics. Color Technol. 2008;124:223–8.
47. Aldib M, Christie RM. Textile applications of photochromic dyes. Part 4: application of commercial photochromic dyes as disperse dyes to polyester by exhaust dyeing. Color Technol. 2011;127:282–7.
48. Billah SM, Christie RM, Morgan KM. Direct coloration of textiles with photochromic dyes. Part 2: the effect of solvents on the colour change of photochromic textiles. Color Technol. 2008;124:229–33.
49. Aldib M, Christie RM. Textile applications of photochromic dyes. Part 5: application of commercial photochromic dyes to polyester fabric by a solvent-based dyeing method. Color Technol. 2013;129:131–43.
50. Aldib M. Photochromic ink formulation for digital inkjet printing and colour measurement of printed polyester fabrics. Color Technol. 2015;131:172–82.
51. Coimbra P, Gil MH, Duarte CM, Heron BM, De Sousa HC. Solubility of a spiroindolinonaphthoxazine photochromic dye in supercritical carbon dioxide: experimental determination and correlation. Fluid Phase Equilib. 2005;238:120–8.
52. Abate MT, Seipel S, Viková M, Vik M, Ferri A, Jinping G, et al. Comparison of the photochromic behaviour of dyes in solution and on polyester fabric applied by supercritical carbon dioxide. IOP Conf Ser: Mat Sci Eng. 2019;459:012026. DOI: 10.1088/1757-899X/459/1/012026.
53. Billah SM, Christie RM, Shamey R. Direct coloration of textiles with photochromic dyes. Part 3: dyeing of wool with photochromic acid dyes. Color Technol. 2012;128:488–92.
54. Shah MR, Christie RM, Morgan KM, Shamey R. Photochromic protein substrates. Mol Cryst Liq Cryst. 2005;431:535–9.
55. Son YA, Park YM, Park SY, Shin CJ, Kim SH. Exhaustion studies of spirooxazine dye having reactive anchor on polyamide fibers and its photochromic properties. Dyes Pigm. 2007;73:76–80.
56. Peng L, Guo R, Jiang S, Lan J, He Y, Huang X. Ultrasound-aided dyeing of cotton fabric with spirooxazines and photochromic properties. Fibers Polym. 2015;16:1312–15.

57. Bertarelli C, Bianco A, Castagna R, Pariani G. Photochromism into optics: opportunities to develop light-triggered optical elements. J Photochem Photobiol C: Photochem Rev. 2011;12:106–25.
58. Xiong Y, Vargas Jentzsch A, Osterrieth JW, Sezgin E, Sazanovich IV, Reglinski K, et al. Spironaphthoxazine switchable dyes for biological imaging. Chem Sci. 2018;9:3029–40.
59. Paramonov SV, Lokshin V, Fedorova OA. Spiropyran, chromene or spirooxazine ligands: insights into mutual relations between complexing and photochromic properties. J Photochem Photobiol C: Photochem Rev. 2011;12:209–36.
60. Gentili PL, Giubila MS, Germania R, Heron BM. Photochromic and luminescent compounds as artificial neuron models. Dyes Pigm. 2018;156:149–59.

Andrew Towns

63 Spiropyran dyes

Abstract: This article furnishes an introduction to one of the most well-known classes of photochromic colorant. While the properties of spiropyran dyes inspired pioneering efforts to exploit photochromism for industrial applications, their lack of robustness held them back from commercialization. Nevertheless, this type of dye remains at the heart of much of the work to develop light-responsive materials upon which many potential applications in different fields of scientific and technological endeavor depend. The article describes the photochromism, synthesis, and applications of spiropyran colorants with an emphasis on the structural subtype that has attracted the greatest scrutiny. It also acts as a springboard to sources of more detail on these aspects.

Keywords: photochromism, photochromic, colorant, dye, functional, spiropyran

The spiropyran (SP) class was the first major family of light-responsive colorants whose photochromism was explored systematically. Investigation of this facet began in 1951 [1]. Thanks to the synthetic accessibility of the SP unit, it remains a popular choice of motif with researchers from many and varied fields who are seeking to exploit photochromism, despite SP photostability being inferior to that of related types of dye. The last seven decades have witnessed the appearance of an enormous amount of literature relating to the photochromic behavior and potential applications of SP colorants, although no major commercial usage has yet emerged. This Chapter introduces the chemistry behind their photochromism and gives a flavor of the diversity of disciplines in which these dyes constitute the crucial photoactive element. It also points the reader in the direction of sources of more detailed information relating to these aspects. But first things first: the following section defines what comprises a member of the SP class.

63.1 Spiropyran dye characteristics and molecular structure

The scientific literature contains reports of an enormous number of photochromic SP-derived substances. They range from 'small molecule' dyes to high molecular weight polymers and biomolecules. All share the structural elements depicted in Figure 63.1. One half of the SP colorant is an aliphatic heterocycle (shown in green), usually containing a nitrogen atom, but which may for instance also comprise oxygen

This article has previously been published in the journal Physical Sciences Reviews. Please cite as: A. Towns, Spiropyran Dyes *Physical Sciences Reviews* [Online] 2021, 6. DOI: 10.1515/psr-2020-0197

https://doi.org/10.1515/9783110587128-063

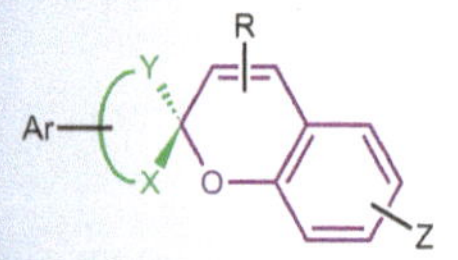

Figure 63.1: Generalized representation of SP dye structure.

and/or sulfur atoms. The heterocycle is connected to a 2*H*-pyran ring (depicted in purple) at the 2-position of the latter to create a spiro link. An aromatic system (usually phenyl, also shown in purple) must be fused to the pyran ring and at least one is almost always joined to the aliphatic heterocycle.

Amongst all this diversity, members of the SP dye class tend to exhibit photochromism that is predominantly T-type, i.e. the color change is driven in one direction by light and by thermal energy in the other [2]. The photochromic switching is usually of a positive nature: irradiation with light converts the dye from a colorless (or weakly colored) state to one which is intensely colored. It fades back in the absence of light to the original color-free form. The transformation is reversible: the rate of thermal reversion from colored to colorless states slows as temperature decreases.

Typical commercial SP dyes respond to radiation within the near ultraviolet (UV) and UV-A, most effectively in the region of 330–370 nm, leading to rapid and intense photocoloration. They may have an element of P-type photochromic character, whereby visible light of appropriate wavelengths (generally >450 nm) aids return to the colorless state. The quantum yields for photobleaching in such dyes are far lower than those for photocoloration. The spectral power distribution of unfiltered sunlight means that irradiation at around 20 °C leads to the rate of coloration far exceeding the rates of thermal or photochemical reversion. While distinction in the character of their photochromism can therefore seem slightly blurred, for practical purposes at usual ambient temperatures they may be considered heliochromic – in predominantly T-type fashion, they rapidly become colored upon exposure to strong sunshine, but fade to colorless in the dark. (Weak P-type behavior of SPs is still sufficient for the purposes of some potential applications as outlined in Section 4.) There are many influences on the rate of development of an SP dye's photocoloration, the intensity attained, and resilience of the effect, plus the rate of thermal fade: these include the nature of the incident light, the environment of the dye, as well as colorant molecular structure – Section 63.3 provides a taste of this last aspect.

Examples are known of SPs whose constitution makes them intrinsically negatively photochromic: these strongly colored materials photobleach upon exposure to light and return to a colored state when this stimulus is removed [3]. Certain circumstances can lead to a switch in photochromism from positive to negative, such as a change in the pH or polarity of the solvent in which an SP dye is dissolved, or complexation with metal ions [4]. (Note that the word "negative" in this context does not equate to a lack photochromism!) The rest of this Chapter will focus almost entirely on SPs which exhibit positive photochromic properties.

Known photochromic SP dyes number in the thousands and exhibit considerable variation in structure [5–8]. Their behavior, along with relationships between structure and properties, varies markedly. Consequently, the remainder of this Chapter concentrates primarily on heliochromic SPs of the kind that have received the most attention from researchers, and which are available commercially off-the-shelf – they possess the generic molecular structure shown in Figure 63.2.

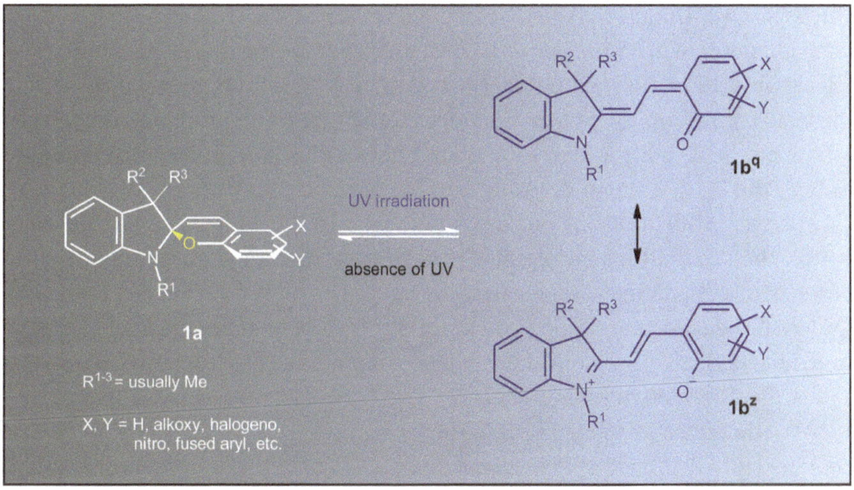

Figure 63.2: Generalized structure **1a** and photochromism of indolinospiropyran dye, showing just one of the isomers of the ring-opened species **1b** produced by irradiation.

Informal terms for this subtype **1a** of SP include indolinospirobenzopyrans and even just indolinospiropyrans. The defining structural motif is an indoline residue bonded in spiro fashion to the 2-position of a 2*H*-pyran ring that forms part of a benzopyran system. Since the last of these structural features also goes by the name 'chromene', such SPs may be encountered with the alternative designation of 'spirochromene'. Given that 'chromene' is a label for members of the naphthopyran dye class, which generally do not feature a spiro linkage on their pyran rings, care must be taken to avoid confusing the two colorant types since their properties tend to be rather different. (Dye **1a** and its derivatives are often referred to within the literature as 'BIPS' when R^{1-3} = Me, the acronym being derived from the name 'BenzIndolinoPyryloSpiran' [9].)

Figure 63.2 also illustrates the transformations that ultimately occur upon photoexcitation and thermal reversion. Prior to irradiation, the dye exists largely as the colorless (or only very weakly colored) compound **1a**. It is often referred to as the 'ring-closed' form, signifying that it comprises an intact pyran ring. Its C_{spiro}–O bond (shown in yellow in Figure 63.2) is relatively long in terms of ether linkages: typically 146–150pm compared to 141–143pm for six-membered oxygen-containing rings [10]. The increased length, signifying weakness and a propensity to be readily cleaved, has been explained as arising through n–σ* interaction between the lone pair orbital of the indoline nitrogen atom and the antibonding orbital of the C_{spiro}–O bond [10–12]. It has also been argued that this interaction becomes more pronounced when SPs are in an electronically excited state, reducing ether bond strength in comparison to that in the corresponding ground state, which further favors photodissociation [10, 13].

Since the two aromatic halves of the dye are orthogonal to each other, their π-systems do not interact strongly. As a consequence, the ring-closed substance typically either absorbs entirely within the UV region, or its spectrum bleeds only weakly into the visible. (See [14, 15] for some examples of SP dye configuration in ring-closed form obtained through X-ray crystallography.) Photoexcitation through absorption of UV leads to the C_{spiro}–O bond breaking and the formation of intensely-colored ring-opened species. While all these processes take place on pico- or nano-second timescales [16], the rate and extent of color development is strongly dependent on the flux of incident UV and may be perceived as building up over the course of seconds or minutes. The destruction of the spiro link sets off a chain of photochemical events which generates a mix of more planar photoisomers whose electronic structures correspond to merocyanine dyes [17]. The most energetically stable of those photoisomers present, the *trans-trans-cis* (TTC) form, is depicted as **1b** in Figure 63.2. The polymethine [18] 'photomerocyanines' are strongly colored owing to the conjugation between the once-separate aromatic systems that results in π-electron delocalization across their length. Extinction coefficients of structures **1b** are a few tens of thousands when quantified in units of $M^{-1}\,cm^{-1}$. The intrinsic high absorption intensity of these species means that even if only a small proportion of dye exists in solution as its ring-opened form, the liquid develops a deep color. (The ring-opened and -closed forms also differ significantly in ways other than their color, e.g. dipole moment, luminescence, basicity, etc. Their contrasting properties form the basis for numerous potential applications and are discussed in more detail in Section 4.)

Photoisomers **1b** are metastable. These species spontaneously ring-close back to **1a** with half-lives ($t_{1/2}$, i.e. time taken for absorbance at peak intensity to halve [2]) conveniently specified in seconds at room temperature, which for the purposes of this Chapter is assumed to be in the range 15–30 °C. Consequently, irradiation of SP dyes under the right conditions with an intense source of UV generates deep and reversible photocoloration. Eventually a dynamic equilibrium of constant concentrations of interchanging ring-opened and -closed isomers is set up [2]. A decrease in flux of UV light causes the proportion of photomerocyanine to diminish, and thus color to weaken, as the rate of ring-closure grows relative to ring-opening. In the dark as solutions in non-polar solvents at room temperature, SP dyes thus essentially exist in ring-closed form. Removal of the UV source leads to them thermally bleaching to a colorless state. The rate of thermal fade in dilute solution often approximates first order decay, i.e. can be defined by a rate constant that depends on SP structure and environment. However, circumstances may dictate that the rate of decay is more complex owing to the presence of photomerocyanine aggregates or isomers with greatly different stability [19]. The intricacies of the impact of placing an SP dye in a polymeric matrix on its photochromism are well covered in [20].

Many studies have examined in detail the mechanistic steps involved in the photochromic transformations of numerous SPs. The body of work is too large to summarize here other than to state that the findings are complex, and the photochemistry is heavily dependent on colorant structure and environment [21]. For a recent review of the mechanism of SP photochromism, see [22].

Members of the SP class tend not to exhibit photochromism in crystalline solid form at room temperature [6]. (Note that this statement is predicated on a definition of photochromism as a reversible visible color change in contrast to IUPAC's definition which includes switching which is so weak and fast that it is not perceptible to the human eye [2, 23].) To make the most of their photochromic behavior, SPs usually must be placed in environments which afford sufficient free volume to allow the necessary change in geometry associated with photocoloration and thermal fade, such as solution in organic solvents or flexible polymer matrices. A few exceptions to this general rule are known. The first examples of SPs that are photochromic in crystalline form at around 20 °C were reported in 2000, reversibly switching from yellow to deep red upon exposure to near UV [24]. Other SP colorants described since then that display room-temperature solid-state photochromism share a molecular motif: a quaternized azaheterocycle atom whose positive charge is associated with a bulky counterion, e.g. iodide [25]. The anion of the appropriate size leads to crystal packing which provides a reaction cavity of sufficient volume for transformation between ring-closed and photomerocyanine forms [26]. However, a special case of photochromism in a crystalline non quaternized SP colorant is also known [27]. The discovery that even crystals of simple SPs, e.g. **1a** (R^{1-3} = Me; XY = benzo), are photochromic at −180 °C suggests that suppression of solid-state photochromism in crystals of commercially-available indolino-derivatives at *ca.* 20 °C relates instead to very rapid thermal reversion rather inhibition of ring-opening [28]. (Thin films of simple SPs have also been reported to exhibit photochromism at room temperature [8, 29] but, while often not well characterized, they lack a highly crystalline nature. A recent report of SP photochromism in ultrahigh molecular weight polyethylene films that have been drawn to give a highly oriented and crystalline structure demonstrates how one must be careful when making generalizations about the behavior of SPs [30].)

Many members of the SP class exhibit color changes triggered by stimuli other than light [11, 22, 31]. However, their photochromism accounts for the majority of academic interest in them. Indolinospiropyrans continue be at the center of most of the work published in the literature involving SP colorants. As we shall see, one motif in particular within this subset of dyes dominates research seeking to exploit the chromism of SPs: it does so to the extent that one might be forgiven for thinking that there is little more to this class. Nevertheless, numerous examples exist of derivatives comprising alternative (hetero)aromatic ring structures to indoline. Part of the lasting technical appeal of the SP photochromic unit is its synthetic accessibility. The next section briefly touches on this aspect.

63.2 Synthesis of spiropyrans

Research quantities of SPs are marketed by laboratory reagent suppliers [32] at prices which are reflective of their small production volumes, i.e. in the region of tens of dollars per gram. They are usually indolinospiropyrans synthesized by reaction of Fischer's base **2** (R = Me, X = H; 2-methylidene-1,3,3-trimethylindoline) or its close analogues with salicylaldehyde **3** (Y = Z = H; *o*-hydroxybenzaldehyde) or its simple derivatives (see Figure 63.3). A commercial SP of this kind is typically supplied in a purity of at least 98% area (determined by HPLC) as an off-white or lightly colored solid. Appearance strongly depends on factors such as impurity profile and process of isolation in addition to its molecular structure.

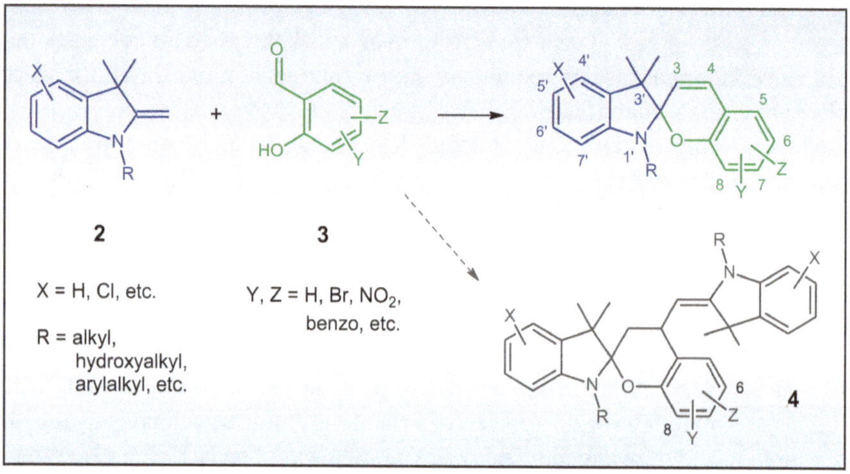

Figure 63.3: Generalized synthetic pathway to indolinospiropyran dyes.

Commercially available indolinospiropyrans readily dissolve in organic solvents of low to medium polarity. Aromatic hydrocarbons as well as common ethers and ketones, e.g. toluene, THF and acetone, act as good solvents. Their solubility in alcohols is much lower, while they are essentially insoluble in water. (However, the diversity in molecular structure of SPs reported in the literature means that there are plenty of exceptions to these generalizations concerning appearance and solubility.)

Provided that care is taken over quality of raw materials and selection of conditions, synthetic yields of indolinospiropyrans tend to be good: typically at least 70% and sometimes being near quantitative. Synthesis often simply involves condensation of **2** and **3** through reflux in an alcohol, then collection of the SP dye by filtration from the mother liquor after cooling. The approach relies on the relatively low solubility of the product in alcohols relative to residual starting materials and side-products. Aside from washing of the isolated product with fresh solvent to displace the remaining impurity-laden mother liquor, further purification may be required to remove colored side-products and residual starting material. Usually recrystallization is sufficient. Commercially at least, column chromatography remains very much a last resort owing to the expense it entails, particularly in terms of solvent use and time.

To avoid carrying through into product traces of strongly colored impurities that tend to form in Fischer's base derivatives **2** during storage, vacuum distillation of these starting materials prior to use is often performed. In instances where such bases readily precipitate as hydrohalide or perchlorate salts, it is often advantageous to isolate them in these forms. Their solubilities as these kinds of salts in organic solvents are low, enabling ready removal of contaminants without resorting to distillation, since the impurities tend to remain in the mother liquor. In addition, the resultant pure dried solid salts are much more stable and easier to handle than the liquid free bases. Rather than convert them back to free base form immediately before use, the acid salts may be employed directly. When doing so, a solvent-miscible organic base is also added to scavenge acidity. Trialkylamines or cycloaliphatic amines, like triethylamine and piperidine, respectively, work well in this regard. The combination of salt and amine is claimed to be beneficial by minimizing generation of side-products **4** [8]. The formation of 'di-condensed' non-photochromic compounds **4** (R = Me; X = Y = H; Z = H, 8-OMe, 6-Br) during SP synthesis was first reported in 1952 [33], but speculation about their exact structures was not ended until over half a century later by X-ray crystallography of **4** (R = Me; X = Y = H; Z = 6-CO_2H) [34]. These side-products are not generated by reaction of SP with **2**, but through trapping of an intermediate on the pathway to SP formation [35]. One strategy to minimize formation of **4** is slow addition of **2** to a small excess (5–10mol%) of **3** [36]. Further information on this type of side-product can be found in [5]. (Note that nomenclature denoting positioning of substituents varies in the literature; Figure 63.2 shows the numbering system used in this Chapter with the prime notation employed to demarcate locations on the indoline system, but some reports reverse this usage and employ it to indicate placement on the benzopyran moiety.)

In cases where the requisite salicylaldehyde precursor is not available, a pathway starting from Fischer's aldehyde and its derivatives may serve as a useful alternative. For further details as well as an excellent overview of the practicalities involved in SP dye synthesis, the reader is referred to [36] – although published over twenty years ago, it features a perspective gained in industry and also discusses the preparation and quality of indoline- and salicylaldehyde-based intermediates in depth.

The Fischer's base moiety **2** is readily replaced with other heterocycles, producing considerable variation in the kinetics and resilience of photochromism. Furthermore, subsequent to its synthesis, the SP skeleton is sufficiently tolerant in certain instances to functional group transformation for property modification or the attachment of reactive motifs to enable inclusion of the photochromic unit into/onto polymeric chains and other substrates. Overviews of such synthetic possibilities are given in [5–8, 12]. A wide range of behavior can be accessed just by altering substituents of indolinospiropyrans **1**. The following section illustrates some of these possibilities.

63.3 Color and constitution of spiropyran colorants

Relationships across the SP class between structure and aspects of photochromic behavior (e.g. hue, intensity, thermal reversion rate, robustness) are complicated. In some cases, general rules linking constitution to photocoloration properties have been formulated and rationalized in terms of intra- and inter-molecular electronic and steric effects of dye substituents. Nevertheless, researchers have long bemoaned the many exceptions to trends, or even the absence of correlations, that become apparent not just when comparing families of SP dyes, but also within the indolinospiropyran series [7, 37]. As is the case with non-SP photochromic colorants, the complex interplay between geometric, electronic, and environmental influences means that there is often no substitute for synthesis and testing of colorants when assessing the impact of making a structural change on photochromism at a fine-grained level.

Commercial indolinospiropyrans furnish purplish-red to blue photocoloration in non-polar solvents and polymers. Environment strongly affects photochromism of this kind of SP. The hues of photomerocyanine species are sensitive to solvent polarity, which also markedly influences kinetics and thus intensity of color before, during and after exposure to UV. Depending upon SP structure, absorption maxima (λ_{max}) either become longer as solvent polarity increases (positive solvatochromism) or shorter (negative solvatochromism). Tables 63.1 and 63.2 illustrate this contrast in terms of the λ_{max} of ring-opened photoisomers of two indolinospiropyrans, $\mathbf{5}^{\mathbf{SP}}$ and $\mathbf{6}^{\mathbf{SP}}$, that are often encountered in the literature.

The electronic structures of the merocyanine products of photoisomerization can be regarded as hybrids of the extreme quinonoid and zwitterionic resonance forms, $\mathbf{1b}^{\mathbf{q}}$ and $\mathbf{1b}^{\mathbf{z}}$, respectively (see Figure 63.2). The degree of contribution from each form is influenced by dye structure and the polarity of the colorant's environment. The significant quinonoid character of the photomerocyanine $\mathbf{5}^{\mathbf{MC}}$ (X = CH) in the case of indolinospironaphthopyran $\mathbf{5}^{\mathbf{SP}}$ leads to mild positive solvatochromism (see Table 63.1). A non-polar solvent like toluene widens the energy gap between the relatively non-polar quinonoid ground states of the photoisomers, e.g. $\mathbf{1b}^{\mathbf{q}}$, and their more polar

Table 63.1: Solvatochromism and thermal fade of merocyanine forms 5^{MC} of indolinospironaphthopyran and -oxazine dyes 5^{SP} and 5^{SO}, respectively, at *ca.* 25 °C.

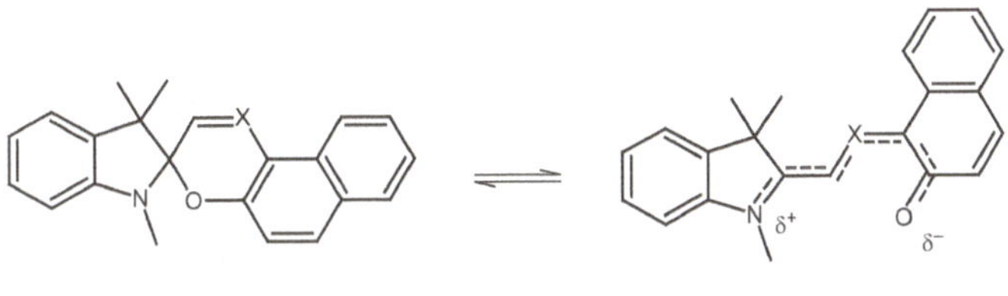

X = CH, 5^{SP} ; X = N, 5^{SO} 5^{MC}

Solvent		5^{MC} (X = CH)			5^{MC} (X = N)		
		λ_{max} (nm)		$t_{1/2}$ (s)*	λ_{max} (nm)		$t_{1/2}$ (s)*
		[38]	Other references	[38]	[38]	Other references	[38]
Toluene	polarity	558	530 [39], 563 [37], 571 [40]	0.2	594	588 [39], 590 [37], 593 [41]	1.3 (2.6 [41])
1,4-Dioxane		558	530 [39]	<0.02	591	578 [39]	1.0
Acetonitrile		561	562 [39]	0.02	597		0.4
Ethanol		565	568 [37, 39]		607	608 [39], 612 [42]	

*calculated from literature rate constants assuming first-order decay

excited states by lowering the energy of the former relative to the latter. Conversely, this difference narrows in a polar solvent, through greater stabilization of the excited states, so the violet-red photocoloration produced by irradiation of 5^{SP} in ethanol is bathochromic relative to that observed in solvents of lower polarity. Such behavior parallels that of members of the structurally analogous indolinospironaphthoxazine class, i.e. those based on 5^{SO} [51]. Photomerocyanines of SPs are intrinsically more hypsochromic than their direct spirooxazine counterparts, e.g. photocoloration produced from 5^{SP} is purplish in non-polar solvents, whereas that from 5^{SO} is blue (see Table 63.1).

Care must be taken when making comparisons based solely on λ_{max}. The spectra of ring-opened SPs **1b** often consist of an irregularly shaped band in the visible region that may possess a shoulder or more than one peak. These features arise through the presence of a mixture of photomerocyanine geometric isomers whose components possess different properties [35, 52] irrespective of irregularities in shape caused by vibronic transitions [53]. The technique and conditions of measurement (e.g. temperature, illuminant spectral power distribution, SP concentration, purity of dye and solvent, etc.) influence the spectra and kinetics, so that often there is considerable variation in results reported for a particular SP derivative by different research groups. One must be wary when trying to draw conclusions using data that

Table 63.2: Solvatochromism and half-life of photocoloration of an important and commonly used type of indolinospirobenzopyran dye **6** possessing 6-nitro substitution.

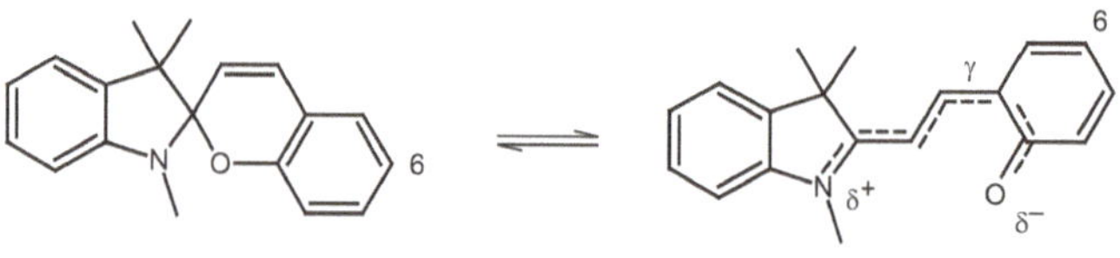

6^{SP} 6^{MC}

Solvent		6^{MC} (6-NO_2) λ_{max} (nm)		6^{MC} (6-NO_2) $t_{1/2}$ (s)*	
		[43]	Other references	[43, 44]	Other references
Toluene		600	596 [45], 597 [40], 598 [37], 600 [46], 602 [39], 603 [44]	5.7	
1,4-Dioxane		590	590 [39, 44, 47]	11	12 [47]
Tetrahydrofuran		584	574 [45], 583 [39], 584 [48], 585 [44]	26	
Acetone	Polarity** ↓	568	561 [44], 562 [45], 565 [46, 47]	95	95 [48], 96 [47]
N,N-Dimethylformamide		562	558 [44], 560 [47]	350	450 [47]
Acetonitrile			550 [39], 555 [47], 557 [44], 560 [46]		240 [48]
Ethanol		540	532 [37, 40], 534 [45], 535 [46, 49], 537 [48], 538 [44], 545 [39]	1000	1130 [48]

*calculated from literature rate constants obtained at 25 °C assuming first-order decay
**ranking based on E_T scale in [50]

originate from more than one source! The “Other references” columns of Table 63.1 demonstrates how literature results can vary markedly by listing values for λ_{max} of $\mathbf{5}^{MC}$ (X = CH and N) gathered from just a few of the many sources of spectroscopic data relating to the photochromism of $\mathbf{5}^{SP}$ and $\mathbf{5}^{SO}$.

Simple SPs based on $\mathbf{5}^{SP}$ tend to furnish systems that exhibit significantly higher rates of thermally-driven ring-closure than their spirooxazine analogues $\mathbf{5}^{SO}$ [38]. Thermal decay accelerates as solvent polarity increases (see Table 63.1). However, replacing the fused benzo ring of indolinospironaphthopyran $\mathbf{5}^{SP}$ with a 6-nitro group to give $\mathbf{6}^{SP}$ greatly slows down the rate of thermal ring-closure (see Table 63.2). This retardation of reversion contributes to a huge increase in intensity of photochromic response. Consequently, the 6-nitrobenzopyran motif is very commonly encountered in the SP literature. A large proportion of studies adapt this unit, adding points for attachment to other materials, often at the 1′-position. As will be discussed below, the incorporation of the 6-nitro function leads to considerable changes in other aspects of photochromism. (Be alive to this influence: often in papers reporting exploitation of

$\mathbf{6^{SP}}$ or its close derivatives, authors will make generalizations in explaining the behavior of such dyes, writing in a way that implies they apply to the whole class of SP colorants, which is definitely not the case.)

In contrast to derivatives **5**, the presence of the strong electron-accepting nitro function at the 6-position of the benzopyran moiety of $\mathbf{6^{SP}}$ confers much more pronounced zwitterionic character on the merocyanine species $\mathbf{6^{MC}}$ (6-NO_2) produced by UV-induced ring-opening. The nitro group stabilizes separation of charge, thereby promoting the contribution of $\mathbf{1b^{z}}$ to photoisomer ground state electronic structure. The greater zwitterionic character turns its solvatochromism negative, i.e. as solvent polarity is adjusted upwards, maximal absorbance of $\mathbf{6^{MC}}$ (6-NO_2) shifts to shorter wavelengths (see Table 63.2) in contrast to $\mathbf{5^{MC}}$. The violet photocoloration produced on irradiation of $\mathbf{6^{SP}}$ (6-NO_2) in solvents of low polarity becomes much redder when the environment grows more polar. The TTC photoisomer (depicted as part of Table 63.2 by $\mathbf{6^{MC}}$) is the dominant product of ring-opening. Irradiation of $\mathbf{6^{SP}}$ (6-NO_2) in acetonitrile solution to a steady state furnishes TTC:TTT photoisomers (TTT = *trans-trans-trans* where the phenoxy ring is rotated 180° about the γ-bond of $\mathbf{6^{MC}}$ relative to the TTC isomer) in a ratio of ~ 9:1, which absorb maximally at 557 nm and 595 nm, respectively [54]. The behavior of **6** (6-NO_2) provides a good example of how sensitive SP photochromism is towards environment: for a detailed account of the impact of solvent on the spectroscopic properties, thermodynamics, and kinetics of the photocoloration of $\mathbf{6^{SP}}$ (6-NO_2), see [55].

The very pronounced impact of 6-nitro substitution on SP photochromism is further illustrated by Table 63.3. It leads to much more prominent photochromism compared to that of other simple indolinobenzopyran analogues that do not bear nitro groups. At room temperature, the photochromism of the parent SP **6** (6-H), as well as 6-substituted derivatives bearing moderate electron donors (6-OCH_3, 6-CH_3) or acceptors (6-Br, 6-CO_2CH_3), is too weak to be of much practical use for the production of visual effects. In marked contrast, the 6-nitro derivative **6** (6-NO_2) is strongly photochromic. It colors up far more deeply as indicated by intensity of photocoloration being at least an order of magnitude greater than that of the other derivatives in Table 63.3. An increased quantum yield for ring-opening, as well as a lengthening of half-life, account at least in part for the much more pronounced photochromism [22]. In other words, the proportion of photons absorbed that result in photoisomerization of $\mathbf{6^{SP}}$ to colored species is enhanced, while the rate at which the photomerocyanines revert to the ring-closed form is retarded. Consequently, in non-polar polymers and solutions, the rate of thermal fade is slowed sufficiently to result in rapid and intense photocoloration upon irradiation with UV-A sources from an initial state of low residual color. Such systems can return to this state within a few minutes.

The ease of synthesis of $\mathbf{6^{SP}}$ (6-NO_2) by condensation of the readily available Fischer's base **2** (R = Me, X = H) and 5-nitrosalicylaldehyde **3** (Y = 5-NO_2, Z = H), in addition to its good photoresponsivity and tendency to color up deeply, explain its status as the most intensely studied SP derivative. The colorant does not just serve as a model in many mechanistic investigations of SP photochromism: it continues

Table 63.3: Influence of benzopyran moiety substituents on photochromic properties in 1,4-dioxane solution at 22 °C (data adapted from [56]).

6-substituent of dye 6	$t_{1/2}$ (s)*	Intensity (M^{-1} cm^{-1})**	Durability**
OCH_3	0.2	500	**8000**
CH_3	0.3	250	**17,000**
H	0.7	200	**9000**
Br	2	200	900
CO_2CH_3	**23**	100	420
$COCH_3$	**20**	1000	2350
NO_2	**18**	**>28000*****	<100***

* $t_{1/2}$ calculated from rate of thermal fade assuming first order decay
** "Intensity" of photocoloration is apparent extinction coefficient and "Durability" is the number of UV lamp flashes of 450 J input until photocoloration intensity falls to half of its original value for 4 mM dye solution
*** figures assumed based on values of 28,000 and 100 being obtained with flashes of 300 J input in case of 0.1 mM dye solution

to attract the greatest attention of efforts to exploit SP-based technology through use in its own right or following adaptation of its structure to facilitate adsorption or covalent linkage to other materials. However, the convenience and responsive nature of $\mathbf{6}^{\mathbf{SP}}$ (6-NO_2) comes at a cost: the dye readily fatigues, losing its activity far more quickly than non-nitro substituted SP colorants (see "Durability" column of Table 63.3). Irreversible side-reactions at the expense of desired photoisomerization tend to occur more readily, so that the concentration of dye decreases faster and the photoactivity of what remains is reduced owing to degradation products competing for UV. The low photostability of the dye is attributed to its nitro function dictating that photoisomerization takes place *via* a triplet excited state pathway. Such a route leads to a greater propensity for photochemical side reactions than is the case with non-nitro substituted SP dyes, like **6** (6-substituent ≠ NO_2), which photoisomerise by means of excited singlet states [22].

Environment plays a major role in determining the longevity of photochromic performance, e.g. by influencing rate of photooxidation in addition to the prevalence of irreversible side-reactions competing with photoisomerization. Breakdown products, many of which have been identified [57], may cause increased residual color in a non-activated state in addition to soaking up incident UV instead of the colorant. The net result is weakening and disappearance of photochromism. Generally, SPs may be cycled a few hundred times before their photochromic behavior is substantially lost, although some do not last more than five cycles, while others may prove much more robust (>10,000 cycles) [6]. Table 63.3 not only demonstrates that rate of fatigue varies by orders of magnitude between SPs of similar structure, but it also reveals the

Achilles' heel of the photochromic unit based on the popular 6-nitroindolinospirobenzopyran dye **6**^{SP} (6-NO_2). We will return to this topic in the next section.

The ready commercial availability of simple indolinospiropyrans, and their continued domination of the academic literature, mean that one might be led into thinking there is not much more to the SP class beyond dye **6**^{SP} (6-NO_2) and **5**^{SP} as well as their close analogues. However, reports abound from researchers around the world that relate to the impact of myriad structural variations on photocoloration, for example, replacing the indoline moiety with other heterocycles [6, 7, 37]. Many such modifications simultaneously change color and kinetics in a substantial manner as do alterations to substitution in the benzopyran moiety. The thiopyran analogue of **6**^{SP} (6-NO_2), **7a** (see Figure 63.4), gives photocoloration that extends into the near infrared (NIR) owing to the presence of the sulfur atom shifting absorption bathochromically by around 100 nm in polymers and polar solvents [58]. Fatigue resistance remains poor. Another means of pushing absorption into the NIR involves appropriately loading the benzopyran moiety with electron-withdrawing functions as in the case of **7c**. This recently reported colorant furnishes photocoloration and fluorescence that bleeds into the NIR (λ_{max} 738 nm and λ_{em} 791 nm in acetonitrile) [59]. Its merocyanine photoproducts are some of the most bathochromic known for SPs. The absorption of the ring-closed form of **7c** is red-shifted sufficiently into the visible region that it confers a deep reddish-yellow color. While photosensitive materials exhibiting NIR absorption and fluorescence were originally of interest for data storage system development that relied on NIR lasers, the focus of attempts to exploit them has shifted to the biotechnology sector (see Section 4).

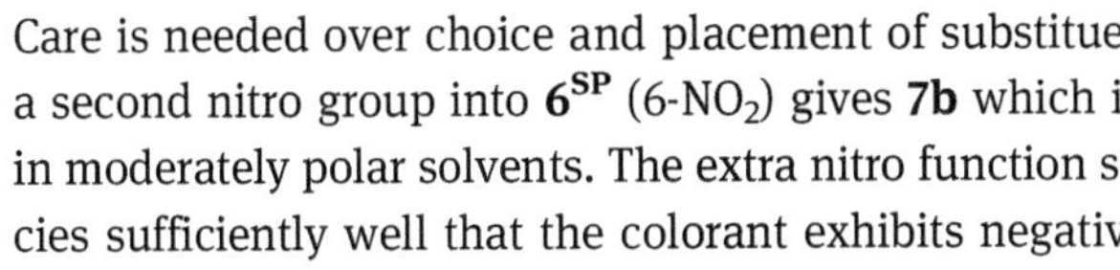

Figure 63.4: Some variants of indolinospiropyran colorants.

Care is needed over choice and placement of substituents. For example, introducing a second nitro group into **6**^{SP} (6-NO_2) gives **7b** which is not positively photochromic in moderately polar solvents. The extra nitro function stabilizes the merocyanine species sufficiently well that the colorant exhibits negative photochromism: the red-orange coloration of a solution of **7b** in acetone at room temperature protected from strong daylight reversibly photobleaches to a near colorless state when irradiated with white light [3, 4].

For all the structural diversity encountered in the academic literature and the huge number of studies attempting to exploit the SP unit, some aspects of design are underutilized by academia. For example, locating bulky β-branched *N*-alkyl substituents on the indoline moiety of spirooxazine dyes is a tactic employed industrially for many years to inhibit thermal ring-closure and thus boost photocoloration intensity [51]. However, this approach and others remain surprisingly overlooked as a means of enhancing SP photocoloration. Replacement of the *N*-methyl function of Fischer's base with a bulky alkyl group (neopentyl), for instance, was only reported in this context a few years ago [60].

Before leaving this subject, it is worth reiterating that comparisons of literature data published in different studies that relate to the photochromic properties of SPs in solution must be undertaken with caution. They are highly sensitive to factors which include the power, spectra distribution and filtering of stimulating light source as well as dye concentration, temperature, and presence of oxygen. Nevertheless, technologists attempting to exploit the photochromism of the SP unit can make use of design rules like those outlined here to guide their efforts. Lots of potential uses are still being explored in connection with the switching of SP colorants, many years after their chromism was discovered. There is insufficient space to cover them all here, but the next section looks at some of the major applications.

63.4 Applications of spiropyran dyes

The observation of thermochromism in SP derivatives nearly a century ago first sparked interest in them [61, 62]. In 1940, indolinospiropyrans were reported to be thermochromic [63]. However, it was not until a decade later that the photochromism of these derivatives was discovered and their light sensitivity became the subject of detailed investigation. SPs were the first of the major types of photochromic dye to be scrutinized by investigators equipped with spectroscopic instrumentation. Their photochromic properties – and the mechanisms behind them – were studied in depth during the 1950s and 1960s by commercial, academic, and military organizations [9, 37, 64].

The initial discovery of photochromism in SPs came in 1951 [1]. UV-responsive color changes were observed in the case of two dyes, one of which was $\mathbf{5^{SP}}$. When their solutions were chilled below −30 °C, the rate of thermal reversion became so slow that their P-type photochromic character could be brought to the fore, i.e. it became possible to effect the transformation back to the colorless state with visible light [66]. This kind of photoswitching of SPs was of interest for one of the first functional uses of photochromic dyes to be advocated: the construction of high-density memory based on optical properties was proposed by Dr EH Frei [1] and outlined in terms of an SP-based system as early as 1955–1956 [67]. While examples of SPs were very soon found that gave photochromic solutions at ambient temperatures and prototypes constructed, following extensive investigations, it was concluded half a century ago that SP dyes were too fatigue-prone to be useful in this capacity [64].

The lack of stability of SP dyes to repeated or continuous exposure to light has long been known [37]. They are susceptible to photochemical fatigue and attack by environmental agents, including oxygen and humidity. As was briefly mentioned in the preceding section, the nature of the moiety on the pyran side of the spiro linkage exerts a strong influence, although relationships tend to be complex [57]. Irrespective of the fugacity of the commonly used 6-nitro derivative of indolinospiropyran $\mathbf{6}^{\mathbf{SP}}$ (6-NO_2), SPs are very delicate compared to conventional non-chromic colorants: their degradation in certain environments is marked enough for switching performance to drop off markedly within a few cycles. Consequently, combatting loss of reversibility remains an open problem, worthy of a Nobel Laureate's attention [68]. Immobilization of SP units by covalent attachment to other materials, e.g. grafting them onto macromolecule chains, is a strategy for increasing the robustness of their photochromism [69, 70]. Despite such efforts, SP-based systems will never be sufficiently resilient for certain applications. SP dyes still sometimes get selected as the central component of misguided studies exploring the creation of light-responsive structures and devices subject to prolonged direct exposure to strong sunshine, e.g. architectural glazing and even self-regulating solar cells [71]: they are fundamentally unsuited to such uses.

SP dyes continue to be examined as colorants for fibers and textiles, e.g. poly (methyl methacrylate) nanofibers [72], despite the invariably poor photostability of the coloration produced. Unfortunately, in many studies, authors do not address the lack of robustness, or even completely ignore it. For example, a photochromic cotton was produced by treatment with a polyurethane endcapped through covalent binding with SP dyes [73]. The material showed a steady loss in performance over just twenty cycles of irradiation–thermal reversion. Resilience to sunshine will thus be orders of magnitude lower than what is typically acceptable for commercial textiles. As a source of garment coloration, the use of SPs is essentially a dead-end.

More suitable applications for off-the-shelf commercial SPs are those where exposure to light is limited. Security printing of optically variable overt or covert marks with SP-containing anti-counterfeiting inks is one such avenue [74]. The simplest means involves production of a color change upon exposure to UV. Interest continues in more sophisticated systems, e.g. whereby the properties of liquid crystals films are modified by doping them with SP dyes and then irradiating.

The lack of robustness of SP colorants, as well as the limited color gamut of derivatives with the desired rate of thermal reversion, proved to be key obstacles to the production of sun-light responsive ophthalmic lenses of commercially acceptable longevity [9]. The realization that structurally similar spirooxazines [51] possessed greater robustness [42] led to a switch in focus concerning this application. Along with the naphthopyran class [75], they went on to taste long-lasting industrial success instead of SPs. Nevertheless, academia's appetite for SP photoswitches remains undimmed. Research is now directed at other fields, many of which do not depend directly on a color change, nor count photostability as a concern.

The continued broad interest in SP photochromism stems from the ring-closed and -opened forms differing not just in their color, but also in numerous other ways as exemplified by Figure 63.5 for SP dye **6** (6-NO_2). Irradiation with UV brings about increases in polarity as well as tendencies to luminesce, protonate, complex with metal ions, and self-associate. Part of the allure of SPs to some researchers lies in the driving force behind these substantial property changes not being restricted to light: as well as temperature, reversible switching in SP dyes may be induced depending on the system by changes to polarity or pH of the medium, and even mechanical stress [22, 69]. Conversely, substantial changes in SP photochromism, including the direction, kinetics, and efficiency of light-driven switching, as well as the colors involved, can be produced under the right conditions by complexation [76] or protonation.

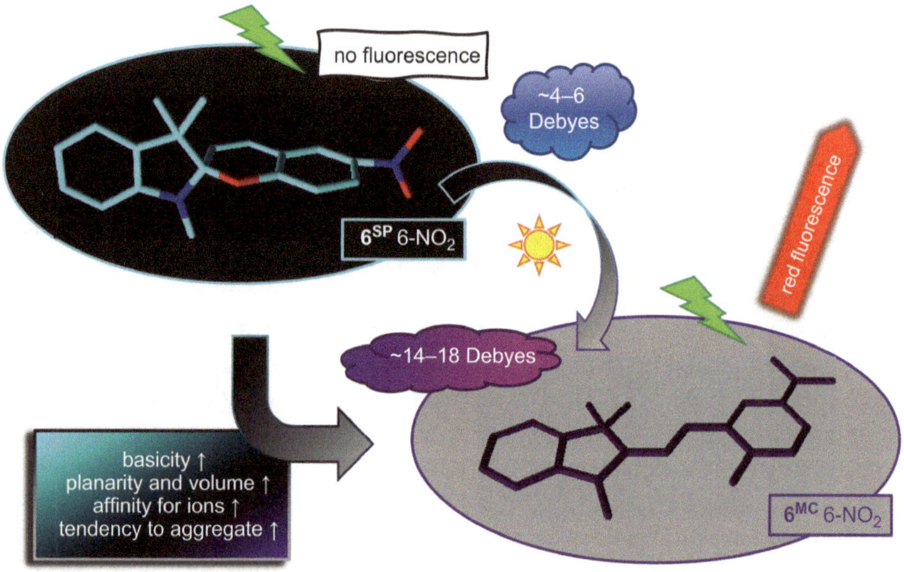

Figure 63.5: Differences in properties of ring-closed (^{SP}) and ring-opened (^{MC}) forms of 6-nitro-substituted indolinospirobenzopyran **6** (based on data in [69]).

Very early on in the study of the photochromism of SPs, their potential utility as functional colorants was foreseen. While decades later they continue to receive attention, much remains to be done before a major new application becomes a commercial reality [77]. Just a few of examples of the many avenues of research in which SP derivatives can be found are tabulated below (see Table 63.4).

SP photochromism has fired the imaginations of scientists and technologists as a facilitator in a wide variety of fields. The overarching theme is one of exploiting the changes in one or more properties that occur upon irradiation with light. The multidisciplinary nature of much modern research is reflected by the entries in Table 63.4: several areas bleed into each other, such as nanotechnology and biomedicine, which in turn overlaps with imaging. Examples of SP use have been roughly divided up into the following subsections, which give a flavor of the ways in which researchers are seeking to exploit them. Many of the fields are very active and fast changing, but

Table 63.4: Some examples of switchable properties of SPs and instances of applications in which their exploitation has been attempted.

Photoreversible property	Example application
Absorbance	Memory element for data storage [78], super-resolution lithography [79], photonic circuitry component [80], security printing, dosimetry
Geometry	Photocontrol of biomolecule activity [81, 82], chemotherapeutics, photoextraction [83], security printing [84]
Ion affinity	Imaging of biological structures [85, 86], colorimetric analysis [83, 86]
Luminescence [87]	Memory element [78], super-resolution microscopy [87]
Optical non-linearity	Memory element [78], optoelectronics
pKa	Microfluidic device photovalve [88, 89], photonics, photoregulation of polymerization [90]
Polarity	Photomodulation of adhesion or wettability [91], reaction control [92], optoelectronics [93]
Refractive index	Memory element [78], holography [78, 94]

reviews often appear that summarize developments, for example, nanotechnology [95]. Table 63.4 is by no means exhaustive!

63.4.1 Information technology and optoelectronics

The development thread relating to the creation of high-density memory devices which began with the discovery of SP photochromism continues to evolve. It shifted long ago from involving simple switching of a matrix of 2D points between states of different absorbance. Other properties have been used to read the state of SPs without altering that state, e.g. fluorescence, refractive index, non-linear optical response, etc. Time will tell whether the strategy of immobilization of SP units in macromolecules or onto surfaces means that the optimism of the recent statement "properly immobilized spiropyrans can enable realization of information storage devices" [69] is well-placed. Over six decades since the idea was conceived, an SP-based technology that finally supplants commercial storage methods still appears remote. The potential of SPs in the construction of all-optical devices to process information digitized as light signals faster than electronic circuitry has long been recognized [80]. They continue to be the subject of research as functional dyes in components such as logic gates and switches in photonic devices. SP colorants also lie at the center of related work that seeks to manipulate light for recording or processing of digital data, e.g. the modulation of refractive index in holography [94] and controlling the optical properties of liquid crystalline materials [96].

63.4.2 Imaging and sensing

Various research activities center on SP-related materials to construct alternative views of the world by relying on the contrast in properties of ring-opened and -closed forms. As is the case with other photochromic dye classes, the SP-type of colorant is also of interest in super-resolution microscopy, which is capable of imaging detail in structures whose nanoscale dimensions lie below the diffraction limit of visible light [87]. Photoswitching the fluorescence emission of SP dyes provides the basis for image build-up with resolution in the realm of tens of nanometers. Means of improving image quality, for example, by reducing noise, include developing SP-based systems that switch between multiple luminescence colors [87]. One strategy modifies the picture presented in Figure 63.4 by adjusting the structure of $\mathbf{6}^{\mathbf{SP}}$ (6-NO_2) so that it fluoresces, emitting green light.

SP units have been explored as chemosensors to detect the presence of anions, metal cations, gases, and biomolecules [31, 86]. For example, the contrast between the abilities of ring-opened and -closed isomers to complex with ions and luminesce is at the heart of an approach to visualize lithium ion flow in cancer cells through fluorescence microscopy. UV irradiation of a non-emissive SP dye within a biological sample generates merocyanine species that complex with lithium ions in preference to other metal cations common in cellular systems; the fluorescence of the complex allows mapping of lithium distribution [85]. The dye can be partially "reset" by irradiation with white light, causing much reversion to the ring-closed form; the dye also gradually forms the fluorescent merocyanine–lithium complex in the dark. A substantial amount of work has been conducted into the complexation of SP-based species with metal ions [83, 97], although such studies remain in the realm of academia.

Development work on another form of imaging, super-resolution lithography, has also involved use of SP dye chemistry. This technique is of interest as a means of 'writing' photoresists of nanoscale detail in three dimensions. The concept has been demonstrated using a co-polymer produced from spirothiopyran monomer closely related in structure to **7a** [79]. In the presence of multifunctional maleimides, light is employed to selectively cross-link it where desired. Only the photomerocyanine species, and not the ring-closed form, undergoes chemical reaction (Michael addition) with the maleimido cross-linker. The process relies on simultaneous irradiation of SP units with UV and green light, which, respectively, increases and decreases merocyanine concentration. By varying the intensity of visible light, a pattern of cross-linked material is created; unpolymerized regions are then washed away to reveal the desired nanostructure. The technique relies on the visible radiation inhibiting formation of the reactive photomerocyanine, so cross-linking occurs only where its intensity is low.

Prototype dosimeters which visualize exposure to hazardous forms of radiation have been constructed with SPs. Real-time indication of X-ray exposure was made possible using dye $\mathbf{6}^{\mathbf{SP}}$ (6-NO_2) as an alternative to commercial dosimeter systems. The response depended upon the interaction in a polymeric matrix between the colorant and an inorganic phosphor that emits UV following absorption of X-ray radiation energy. Exposure of this combination in poly(lactic acid) [98] or polystyrene [99] to X-rays led to the reversible development of light purple and deep blue coloration, respectively, owing to the phosphor emitting UV of around 380 nm which stimulated ring-opening of the SP derivative. The same dye was also used in poly(methyl methacrylate) that was free of inorganic phosphor, giving weak direct response to irradiation. An enhanced effect was obtained when an organic phosphor was present, with the assembly having the advantage of being more “tissue-equivalent” with respect to energy absorption [100]. The approach was ultimately unsuccessful owing to the impermanence of the coloration.

In contrast, the T-type behavior of SP colorant chemistry served as the crucial mechanism for the commercially piloted OnVu™ temperature–time indicator (TTI) labelling system [101, 102]. The TTI reveals when refrigeration of perishable foodstuffs has been disrupted, and thus the shelf-life date rendered invalid, or serves directly as an indicator of shelf-life. The technology depended on solid-state SP photochromism [103] although an alternative system reliant on photochromic crystals of benzylpyridine derivative was also explored [104]. Immediately before use, the photochromic ink of the TTI label is activated by UV from LEDs to ring-open the dye, developing a blue color. The extent of photoisomerization, and thus depth of coloration, depends on irradiation time [105]. Post-activation, the coloration slowly fades. A film, laminated on top of the ink directly following activation, acts as a filter to block any further photochromic response to accidental or deliberate exposure to UV that could reactivate the colorant. As would be expected from the T-type nature of the dye, the rate of bleaching is dependent on temperature, so the depth of color of the activated ink acts as a record of its thermal history. The light bluish color of a permanent reference also printed on the label serves as a marker for too lengthy exposure to insufficiently cold temperatures.

Research is targeting the development of another form of sensing with SP colorants that does not rely on their light sensitivity, but in which they play the role of mechanophores. These chromic substances undergo chemical transformation upon application of a mechanical force [106]; SP-based materials whose color changes in response to stress, i.e. exhibit mechanochromism, were first reported in 2007 [107]. The model involves covalently incorporating the SP unit into a polymeric matrix, linking the moieties on each side of the spiro junction to different chains of the network. Application of mechanical force to the bulk material may thus lead to rupturing of the C_{spiro}–O bond as the halves are pulled in different directions by the matrix. Cleavage produces the colored merocyanine form. Stress and strain experienced by the bulk material can thus be detected optically through the presence of the ring-opened form absorbing light or fluorescing. The SP unit is the most well-established reversible type of mechanophore utilized in the development of stress indicators. However, much remains to be done to create practical real-world systems. Targets include not only increasing the degree of SP mechanoactivation but also lowering the threshold at which mechanochromism is observed to usefully low levels. For a helpful recent overviews of SP application in this field, see [31, 108].

63.4.3 Material science and process technology

Much effort has been directed at SP photochromism as a tool to manipulate the properties of materials by means of light. For example, the wettability of surfaces, and thus interaction with other species, can be reversibly photoregulated by introducing SP functionality into polymers or coatings [91, 109]. When the photochromic unit is based upon $\mathbf{6}^{\mathbf{SP}}$ (6-NO_2), irradiation of such materials with UV increases surface hydrophilicity by raising the concentration of polar merocyanine relative to ring-closed colorant, altering surface free energy, which in turn affects ease of wetting. While other stimuli can induce such change, light has the advantage of being non-invasive and precise. This level of control has also driven interest in the design of SP-based components that direct flows of liquids in channels within microfluidic devices [88, 89]. Use of photochromism in this manner could simplify their construction and lower cost, enabling potential mass use in a wide range of applications such as rapid assay of biological samples. Although altering channel wettability has been proposed as one mode of controlling flow ('photocapillarity') [110], research with SP units continues to focus primarily on modulating the swelling of hydrogel-based microvalves into which they have been incorporated. Their presence allows liquid flow within the device to be directed and regulated by localized irradiation with LED light. Without irradiation in the polymer, SP units tend to transform into hydrophilic protonated yellow merocyanines, which encourages the matrix itself to expand by absorbing moisture, thereby leading to restriction of flow. Irradiation with visible light drives the SP unit back to its more hydrophobic (non-protonated) colorless ring-closed form, causing the hydrogel to expel water and contract, thereby opening the microvalve. Proof of concept has also been demonstrated for SP-containing polymers to control degree of mixing of fluids as well as to release or take up metal ions on demand from them.

SP-based systems also offer spatiotemporal control over chemical processes. They enable light to govern when and where polymerization takes place [90]: proof of concept has been demonstrated by means of a negatively photochromic SP whose electrically neutral merocyanine form is ring-closed with visible light, releasing protons to induce polymerization of lactone monomer. Removal of irradiation led to reversion to more basic ring-opened species and slowing of polymerization. Electrochemical systems may also be modulated with light, i.e. photocontrolling redox reactions by light-actuated switching between spiropyran and protonated photomerocyanine forms which in turn influences whether reactants interact with electrode surfaces onto which the colorant has been adsorbed. Conversely biomolecules have been labelled with SP units whose status controls the redox reaction. While proof of concept has been demonstrated with many types of system, they remain "scientific toys" and much work remains to translate them to practical applications [92].

The size of micelles can be controlled by reversible 'photoloading' of SP **6** [111]. Irradiation leads to generation of photomerocyanine species $\mathbf{6}^{\mathbf{MC}}$ with high $\mathbf{1b}^{z}$ character which have affinity with the zwitterionic core of the micelles in the system. They diffuse inwards, swelling micelles up to twice their original diameter. The polar interior stabilizes the ring-opened form so that the colorant can be expelled on demand, and the size of the micelle shrunk, by irradiation with visible light to regenerate the non-polar ring-closed form $\mathbf{6}^{\mathbf{SP}}$. In principle, this strategy could enable the manipulation of properties and transport of materials at the nanometer scale, including in biological systems, which are covered next.

63.4.4 Biotechnology

The photocontrol of polypeptide conformation, enzymatic activity, and transport of bioactive substances through membranes hold great interest to those working towards new means of identifying and treating medical conditions [81, 82]. SP-based photoswitching, like that of other photochromic families, is of interest in facilitating them [112]. Remote regulation with light offers the potential advantages of non-invasiveness and precise spatial control. Biomolecules combined with SP units are also regarded as a resource for applications outside of the life sciences that require light responsive materials. The already large, and rapidly growing, body of published work acts as testament to both approaches. It includes the creation of nanostructures that enable light-actuated release of materials, such as bioactives, from nanoparticles [113], for example by altering membrane permeability. Of particular interest for clinical scenarios is the potential of SP-based systems exhibiting negative photochromism. In principle, they could obviate use of toxic ultraviolet light by enabling visible light to effect photoisomerization. As mentioned in Section 63.4.2, SP colorants form the subject of efforts to develop biosensors. For examples, the propensity of SP photomerocyanine species to intercalate with double-stranded RNA, in contrast to their ring-closed isomers, could serve as a means of viral detection or the determination of the effectiveness of a chemotherapeutic treatment [114].

63.5 Summary

Around seventy years have passed since the scientific community began its investigation into the photochromism of SPs. Its interest was first piqued by their potential as light-actuated functional colorants. However, even now, new insights into the photochromism of this class of dye continue to be gained. Arguably more research exploiting the SP unit as a switch is being conducted than ever. The continuing efforts to develop novel applications stretch across many disciplines. Members of the SP class did not prove robust enough for use in ophthalmic lenses, and thus hold only minor commercial significance. They remain at the forefront of research to develop new technologies in fields ranging from nanomachinery to optoelectronics to biomedicine. Translation of these development strands into products for mass consumption still seems many years away. Nevertheless, if any one of these endeavors bore fruit and made it to a mainstream market, then the SP class would at long last become the most industrially important type of photochromic colorant.

References

1. Hirschberg Y. The photochemical memory. New Scientist. 1960;7:1423–5. https://books.google.co.uk/books?id=RRVRuUO5vFIC&pg=PA1379&source=gbs_toc&cad=2#v=onepage&q&f=false Accessed: 9 Feb 2021.
2. Towns A. Photochromic Dyes. in press (10.1015/psr-2020-0191).
3. Aiken S, Edgar RJ, Gabbutt CD, Heron BM, Hobson PA. Negatively photochromic organic compounds: exploring the dark side. Dyes Pigm. 2018;149:92–121.
4. Barachevsky VA. Negative photochromism in organic systems. Rev J Chem. 2017;7:334–71.
5. Zakhs ÉR, Martynova VM, Éfros LS. Synthesis and Properties of Spiropyrans that are capable of reversible opening of the pyran ring (Review). Chem Het Comp. 1979:351–72.
6. Samat A, De Keukeleire D, Guglielmetti R. Synthesis and spectrokinetic properties of photochromic spiropyrans. Bull Soc Chim Belg. 1991;100:679–700.
7. Guglielmetti R. Chapter 8 ("4n+2 Systems: Spiropyrans"). In: Dürr H, Bouas-Laurent H, editors. Photochromism molecules and systems. Revised ed. Amsterdam: Elsevier, 2003:314–466.
8. Lukyanov BS, Lukyanova MB. Spiropyrans: synthesis, properties and application (review). Chem Het Comp. 2005;41:281–311.
9. Bertelson RC. Reminiscences about organic photochromics. Mol Cryst Liq Cryst. 1994;246:1–8.
10. Aldoshin SM. Spiropyrans: structural features and photochemical properties. Russ Chem Rev. 1990;59:663–85.
11. Kholmanskii AS, Dyumaev KM. The photochemistry and photophysics of spiropyrans. Russ Chem Rev. 1987;56:136–51.
12. Minkin VI. Photo-, Thermo-, Solvato-, and electrochromic spiroheterocyclic compounds. Chem Rev. 2004;104:2751–76.
13. Dietz F, El'tsov AV. Chapter 1 ("Theoretical Studies of the Photochromism of Organic Compounds"). In: El'tsov AV, editor. Organic photochromes. New York: Consultants Bureau/Plenum, 1990:1–44.
14. Aldoshin S. Chapter 7 ("Structural Studies by X-Ray Diffraction"). In: Crano JC, Guglielmetti RJ, editors. Organic photochromic and thermochromic compounds. vol. 2. New York: Plenum, 1998:297–355.
15. Seiler VK, Tumanov N, Robeyns K, Wouters J, Champagne B, Leyssens T. A structural analysis of spiropyran and spirooxazine compounds and their polymorphs. Crystals. 2017;7:84. DOI:10.3390/cryst7030084.
16. Lenoble C, Becker RS. Photophysics, photochemistry and kinetics of indolinospiropyran derivatives and an indolinospiropyran. J Photochem. 1986;34:83–8.
17. Mustroph H. Cyanine dyes. Phys Sci Rev. 2020;5:20190145. DOI:10.1515/psr-2019-0145.

18. Mustroph H. Polymethine dyes. Phys Sci Rev. 2020;5:20190084. DOI:10.1515/psr-2019-0084.
19. Krongauz VA. Chapter 21 ("Environmental Effects on Organic Photochromic Systems"). In: Dürr H, Bouas-Laurent H, editors. Photochromism molecules and systems. Revised ed. Amsterdam: Elsevier, 2003:793–821.
20. Such G, Evans RA, Yee LH, Davis TP. Factors influencing photochromism of spiro-compounds within polymeric matrices. J Macromol Sci C Polym Rev. 2003;43:547–79.
21. Hobley J, Lear MJ, Fukumura H. Chapter 8 ("Photo-Switching Spiropyrans and Related Compounds"). In: Ramamurthy V, Schanze KS, editors. Photochemistry of organic molecules in isotropic and anisotropic media. New York: Marcel Dekker, 2003:353–404.
22. Kortekaas L, Browne WR. The evolution of spiropyran: fundamentals and progress of an extraordinarily versatile photochrome. Chem Soc Rev. 2019;48:3406–24.
23. International Union of Pure and Applied Chemistry Organic and Biomolecular Chemistry Division. Glossary of terms used in photochemistry, 3rd edition (IUPAC Recommendations 2006). Pure Appl Chem. 2007;79:293–465.
24. Bénard S, Yu P. New spiropyrans showing crystalline-state photochromism. Adv Mater. 2000;12:48–50.
25. Yurieva EA, Aldoshin SM. Spiropyran salts and their neutral precursors: synthesis, crystal structure, photochromic transformations in solutions and solid state. Org Photonics Photovolt. 2015;3:42–53.
26. Funasako Y, Ason M, Takebayashi J-I, Inokuchi M. Solid-state photochromism of salts of cationic spiropyran with various anions: a correlation between reaction cavity volumes and reactivity. Cryst Growth Des. 2019;19:7308–14.
27. Godsi O, Peskin U, Kapon M, Natan E, Eichen Y. Site effects in controlling the chemical reactivity in crystals: solid-state photochromism of N-(n-propyl)nitrospiropyrane. Chem Commun. 2001;2132–3.
28. Harada J, Kawazoe Y, Ogawa K. Photochromism of spiropyrans and spirooxazines in the solid state: low temperature enhances photocoloration. Chem Commun. 2010;46:2593–5.
29. Lukyanov BS, Metelitsa AV, Voloshin NA, Alexeenko YS, Lukyanova MB, Vasilyuk GT, et al. Solid state photochromism of spiropyrans. Int J Photoenerg. 2005;7:17–22.
30. Lafleur SSD, Severn JR, Verpaalen RC, Schenning AP, Bastiaansen CW. Rewritable optical patterns in light-responsive ultrahigh molecular weight polyethylene. ACS Appl Polym Mater. 2019;1:392–6.
31. Abdollahi A, Roghani-Mamaqani H, Razavia B. Stimuli-chromism of photoswitches in smart polymers: recent advances and applications as chemosensors. Prog Polym Sci. 2019;98:101149.
32. For example: Tokyo Chemical Industry, Photochromic Dyes. www.tcichemicals.com/eshop/en/gb/category_index/12995/ Accessed: 9 Feb 2021. Merck, Photochromic and Thermochromic Dyes, www.sigmaaldrich.com/materials-science/material-science-products.html?TablePage=9541081 Accessed: 9 Feb 2021.
33. Koelsch CF, Workman WR. Some Thermochromic Spiropyrans. J Am Chem Soc. 1952;74:6288–9.
34. Keum S-R, Ku B-S, Shin J-T, Ko JJ, Buncel E. Stereoselective formation of diconsdensed spiropyran product obtained from the reaction of excess Fischer base with salicaldehydes: first full characterization of X-ray crystal structure analysis of a DC·acetone crystal. Tetrahedron. 2005;61:6720–5.
35. Keum S-R, Kazmaier PM, Cheon K-S, Manderville RA, Buncel E. The structural identification of dicondensed products derived from the reaction of excess Fischer's base with salicylaldehydes. Bull Korean Chem Soc. 1996;17:391–3.

36. Bertelson RC. Chapter 1 ("Spiropyrans"). In: Crano JC, Guglielmetti RJ, editors. Organic photochromic and thermochromic compounds. vol. 1. New York: Plenum, 1998:11–83.
37. Bertelson RC. Chapter 3 ("Photochromic process involving heterolytic cleavage"). In: Brown GH, editor. Photochromism – Techniques of chemistry. vol. III. London: Wiley, 1971:45–431.
38. Kellmann A, Tfibel F, Dubest R, Levoir P, Aubard J, Pottier E, et al. Photophysics and kinetics of two photochromic indolinospirooxazines and one indolinospironaphthopyran. J Photochem Photobiol A Chem. 1989;49:63–73.
39. Lee YS, Kim JG, Huh YD, Kim MK. Thermochromism of spiropyran and spirooxazine derivatives. J Korean Chem Soc. 1994;38:864–72.
40. Feng K-C, Griffiths J. Thermochromic and photochromic properties of some new spiropyran systems. Adv Col Sci Technol. 2001;4:12–20.
41. Vázquez-Mera N, Roscini C, Hernando J, Ruiz-Molina D. Liquid-filled capsules as fast responsive photochromic materials. Adv Optical Mater. 2013;1:631–6.
42. Chu NY. Chapter 10 ("4n+2 Systems: Spirooxazines"). In: Dürr H, Bouas-Laurent H, editors. Photochromism molecules and systems. Revised ed. Amsterdam: Elsevier, 2003:493–509.
43. Wojtyk JT, Wasey A, Kazmaier PM, Hoz S, Buncel E. Thermal reversion mechanism of N-functionalized merocyanines to spiropyrans: a solvatochromic, solvatokinetic, and semiempirical study. J Phys Chem A. 2000;104:9046–55.
44. Keum S-R, Hur M-S, Kazmaier PM, Buncel E. Thermo- and photochromic dyes: indoline-benzospiropyrans. Part 1. UV-VIS spectroscopic studies of 1,3,3-spiro(2H-1-benzopyran-2,2′-indolines) and the open-chain merocyanine forms; solvatochromism and medium effects on spiro ring formation. Can J Chem. 1991;69:1940–7.
45. Rosario R, Gust D, Hayes M, Springer J, Garcia AA. Solvatochromic study of the microenvironment of surface-bound spiropyrans. Langmuir. 2003;19:8801–6.
46. Chibisov AK, Görner H. Photoprocesses in spiropyran-derived merocyanines. J Phys Chem A. 1997;101:4305–12.
47. Sueishi Y, Ohcho M, Nishimura N. Kinetic studies of solvent and pressure effects on thermochromic behavior of 6-nitrospiropyran. Bull Soc Chem Jpn. 1985;58:2608–13.
48. Zhang S, Zhang Q, Ye B, Li X, Zhang X, Deng Y. Photochromism of spiropyran in ionic liquids: enhanced fluorescence and delayed thermal reversion. J Phys Chem B. 2009;113:6012–9.
49. Görner H. Photoprocesses in spiropyrans and their merocyanine isomers: Effects of temperature and viscosity. Chem Phys. 1997;222:315–29.
50. Reichardt C. Solvents and solvent effects in organic chemistry. vols 339–405. 2nd ed. Weinheim: VCH; 1988.
51. Towns A. Spirooxazine Dyes. Phys Sci Rev. 2020;5:20200013. 10.1515/psr-2020-0013.
52. Nuernberger P, Ruetzel S, Brixner T. Multidimensional electronic spectroscopy of photochemical reactions. Angew Chem Int Ed. 2015;54:11368–86.
53. Mustroph H. Dyes: quantum chemical calculation of electronic spectra. Phys Sci Rev. 2019;4:20190040. 10.1515/psr-2019-0040.
54. Ruetzel S, Diekmann M, Nuernberger P, Walter C, Engels B, Brixner T. Multidimensional spectroscopy of photoreactivity. Proc Nat Acad Sci. 2014;111:4764–9.
55. Tian W, Tian J. An insight into the solvent effect on photo-, solvato-chromism of spiropyran through the perspective of intermolecular interactions. Dyes Pigm. 2014;105:66–74.
56. Gautron R. Photochromisme des indolinospiropyranes. IV – Étude de la dégradation par voie physique. Relation avec la structure. Bull Soc Chim Fr. 1968;3190–200.
57. Malatesta V. Chapter 2 ("Photodegradation of Organic Photochromes"). In: Crano JC, Guglielmetti RJ, editors. Organic photochromic and thermochromic compounds. vol. 2. New York: Plenum, 1999:65–166.

58. Seto J. Chapter 7 ("Photochromic Dyes"). In: Matsuoka M, editor. Infrared absorbing dyes. New York: Springer, 1990:71–88.
59. Pugachev AD, Ozhogin IV, Lukyanova MB, Lukyanov BS, Rostovtseva IA, Dorogan IV, et al. Visible to near-IR switches based on photochromic indoline spiropyrans with a conjugated cationic fragment. Spectrochim Acta A Mol Biomol Spectr. 2020;230:118041.
60. Tyurin RV, Lukyanov BS, Chernyshev AV, Malay VI, Kozlenko AS, Tkacheva NS, et al. Effect of bulky substituents on the photochromic properties of indoline spiropyrans containing an annelated aromatic or heteroaromatic fragment. Doklady Chem. 2016;470:268–73.
61. Day JH. Thermochromism. Chem Rev. 1963;63:65–80.
62. Samat A, Lokshin V. Chapter 10 ("Thermochromism of Organic Compounds"). In: Crano JC, Guglielmetti RJ, editors. Organic photochromic and thermochromic compounds. vol. 2. New York: Plenum, 1998:415–66.
63. Wizinger R, Wennig H. Über intramolekulare Ionisation. Helv Chim Acta. 1940;23:247–71.
64. Bertelson RC. Chapter 10 ("Applications of photochromism"). In: Brown GH, editor. Photochromism – techniques of chemistry. vol. III. London: Wiley, 1971:733–840.
65. Jackson G. The Properties of Photochromic Materials. Optica Acta 1969;16:1–16.
66. Fischer E, Hirschberg Y. Formation of coloured forms of spirans by low-temperature irradiation. J Chem Soc. 1952;4522–4.
67. Hirschberg Y. Reversible Formation and Eradication of Colors by Irradiation at Low Temperatures. A Photochemical Memory Model. J Am Chem Soc. 1956;78:2304–12.
68. Kumar S, Soni S, Danowski W, Leach IF, Faraji S, Feringa BL, et al. Eliminating fatigue in surface-bound spiropyrans. J Phys Chem C. 2019;123:25908–14.
69. Florea L, Diamond D, Benito-Lopez F. Photo-responsive polymeric structures based on spiropyran. Macromol Mater Eng. 2012;297:1148–59.
70. Klajn R. Spiropyran-based dynamic materials. Chem Soc Rev. 2014;43:148–84.
71. Ma S, Ting H, Ma Y, Zheng L, Zhang M, Xiao L, et al. Smart photovoltaics based on dye-sensitized solar cells using photochromic spiropyran derivatives as photosensitizers. AIP Adv. 2015;5:057154.
72. Doan HN, Tsuchida H, Iwata T, Kinashi K, Sakai W, Tsutsumi N, et al. Fabrication and photochromic properties of Forcespinning® fibers based on spiropyran-doped poly(methyl methacrylate). RSC Adv. 2017;7:33061.
73. Bao B, Fan J, Wang W, Yu D. Photochromic cotton fabric prepared by spiropyran-terminated water polyurethane coating. Fibers Polym. 2020;21:733–42.
74. Abdollahi A, Sahandi-Zangabad K, Roghani-Mamaqani H. Rewritable anticounterfeiting polymer inks based on functionalized stimuli-responsive latex particles containing spiropyran photoswitches: reversible photopatterning and security marking. ACS Appl Mater Interfaces. 2018;10:39279–92.
75. Towns A. Naphthopyran Dyes. Phys Sci Rev. 2020;5:20190085. DOI:10.1515/psr-2019-0085.
76. Barachevsky VA. Photochromic spirocompounds and chromenes for sensing metal ions. Rev J Chem. 2013;3:52–94.
77. Zhang X, Hou L, Samorì P. Coupling carbon nanomaterials with photochromic molecules for the generation of optically responsive materials. Nat Commun. 2016;7:11118.
78. Berkovic G, Krongauz V, Weiss V. Spiropyrans and spirooxazines for memories and switches. Chem Rev. 2000;100:1741–53.
79. Vijayamohanan H, Palermo EF, Ullal CK. Spirothiopyran-based reversibly saturable photoresist. Chem Mater. 2017;29:4754–60.
80. Raymo FM, Giordani S. All-optical processing with molecular switches. Proc Nat Acad Sci. 2002;99:4941–4.

81. Hao Y, Meng J, Wang S. Photo-responsive polymer materials for biological applications. Chin Chem Lett. 2017;28:2085–91.
82. Szymański W, Beierle JM, Kistemaker HA, Velema WA, Feringa BL. Reversible photocontrol of biological systems by the incorporation of molecular photoswitches. Chem Rev. 2013;113:6114–78.
83. Paramonova SV, Lokshin V, Fedorova OA. Spiropyran, chromene or spirooxazine ligands: Insights into mutual relations between complexing and photochromic properties. J Photochem Photobiol C Photochem Rev. 2011;12:209–36.
84. Petriashvili G, Devadze L, Chanishvili A, Zurabishvili C, Sepashvili N, Ponjavidze N, et al. Spiropyran doped rewritable cholesteric liquid crystal polymer film for the generation of quick response codes. Optical Mater Express. 2018;8:3708–16.
85. Pei JV, Heng S, De Ieso ML, Sylvia G, Kourghi M, Nourmohammadi S, et al. Development of a Photoswitchable Lithium-Sensitive probe to analyze nonselective cation channel activity in migrating cancer cells. Mol Pharmacol. 2019;95:573–83.
86. Ali AA, Kharbash R, Kim Y. Chemo- and biosensing applications of spiropyran and its derivatives – A review. Analytica Chim Acta. 2020;1110:199–223.
87. Kim D, Park SY. Multicolor fluorescence photoswitching: color-correlated versus color-specific switching. Adv Optical Mater. 2018;6:1800678.
88. Ter Schiphorst J, Saez J, Diamond D, Benito-Lopez F, Schenning AP. Light-responsive polymers for microfluidic applications. Lab Chip. 2018;18:699–709.
89. Dunne A, Francis W, Delaney C, Florea L, Diamond D. Stimuli-controlled fluid control and microvehicle movement in microfluidic channels. In: Hashmi S, editor. Reference module in materials science and materials engineering. Amsterdam: Elsevier, 2017.
90. Zayas MS, Dolinski ND, Self JL, Abdilla A, Hawker CJ, Bates CM, Read de Alaniz J. Tuning merocyanine photoacid structure to enhance solubility and temporal control: application in ring opening polymerization. ChemPhotoChem. 2019;3:467–72.
91. Wagner N, Theato P. Light-induced wettability changes on polymer surfaces. Polymer. 2014;55:3436–53.
92. Katz E. Modified electrodes and electrochemical systems switchable by light signals. Electroanalysis. 2018;30:759–97.
93. Fu L-N, Leng B, Li Y-S, Gao X-K. Photoresponsive organic field-effect transistors involving photochromic molecules. Chin Chem Lett. 2016;27:1319–29.
94. Bertarelli C, Bianco A, Castagna R, Pariani G. Photochromism into optics: opportunities to develop light-triggered optical elements. J Photochem Photobiol C Photochem Rev. 2011;12:106–25.
95. Wang L, Li Q. Photochromism into nanosystems: towards lighting up the future nanoworld. Chem Soc Rev. 2018;47:1044–97.
96. Bisoyi HK, Li Q. Light-driven liquid crystalline materials: from photo-induced phase transitions and property modulations to applications. Chem Rev. 2016;116:15089–166.
97. Sahoo PR, Prakash K, Kumar S. Light controlled receptors for heavy metal ions. Coord Chem Rev. 2018;357:18–49.
98. Kinashi K, Miyamae Y, Nakamura R, Sakai W, Tsutsumi N, Yamane H, et al. A spiropyran-based X-ray sensitive fiber. Chem Commun. 2015;51:11170–3.
99. Tsuchida H, Nakamura R, Kinashi K, Sakai W, Tsutsumi N, Ozaki M, et al. Radiation-induced colour changes in a spiropyran/BaFCl:Eu2+/polystyrene composite film and nonwoven fabric. New J Chem. 2016;40:8658–63.
100. Asai K, Koshimizu M, Fujimoto Y, Asai K. Isomerization behavior of spiropyran-based compounds upon X-ray irradiation. Radiat Meas. 2017;106:166–9.

101. Sadeghi K, Yoon J-Y, Seo J. Chromogenic polymers and their packaging applications: a review. Polym Rev. 2020;60:442–92.
102. Wang S, Liu X, Yang M, Zhang Y, Xiang K, Tang R. Review of time temperature indicators as quality monitors in food packaging. Packag Technol Sci. 2015;28:839–67.
103. Kreyenschmidt J, Christiansen H, Hübner A, Raab V, Petersen B. A novel photochromic time–temperature indicator to support cold chain management. Int J Food Sci Technol. 2010;45:208–15.
104. Biegańska M. Shelf-life monitoring of food using time-temperature indicators (TTI) for application in intelligent packaging. Pol J Commodity Sci. 2017;51:75–85.
105. Brizio AP, Prentice C. Use of smart photochromic indicator for dynamic monitoring of the shelf life of chilled chicken based products. Meat Sci. 2014;96:1219–26.
106. Beyer KB, Clausen-Schaumann H. Mechanochemistry: the mechanical activation of covalent bonds. Chem Rev. 2005;105:2921–48.
107. Potisek SL, Davis DA, Sottos NR, White SR, Moore JS. Mechanophore-Linked Addition Polymers. J Am Chem Soc. 2007;129:13808–9.
108. Li M, Zhang Q, Zhou Y-N, Zhu S. Let spiropyran help polymers feel force! Prog Polym Sci. 2018;79:26–39.
109. Katsonis N, Lubomska M, Pollard MM, Feringa BL, Rudolf P. Synthetic light-activated molecular switches and motors on surfaces. Prog Surf Sci. 2007;82:407–34.
110. Rosario R, Gust D, Hayes M, Jahnke F, Springer J, Garcia AA. Photon-Modulated Wettability Changes on Spiropyran-Coated Surfaces. Langmuir. 2002;18:8062–9.
111. Niskanen J, Vapaavuori J. Photobreathing Zwitterionic Micelles. ChemSystemsChem. 2019;1: e1900018.
112. Cheng H, Yoon J, Tian H. Recent advances in the use of photochromic dyes for photocontrol in biomedicine. Coord Chem Rev. 2018;372:66–84.
113. Jia S, Fong W-K, Graham B, Boyd BJ. Photoswitchable molecules in long-wavelength light-responsive drug delivery: from molecular design to applications. Chem Mater. 2018;30:2873–87.
114. Ali AA, Kang M, Kharbash R, Kim Y. Spiropyran as a potential molecular diagnostic tool for double-stranded RNA detection. BMC Biomedical Eng. 2019;1:6.

Heinz Mustroph

64 Streptocyanine dyes

Abstract: The simplest possible type of polymethine dye consists of an odd-numbered polymethine chain sandwiched between two terminal nitrogen atoms, which do not form part of an unsaturated heterocyclic ring. Originally, this type of colorant was termed streptopolymethine dyes. For systematic reasons they are usually called streptocyanine dyes today. Their color is mainly determined by the lengths of their polymethine chains. Owing to their simple structure they are the favorite compounds for investigation of fundamental structure–property relationships. In the form of dianil hydrochlorides mainly streptocyanines are used to prepare cyanine, hemicyanine, merocyanine, oxonol, pyrylium and thiopyrylium dyes.

Keywords: equilibrium bond lengths, electronic structure, polymethine chain, streptocyanine dyes, streptopolymethine dyes, streptostatic chromophore, strepto vinylene group

64.1 Fundamentals

In the early years of synthetic organic dyes, their chemical structures were studied with many experiments. In parallel, empirical theories on structure–color relationships were developed enabling synthesis of new dyes in a more targeted manner. In addition to the known chromophores in *Witt's* chromophore theory like e.g. keto >C = O and azo group –N = N–, Felicia Zwayer and Stanislaus von Kostanecki considered also the vinylene group –CH = CH– as a chromophore. They divided the chromophores into two groups according to their position in "*cyclostatic*" (chromophores are in a ring like e.g. in 1,4-benzoquinone **1**) and "*streptostatic*" (chromophores are in a chain like e.g. in 2-hydroxychalcone **2**) [1].

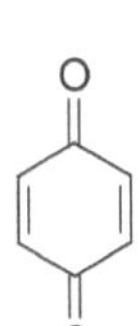

1

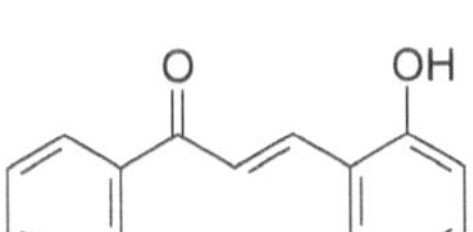

2

This article has previously been published in the journal Physical Sciences Reviews. Please cite as: H. Mustroph, Streptocyanine Dyes *Physical Sciences Reviews* [Online] 2021, 6. DOI: 10.1515/psr-2020-0198

https://doi.org/10.1515/9783110587128-064

Walter König adopted the term “streptostatic” in “streptostatic insertion of vinylene groups into a conjugated system” and labelled the vinylene groups of the polymethine chain “strepto-vinylene groups” [2, 3].

The simplest possible type of polymethine dyes consists of an odd-numbered polymethine chain between two terminal nitrogen atoms, which do not form part of an unsaturated heterocyclic ring and can be described by the general formula **3**,

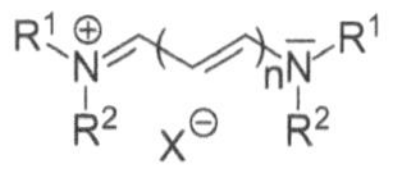

3

with e.g. $N(R^1)R^2 = NH_2$, NHC_2H_5, NHPh, $N(CH_3)_2$, $N(C_2H_5)$,

or saturated heterocyclic rings like e.g.

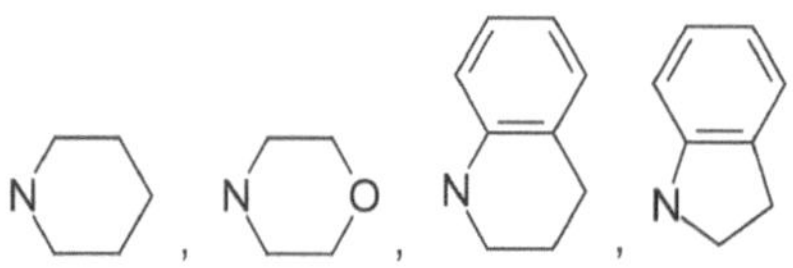

Therefore, König termed this type of colorants *streptopolymethine dyes* [2, 3]. Of course, these dyes are part of the polymethine family. However, they are also the simplest possible type of cyanine dye family. *Cyanine dyes* are defined as compounds that possess two terminal nitrogen atoms which form part of separate unsaturated heterocyclic rings that are in conjugation with a polymethine chain. Therefore, Oskar Riester called the dyes **3** *streptocyanine dyes* [4]. Currently both terms are used [5–9], which can lead to confusion. With regard to the terms *cyanine*, *apocyanine*, *hemicyanine* and *merocyanine dyes*, it would only be logical to use streptocyanine dyes, which is also done in the following text.

The color of these dyes is mainly determined by the lengths of the (strepto)polymethine chain and due to their simple structure they are the favorite compounds of dye chemists who are interested in the fundamental structure–property relationships.

Like all other polymethines the streptocyanine dyes are characterised by an odd number $2n + 3$ of π-centres and $2\,n + 4$ π-electrons (where n is the number of vinylene groups $-CH = CH-$). This special feature has marked impact on their electronic structure and thus their equilibrium structure. Especially series **3** with $N(R^1)R^2 = N(CH_3)_2$, was intensively investigated by ^{13}C NMR chemical shifts (Table 64.1), crystal structure analysis (Table 64.2) and absorption spectra (Table 64.3).

The π-charge density distribution along the polymethine chain alternates markedly. This is evidenced by the ^{13}C NMR chemical shifts δ (^{13}C) along the polymethine chain (Table 64.1).

Table 64.1: ^{13}C NMR chemical shifts δ (^{13}C) in ppm along the polymethine chain of streptocyanine dyes **3** with $N(R^1)R^2 = N(CH_3)_2$, measured in [D6]DMSO [10]. For numbering of the *C*-atoms see **4**.

n	*C*-atom	δ (^{13}C)		
	2	3	4	5
1	163.3	90.5		
2	159.4	106.6	165.3	
3	161.3	105.0	159.7	118.3

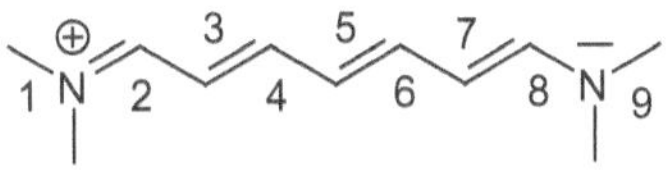

4

In streptocyanine dyes there is no additional influence of heterocyclic rings on the equilibrium structure of the electronic ground state. Therefore, they should come closest to the ideal polymethine model that predicts that all equilibrium bond lengths in symmetrical polymethine dyes along the conjugated chain are equal [6–9]. However, in reality the nitrogen–carbon and all carbon–carbon bond lengths do not have identical values. Also in the simple streptocyanine dyes **3** with $N(R^1)R^2 = N(CH_3)_2$ there is a small difference between the equilibrium bond lengths (Table 64.2). The differences in the carbon–carbon equilibrium bond lengths in symmetrical streptocyanine dyes are small, but they are not completely identical, which is mostly ignored in theoretical considerations.

Table 64.2: Equilibrium bond lengths R_e [pm] of streptocyanine dyes **3** with $N(R^1)R^2 = N(CH_3)_2$ from crystal structure analysis [11–13]. For numbering of the bonds see **4**.

n	Bond	R_e						
	N1-C2	C2-C3	C3-C4	C4-C5	C5-C6	C6-C7	C7-C8	C8-N9
1	132	140	139					C4-(N5) 134
2	131.2	136.5	142.7	138.8	139.9			C6-N(7) 132.3
3	131.2	136.2	137.4	138.2	137.4	140.7	136.4	130.7

Each additional vinylene group in polymethines gives a bathochromic shift of the 0–0 vibronic transition of about 100 nm (the so-called vinylene shift) [6–9]. In symmetrical cyanine and streptocyanine dyes, respectively, the absorption maximum λ_{max} corresponds to the 0–0 vibronic transition and can be used here as value for the 0–0 transition energy [6, 9, 14, 15].

The streptocyanine dyes **3** fit this rule perfectly (Table 64.3). In a logical continuation of the structural principle of streptocyanines, König classified the *diarylmethine dyes* **5** as phenylogous streptocyanine dyes [16].

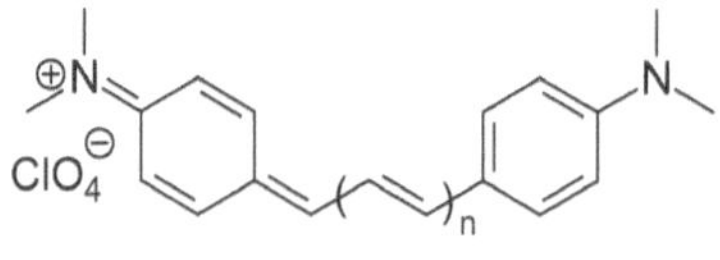

5

Table 64.3: Absorption maxima λ_{max} (nm) and molar absorption coefficients ε_{max} (10^{-3} M^{-1} cm^{-1}) of the streptocyanines **3** $N(R^1)R^2 = N(CH_3)_2$, X = ClO_4 in dichloromethane [17] and the absorption maxima λ_{max} (nm) of diarylmethine dyes **5** in acetic acid [18] with respect to the number of vinylene groups *n*.

n	3		*n*	5
	λ_{max}	ε_{max}		λ_{max}
1	312	64.5		
2	416	119.5		
3	519	207		
4	625	295	0	610
5	734	353	1	693
6	848	(220)	2	790
			3	883

Basically, streptocyanine and diarylmethine dyes with the same lengths of the conjugated system show comparable λ_{max} values (Table 64.3). In addition, the vinylene shift also appears here, but the diarylmethine dyes exhibit a slightly reduced vinylene shift in comparison with the streptocyanine dyes (Table 64.3). König explained this observation by likening part of the chromophore to a "forked conjugated system" due to the phenyl rings [16]. However, cyanine dyes with longer polymethine chain lengths tend to exhibit broadened absorption bands and lower ε values. The reason is the presence of a complex mixture of geometric conformers and isomers, leading to loss of molecular planarity, which influences λ_{max} values. These minor deviations do not undermine the status of the diarylmethine dyes as streptocyanine dyes.

In comparison with **6** a phenyl ring as N-substituent has influence on the conjugated system and causes a bathochromic shift of about 40 nm as exemplified by dye **7** [19]. A (formal) ring closure of the phenyl rings with the polymethine chain, creating aza heterocycles at each end, leads to the cyanine dyes and a substantial bathochromic shift as exemplified by dye **8** [19]. The (formal) ring closure by “cyclostatic” vinylene groups leads to the cyanine dye **9** with an additional bathochromic shift [20].

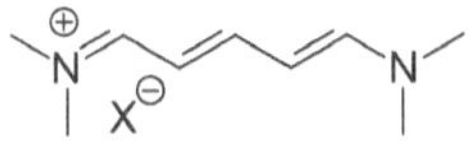

6

λ_{max} = 411 nm
ε_{max} = 93.3 10^{-3} M^{-1} cm^{-1}

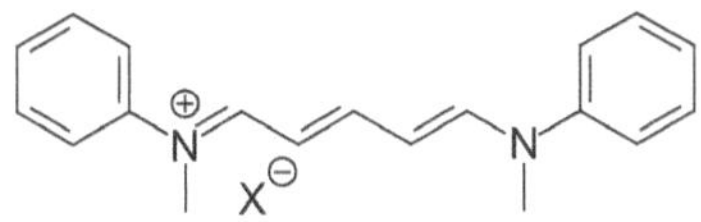

7

λ_{max} = 450 nm
ε_{max} = 91.2 10^{-3} M^{-1} cm^{-1}

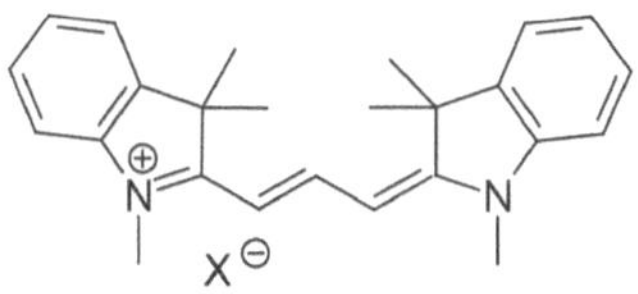

8

λ_{max} = 544 nm
ε_{max} = 131.8 10^{-3} M^{-1} cm^{-1}

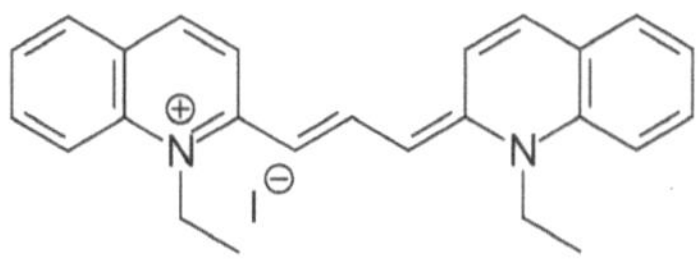

9

λ_{max} = 611 nm
ε_{max} = 197 10^{-3} M^{-1} cm^{-1}

However a ring closure of the substituents $N(R^1)R^2$ to form saturated rings like e.g. in **10** and **11** retains the character of streptocyanine dyes and has little influence on the spectroscopic properties [19].

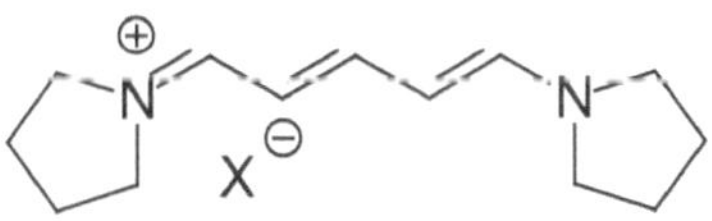

10

λ_{max} = 420 nm
ε_{max} = 114.8 10^{-3} M^{-1} cm^{-1}

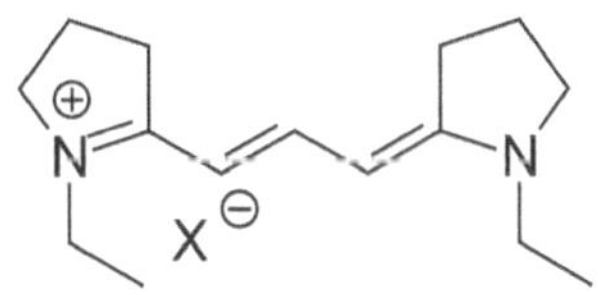

11

λ_{max} = 426 nm
ε_{max} = 125.8 10^{-3} M^{-1} cm^{-1}

64.2 History of the streptocyanine dyes

Ludwig Claisen synthesized the first streptocyanine compound **12** by the reaction of propargyl aldehyde with aniline and subsequent treatment with aniline hydrochloride and in one step by the reaction of propargyl aldehyde diethylacetal with aniline hydrochloride [21].

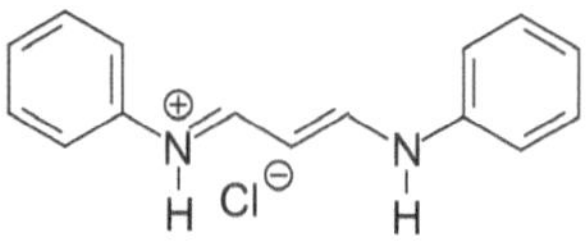

12

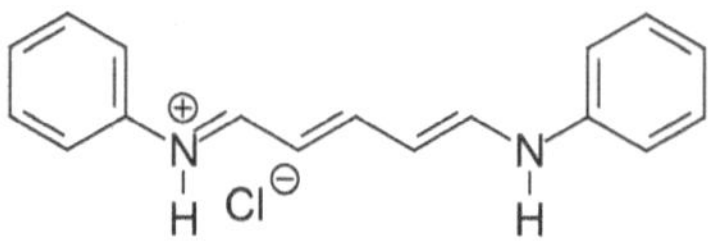

13

However, Claisen did not call **12** "streptocyanine". The crystals show a green-yellow color and the compound was not considered at the time to be a dye.

Theodor Zincke converted N-(2,4-dinitrophenyl)pyridinium chloride (today called Zincke salt) upon treatment with aniline to **13** – the first colored streptocyanine [22].

The glutacondianile hydrochloride **13** was independently synthesized by König using bromcyan for opening the pyridine-ring [23]. He realized that the stability of **13** is too low for application as a dye. Some years later König and Becker found that secondary aromatic amines, as in the case of dye **7**, lead to more stable dyes especially when the side chain forms part of an aliphatic ring, as exemplified by dye **14** [24].

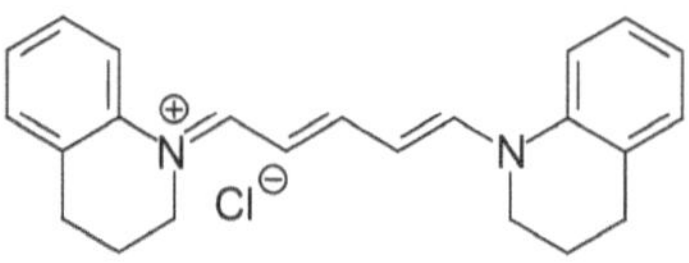

14

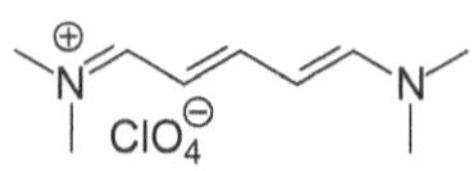

15

Using perchlorate as anion König and Regner were able to synthesize and isolate the first purely aliphatic streptocyanine **15** [25]. With it they opened up the possibility of synthesizing a wide range of purely aliphatic streptocyanine dyes [17].

64.3 General synthetic routes to streptocyanine dyes

There are a lot of synthetic methods to create streptocyanine dyes and here the most common shall be briefly described based on their starting materials.

Usually monomethine streptocyanine compounds **3** ($n = 0$) are prepared by treating triethyl orthoformiate **16** with a mixture of the corresponding amine and amine quaternary salt [5–8, 16, 25, 26]. Instead of triethyl orthoesters the trimethyl orthoesters can be used, too. Furthermore, for chain substituted monomethine streptocyanines the orthoester of the corresponding acids are used like e.g. triethyl orthoacetate, orthopropionate or orthobenzoate.

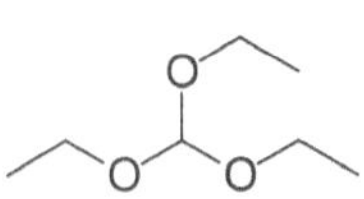

16

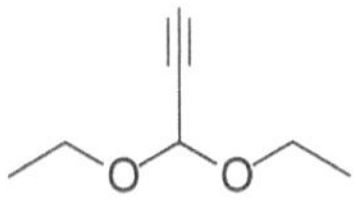

17

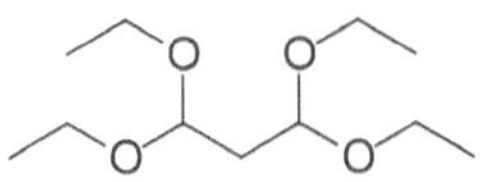

18

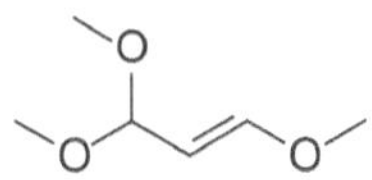

19

For the preparation of trimethine streptocyanine compounds **3** ($n = 1$) Claisen treated propargylaldehyde with aniline and subsequent with aniline hydrochloride. His second described method to use propargylaldehyde diethyl acetal **17** is the more efficient method [21].

Even the alternative starting material malondialdehyde is not very suitable for the preparation of streoptocyanines. It is unstable and condenses with itself. Also here, the acetal malondialdehyde bis (diethyl acetal) **18** is the starting material of the choice [27].

Corresponding to **16**–**18** the dimethyl acetal of 3-methoxy-acrolein (1,3,3-trimethoxy-1-propene) **19** is one possible raw material to synthesize pentamethine streptocyanine compounds **3** ($n = 2$). The more efficient method is the treatment of Zincke salts (N-(2,4- dinitrophenyl)pyridinium salts) with primary or secondary amines [22, 28].

The aminoformylation of cyclopentanone and indan-2-one via Vilsmeier-Haack reaction leads in a direct way to the streptocyanines **20** and **21** [29,30,31].

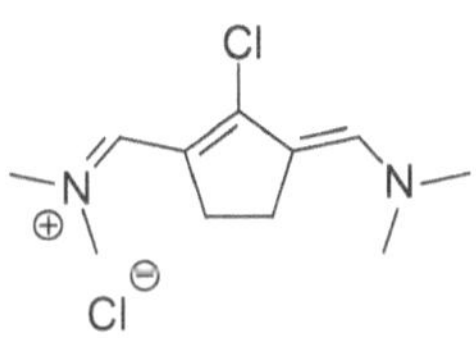

20

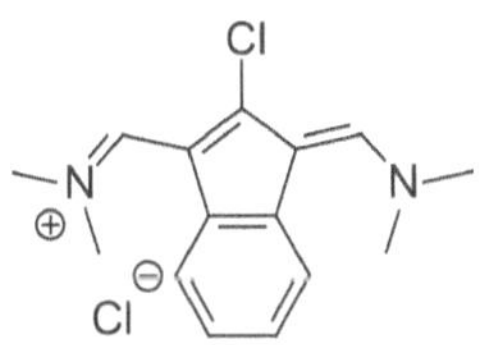

21

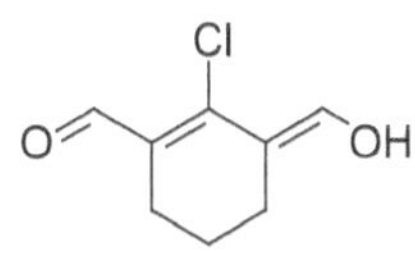

22

But cyclohexanone reacts with the Vilsmeier-Haack reagent to the 2-chloro-3-(hydroxymethylene)-1-cyclohexene-1-carboxaldehyde **22** [31], which can be easily converted into streptocyanine dyes.

64.4 Commercial uses of streptocyanine dyes

Streptocyanines were claimed as spectral sensitizers for silver halide photographic materials [4], but they did not evolve into real technical application as sensitizers, since a large number of cyanine and merocyanine dyes with valuable sensitizing properties were available.

The streptocyanine dyes **7**, **14**, **23** and some derivatives of them play a niche role in paper and textile dyeing.

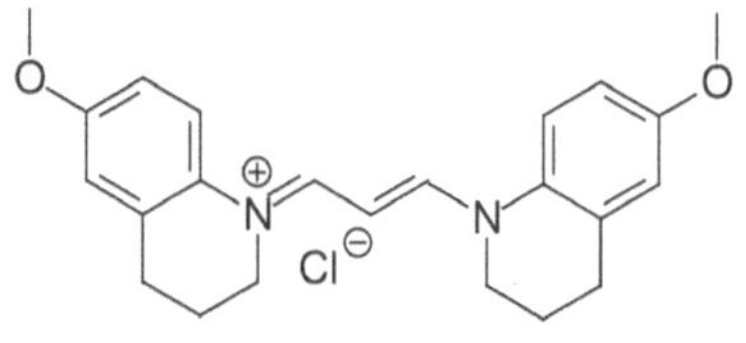

23

Mainly streptocyanines are used to prepare cyanine, hemicyanine, merocyanine, oxonol, pyrylium and thiopyrylium dyes. For this purpose they are used in the form of dianil hydrochlorides as exemplified by **24** (malondialdehyde dianil hydrochloride), **25** (dianil of the aldehyde **22**) and **26** (dianil of **20**).

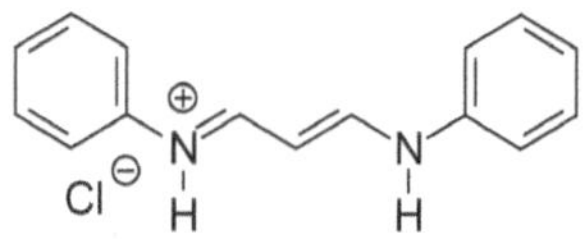

24

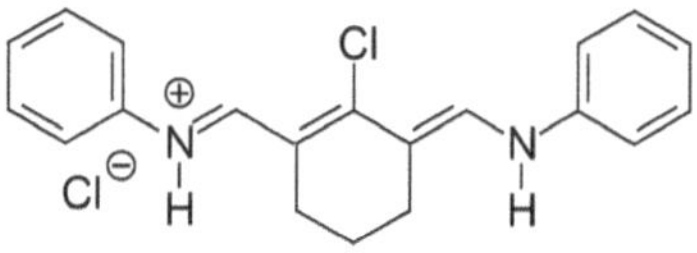

25

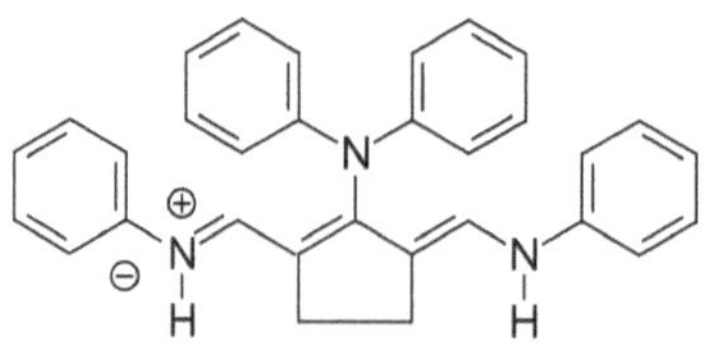

26

One advantage of the dianil hydrochlorides is the controllable reactivity for the synthesis of merocyanine and unsymmetrical cyanine dyes (Figure 1). A 2-methyl heterocyclic quaternary salt **27** is converted by a base into its methylene base which reacts with malondialdehyde dianil hydrochloride **24** to the intermediate **28**. In the next step the anilino group in **28** is converted into the acetanilido group in **29** to increase the reactivity for the following reaction step with the second 2-methyl heterocyclic quaternary salt **30** to the unsymmetrical cyanine dye **31**.

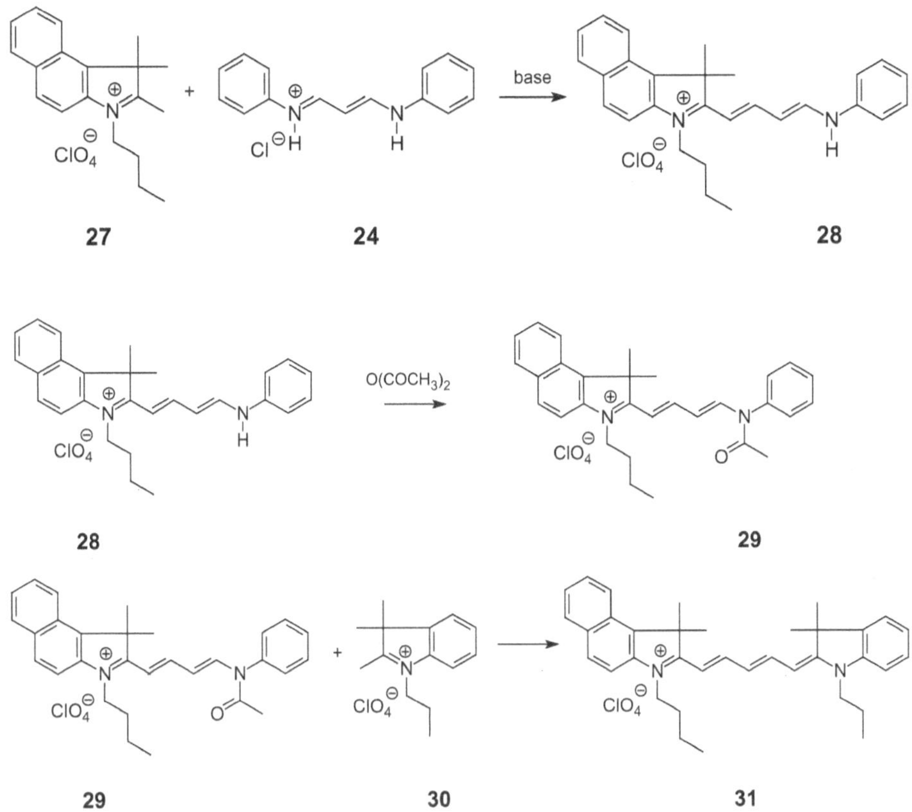

Figure 64.1 Reaction Scheme.

Symmetrical cyanine dyes are synthesized in one step using two mols of a 2-methyl heterocyclic quaternary salt, one mol of the streptocyanine **25**, a base and acetic anhydride in a suitable solvent, to prepare a dye as exemplified by **32** or using another 2-methyl heterocyclic quaternary salt and the streptocyanine **26** to get dye **33.**

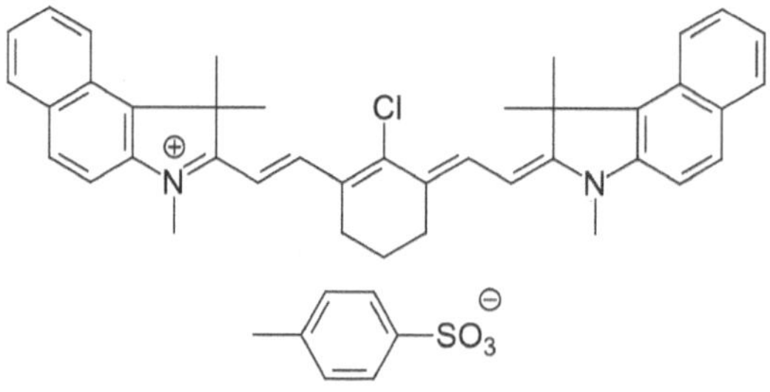

32

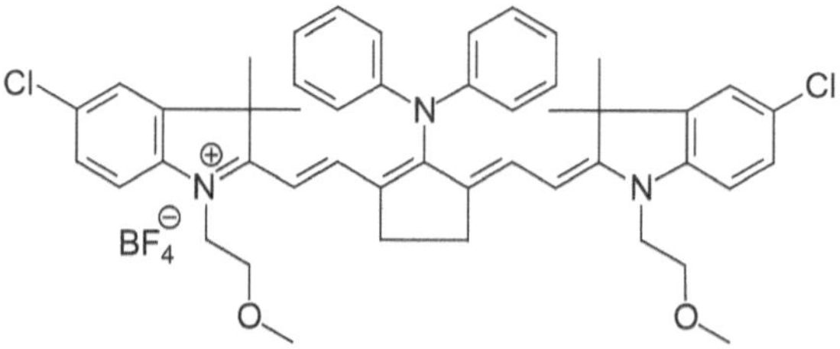

33

To prepare heptamethine cyanine dyes with thermo-sensitive groups it is preferred to place the thermo-sensitive groups already in the streptocyanine dyes as exemplified by **34** and **35** [32, 33]. To increase the stability of the streptocyanine dyes aniline is replaced by the saturated *N*-heterocyclic indoline.

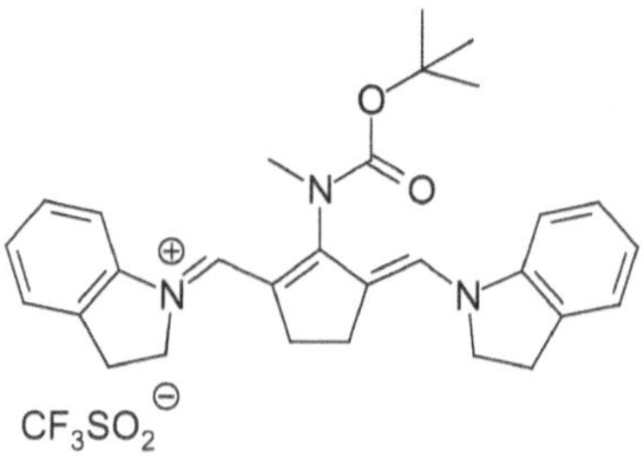

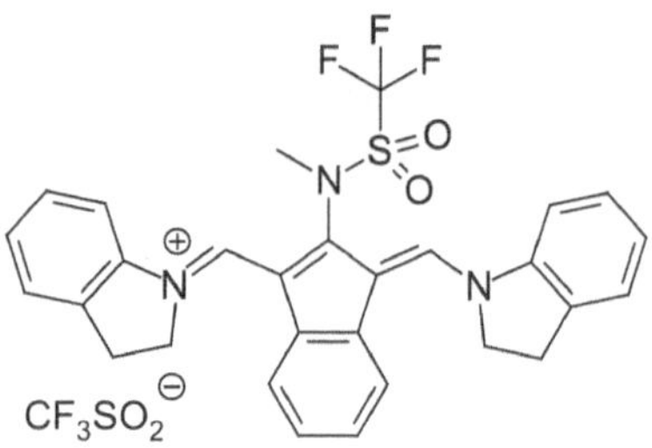

34 **35**

References

1. Zwayer F, Kostanecki SV. Über die Funktion des Doppelchromophors CO.C:C. Ber Dtsch Chem Ges. 1908;41:1335.
2. König W. Über,vinylenhomologe" Indol- und Pyrrol-Farbstoffe. Angew Chem. 1925;38:743.
3. König W. Über den Begriff der „Polymethinfarbstoffe“ und eine davon ableitbare allgemeine Farbstoff-Formel als Grundlage einer neuen Systematik der Farbenchemie. J Prakt Chem. [2]. 1926;112: 1.
4. Riester O. GB 1 229 319. 11 Aug 1967.
5. Berlin L, Riester O. Methoden zur Herstellung von Cyaninen (Polymethinen). In: Müller E, editor. Methoden der organischen Chemie (Houben-Weyl). vol. 5/1d. Stuttgart: Thieme, 1972:231–98.
6. Sturmer DM. Syntheses and properties of cyanine and related dyes. In: Weissberger A, Taylor EC, editors. The chemistry of heterocyclic compounds. vol. 30. New York: Wiley, 1977:441–587.
7. Tyutyulkov N, Fabian J, Mehlhorn A, Dietz F, Tadjer A. Polymethine dyes – structure and properties. Sofia: St Kliment Ohridski University Press; 1991.
8. Berneth H. Methine dyes and pigments. In: Elvers B, editor. Ullmann´s encyclopedia of industrial chemistry. Weinheim: Wiley-VCH, 2012.
9. Mustroph H. Polymethine dyes. Phys Sci Rev. 2020;5. DOI:10.1515/psr-2019-0084.
10. Radeglia R. ^{13}C-NMR-Spektroskopische Untersuchungen der Elektronenstruktur von einfachen Polymethinen. ^{13}C-Chemische Verschiebungen, ^{13}C-H- Kopplungskonstanten und Eu(DPM)$_3$-Verschiebungseffekte. J Prakt Chem. 1973;315:1121.
11. Sieber K, Kutschabsky L, Kulpe S. The molecular and crystal structure of bis- dimethyl trimethine cyanine perchlorate, $C_7H_{15}N_2ClO_4$. Kristall und Technik. 1974;9:1101.
12. Sieber K, Kutschabsky L, Kulpe S. The molecular and crystal structure of bis- dimethyl pentamethine cyanine perchlorate, $C_9H_{17}N_2ClO_4$. Kristall und Technik. 1974;9:1111.
13. Kulpe S, Kuban R-J, Schulz B, Dähne S. Crystal structure determination of bis-dimethyl heptamethine cyanine chloride, $C_{11}N_2H_{19}Cl \cdot 4\ H_2O$. Cryst Res Technol. 1987;22:375.
14. Mustroph H. Cyanine dyes. Phys Sci Rev. 2020;5. DOI:10.1515/psr-2020-0145.
15. Mustroph H, Towns A. Fine structure in electronic spectra of cyanine dyes: Are sub-bands largely determined by a dominant vibration or a collection of singly- excited vibrations? ChemPhysChem. 2018;19:1016.
16. König W, Seifert K. Über das Vinylen-Homologe des „Michlerschen Hydrolblaus“. Ber Dtsch Chem Ges. 1934;67:2112.
17. Malhotra SS, Whiting MC. Researches on polyenes. Part VII. The preparation and electronic absorption spectra of homologous series of simple cyanines, merocyanines, and oxonols. J Chem Soc. 1960;3812.
18. Schmidt H, Wizinger R. Über die Vinylenhomologen der Di- und Triphenylmethanfarbstoffe, III. Liebigs Ann Chem. 1959;623:204.
19. Riester O. Polymethinfarbstoffe. Chimia. 1965;20:330.
20. Mustroph H, Ernst S, Senns B, Towns AD. Molecular electronic spectroscopy: from often neglected fundamental principles to limitations of state- of-the-art computational methods. Color Technol. 2015;131:9.
21. Claisen L. Zur Kenntnis des Propargylaldehyds und des Phenylpropargylaldehyds. Ber Dtsch Chem Ges. 1903;36:3664.
22. Zincke T. Ueber Dinitrophenylpyridiniumchlorid und dessen Umwandlungsproducte. Liebigs Ann Chem. 1904;330:361.
23. König W. Über eine neue vom Pyridin derivierende Klasse von Farbstoffen. J Prakt Chem. [2]. 1904;69: 105.

24. König W, Becker GA. Über den Zusammenhang zwischen Farbe und Konstitution der Pyridinfarbstoffe aus sekundären aromatischen Aminen. J Prakt Chem. [2]. 1912;85: 353.
25. König W, Regner W. Über rein aliphatische Strepto-Pentamethin-Farbstoffe. Ber Dtsch Chem Ges. 1930;63:2823.
26. Hamer FM. The cyanine dyes and related compounds. In: Weissberger A, editor. The chemistry of heterocyclic compounds. vol. 18. New York: Interscience, 1964.
27. Jelinek CF, Kleinschmidt RF. US 2549097. 1 Dec 1948.
28. Štacková L, Štacko P, Klán P. Approach to a substituted heptamethine cyanine chain by the ring opening of Zincke salts. J Am Chem Soc. 2019;141:7155.
29. Zemlicka J, Arnold Z. Synthetic reactions of dimethylformamide. XI. Polyformylation of ketones of other types and the problem of the reaction course. Coll Czech Chem Commun. 1961;26:2852.
30. Arnold Z. Synthetic reactions of dimethylformamide. XXII. Formation and preparation of formyl derivatives of indene. Coll Czech Chem Commun. 1965;30:2783.
31. Reynolds GA, Drexhage KH. Stable heptamethine pyrylium dyes that absorb in the infrared. J Org Chem. 1977;42:885.
32. Callant P, Louwet J. EP 2 234 964. 18 Dec 2008.
33. Callant P. EP 2 072 570. 18 Dec 2008.

Heinz Mustroph

65 Streptomerocyanine dyes

Abstract: Dyes synthesized by condensation of e. g. 4-*N*,*N*-dialkylaminobenzaldehydes, 5-*N*,*N*-dialkylamino-2-thiophenaldehydes or 5-*N*,*N*-dialkylamino-2-furaldehydes, with open chain active methylene compounds have been given the labels merocyanine, neutrocyanine, methine or styryl dyes. All these classifications are misnomers. In terms of systematic nomenclature, their proper classification is as streptomerocyanine dyes. They are used in textile coloration and in dye diffusion thermal transfer printing (D2T2). Furthermore, they have been investigated for their potential as sensitizer dyes in dye sensitized solar cells (DSSC) and for photorefractive applications.

Keywords: benzylidene dyes, merocyanine dyes, neutrocyanine dyes, methinemethine dyes, polymethine dyes, resonance hybrid, streptomerocyanine dyes, styryl dyes

Dyes synthesized by condensation of e. g. 4-*N*,*N*-dialkylaminobenzaldehydes, 5-*N*,*N*-dialkylamino-2-thiophenaldehydes or 5-*N*,*N*-dialkylamino-2-furaldehydes, with open chain active methylene compounds such as malonodinitrile, cyanoacetates, cyanoacetamides, (phenylsulfonyl)acetonitrile or dialkylmalonates have been given the labels merocyanine [1], neutrocyanine [2], methine [3] or styryl dyes [4].

Merocyanine and *neutrocyanine dyes* are defined as *neutral* (zwitterionic – having formal unit electrical charges of opposite sign) *polymethine dyes*, where one terminal component is typically found in *cyanine dyes* (the terminal nitrogen atom in conjugation with the polymethine chain forms part of an unsaturated heterocyclic ring) and the second obtained from an active methylene compound [5, 6]. However, in the condensation products of the above mentioned amino-arylaldehydes with open chain active methylene compounds there is no terminal component typically of cyanine dyes. Therefore, these two classifications are misnomers.

The terms *methine* and *styryl dyes* were derived from individual formulas (contributing structures of Valence Bond theory), which separately, however, do not describe the electronic structure of the molecules [7–9]. Linus Pauling [7] and Theodor Förster [8] had already pointed out that the model with two contributing structures is too simple to describe the electronic structure of cyanine dyes fully. They realized that the resonance of positive charge throughout the whole conjugated system must be considered. Nevertheless, often it is sufficient to describe roughly the electronic structure of such compounds as a *resonance hybrid* between a *charge-separated contributing structure* **S** and a *non-charge-separated contributing structure* **N**.

This article has previously been published in the journal Physical Sciences Reviews. Please cite as: H. Mustroph, Streptomerocyanine Dyes *Physical Sciences Reviews* [Online] 2021, 6. DOI: 10.1515/psr-2020-0211

https://doi.org/10.1515/9783110587128-065

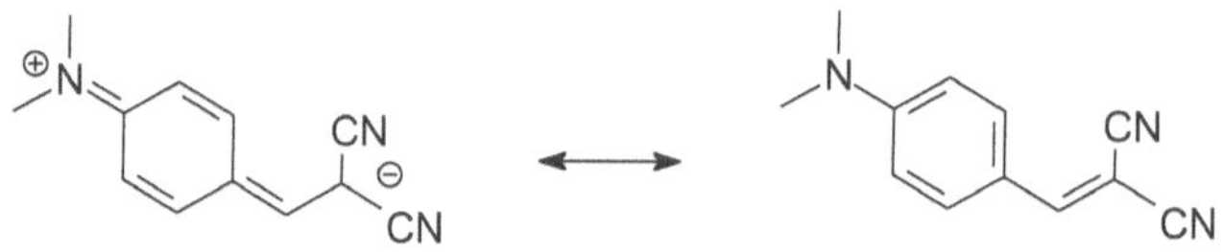

1S **1N**

Considering **1S** the dye could be called methine dye, whereas **1N** would be termed styryl dye. Taking just one of these two narrow views leads sometimes to incorrect conclusions, like e. g.: "Styryl dyes have been a separate classification from polymethine dyes because they are not polymethine dyes." [10].

For simplicity hereafter one contributing structure of the molecule is given only.

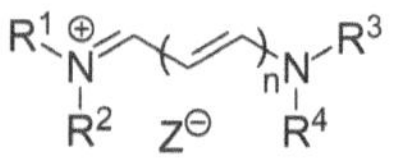

2

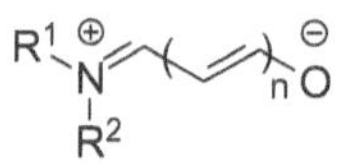

3

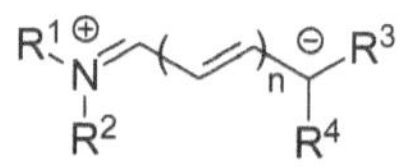

4

The simplest possible type of cyanine dye family are the *streptocyanine dyes* **2**, where the two charge-carrying terminal nitrogen atoms do not form part of an unsaturated heterocyclic ring, but they are part of an open chain $N(R^1)R^2$ group or of saturated heterocyclic rings [5, 6].

As mentioned, for the neutral polymethine dyes, where one terminal component is typically found in cyanine dyes and the second obtained from an active methylene compound, Frances M. Hamer suggested the term *merocyanine* (*μεροσ* = part) *dyes* [11, 12]. Analogously, if the charge-carrying terminal nitrogen atom does not form part of an unsaturated heterocyclic ring like in streptocyanine dyes and the charge-carrying terminal oxygen atom is not constituent of unsaturated heterocycles the dye class **3** should be better named as *streptomerocyanine dyes* [5, 6].

Following this classification in terms of systematic nomenclature, the proper classification of the neutral dyes **4**, where one terminal component is typically found in streptocyanine dyes and the second is an open chain carbanion, substituted with electron acceptors R^3, R^4 like e. g. –CN, $-COOC_2H_5$ or –CONHPh, is also as streptomerocyanine dyes [5, 6].

The simplest model compound is *N,N*-dimethylformamide (DMF) **3** ($n = 0$). In DMF there is no simple single bond between $(CH_3)_2N$ and CHO as it is often drawn. The N–C bond has a high proportion of double bond character, as illustrated by the resonance hybrid between the non-charge-separated contributing structure **5N** and charge-separated contributing structure **5S.** This partial double bond character of the N–C bond causes a hindered internal rotation of the amide group at room temperature which results in a *H* NMR spectrum of DMF at room temperature with two methyl signals [12]

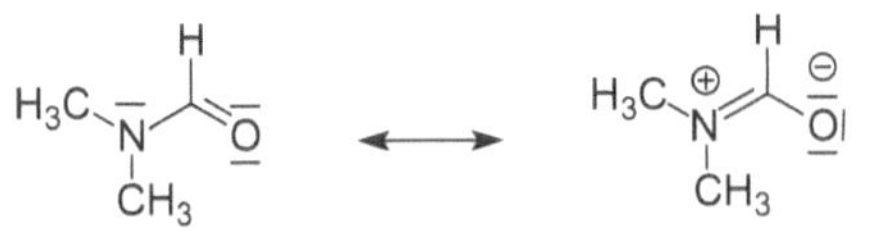

5N **5S**

As model compounds for streptomerocyanines the vinylogous series of DMF was intensively investigated by NMR spectroscopy. A special feature of polymethine dyes is the alternating π-charge density distribution along the polymethine chain, which can be measured by ^{13}C NMR spectroscopy. Also, in this case, the markedly alternating π-charge density along the polymethine chain is evidenced by the ^{13}C NMR chemical shifts δ (^{13}C) (Table 65.1).

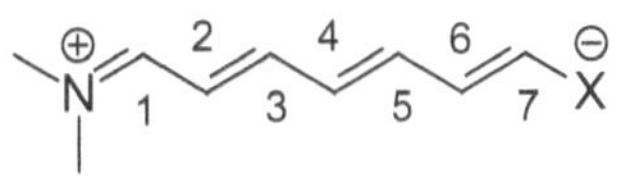

6

Table 65.1: ^{13}C NMR chemical shifts δ (^{13}C) in ppm along the polymethine chain of streptomerocyanine dyes **3** with $N(R^1)R^2 = N(CH_3)_2$, measured in $CDCl_3$ [13]. For numbering of the *C*-atoms see **6** (X = O).

***C*-atom**				δ (^{13}C)			
	1	**2**	**3**	**4**	**5**	**6**	**7**
n							
1	160.7	100.7	187.9				
2	152.6	97.2	156.7	119.6	192.1		
3	149.9	98.7	147.3	117.8	155.8	123.7	192.9

Another special feature of symmetrical (!) polymethine dyes are small differences in the carbon–carbon equilibrium bond lengths in the electronic ground state. In unsymmetrical polymethine dyes (like e. g. hemicyanine, merocyanine or unsymmetrical cyanine dyes) the carbon–carbon equilibrium bond lengths can substantially alternate. To transfer the structural influence in polyatomic molecules to the Franck-Condon principle for diatomic molecules the bond-length alternation (BLA) in the electronic ground state was introduced. BLA is estimated as the average of the absolute difference between adjacent carbon–carbon equilibrium bond lengths in a polymethine chain.

NMR investigations and related quantum chemical calculations on a number of cyanines, merocyanines and polyenes have revealed that there is an almost linear correlation between the ^3J(H,H) coupling constants for *trans* vicinal protons in the polymethine chain and the carbon–carbon equilibrium bond lengths [14–18]. The ^3J (H,H) coupling constants clearly demonstrate that the carbon–carbon equilibrium bond lengths in **3** substantially alternate (Table 65.2). Accordingly, the degree of bond length alternation in solution can be estimated by the absolute difference ΔJ of the ^3J(H,H) coupling constants between adjacent vinylene groups. For illustration, for **3** ($n = 1$) is the value $\Delta J = 4.8$ Hz (Table 65.2). Using a powerful electron donating and electron accepting group like e. g. in **7** and a polar solvent like e. g. DMSO the BLA becomes significantly smaller. For **7** ($n = 1$) is $\Delta J = 1.0$ Hz (Table 65.2). For comparison, for **2** [$n = 2$; $N(R^1)R^2 = N(CH_3)_2$] is $\Delta J = 0.6$ Hz in $CDCl_3$ and $\Delta J = 0.8$ Hz in [D6]DMSO [15].

These experimental results can be explained with the simple model that the contribution of the contributing structures **S** and **N** to the resonance hybrid is determined by the electron donating and accepting abilities of the terminal groups as well the polarity of the solvent [14,15,16,17,18,19]. Increasing contributions from one or the other of **N** or **S** to the ground state electronic structure lead to increasing BLA. This property is important to understand deviations from another characteristic of polymethines.

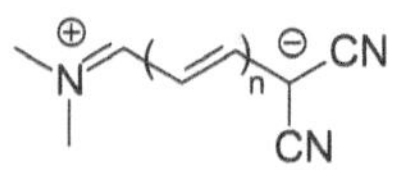

7

Table 65.2: ^3J(H,H) coupling constants in Hz for *trans* vicinal protons in the polymethine chain of streptomerocyanine dyes **3** with $N(R^1)R^2 = N(CH_3)_2$, measured in $CDCl_3$ [15] and **7** measured in [D6] DMSO [16]. For numbering of the *C*-atoms see **6** (**3** X = O, **7** X = $C(CN)_2$).

Bond			**^3J(H,H)**			
	C1–C2	C2–C3	C3–C4	C4–C5	C5–C6	C6–C7
n						
			3			
1	12.9	8.1	14.6	8.5		
2	12.6	11.7	14.0	11.4		
3	12.9	10.7	14.0	11.4	15.0	8.1
			7			
1	11.75	12.75				
2	11.75	12.50	12.50	13.00		
3	12.2	12.3	12.5	12.5	12.5	12.5

A third very characteristic feature of polymethine dyes is the so-called vinylene shift. Each additional vinylene group (–CH = CH–) in polymethines gives a bathochromic shift of the 0–0 vibronic transition of about 100 nm [5, 6, 20–25].

However, a material and widespread error is to discuss this effect in terms of λ_{max} only, without paying attention to the vibronic fine structure of the spectra. The value of λ_{max} is nothing other than the intensity maximum of an absorption band and can correspond to the 0–0, 0–1 or any other vibronic transition. Only in the case that the contribution of both contributing structures **N** and **S** to the electronic ground state is similar, then BLA is small. If BLA in the ground state is small, the difference between the equilibrium bond lengths in the electronic ground and excited states is small and λ_{max} corresponds to the 0–0 vibronic transition [24–26]. The discussion of λ_{max} often leads to wrong conclusions as illustrated with the streptomerocyanines **3** (Table 65.3). Considering λ_{max} only, the conclusion would be, **3** does not belong to the polymethine dyes, because λ_{max} does not exhibit the vinylene shift. Due to the great value for ΔJ (BLA), the difference between the equilibrium bond lengths in the electronic ground and excited states is great, resulting in absorption spectra which do not exhibit a clear vibrational structure [27], and it is highly likely, that λ_{max} does not represent the 0–0 transition. Replacing oxygen in **3** with malodinitrile as in **7** the BLA is reduced (Table 65.2), the absorption spectra exhibit a clear vibrational structure [16], and λ_{max} shows the characteristic vinylene shift (Table 65.3).

Table 65.3: Absorption maxima λ_{max} (nm) of the streptomerocyanines **3** in dichloromethane [27] and **7** in EtOH [16] in dependence on the number of vinylene groups *n*.

n	**3**	**7**
	λ_{max}	λ_{max}
1	283	372
2	361	473
3	421	579
4	462	670

In a logical continuation of the structural principle of streptocyanines, König and Seifert classified the *diarylmethine dyes* as phenylogous streptocyanine dyes [28]. Following this classification in terms of systematic nomenclature, the proper classification of the neutral dye class represented by **1** is as streptomerocyanine dyes, too. All experimental results and theoretical models make clear that the “styryl dyes” should not be considered as a separate class to polymethine dyes. They clearly belong to the large class of polymethine dyes.

In addition, the phenylogous dyes with a 4-R_2N-aryl group and a heterocycle with a terminal open-chain carbanion like e. g. **8** are part of the polymethine dyes.

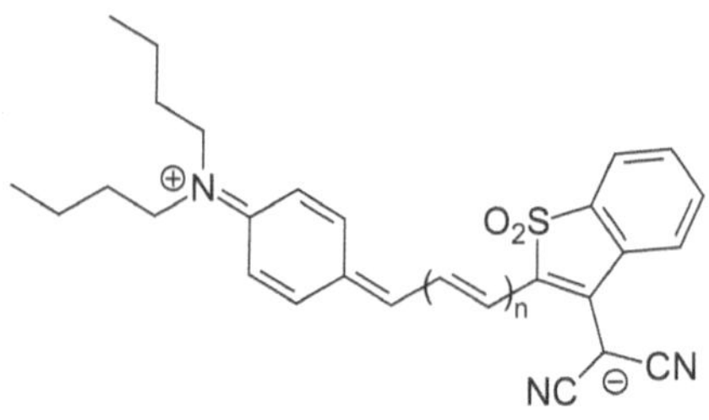

8

The dyes **8** are interesting model compounds, because the polarity of the solvent influences the balance of contributions of **N** and **S** to the resonance hybrid of **N** and **S** substantially. This is illustrated with ΔJ (BLA). For **8** (n = 1) in [D8]dioxane is $\Delta J = 2.5$ Hz and in [D6]DMSO is $\Delta J = 1.2$ Hz [17] – similar to the value for **7** (n = 1), where it is $\Delta J = 1.0$ Hz.

In dioxane λ_{max} does not reflect the vinylene shift and due to $\Delta J = 2.5$ Hz it is again highly likely, that λ_{max} does not represent the 0–0 transition. In addition, the comparison of λ_{max} **8** ($n = 2$) and **8** ($n = 3$) in dioxane is a very nice illustration of just how big an impact vibronic structure can make on λ_{max} (Table 65.4)! On the other hand one can expect with $\Delta J = 1.2$ Hz in DMSO that the difference between the equilibrium bond lengths in the electronic ground and excited states is small and λ_{max} will represent the 0–0 transition and reflects the vinylene shift (Table 65.4).

Table 65.4: Absorption maxima λ_{max} (nm) of the streptomerocyanines **8** in dioxane and in DMSO in dependence on the number of vinylene groups n [17].

n	Dioxane	DMSO
	λ_{max}	λ_{max}
0	581	609
1	657	709
2	721	805
3	712	886

The absorption maxima of streptomerocyanine dyes formed from the condensation of 4-*N,N*-dialkylaminobenzaldehydes with open chain active methylene compounds lie below 450 nm and so only yellow color shades are obtainable. Due to their brightness and excellent lightfastness they are extensively used commercially in the coloration of synthetic fibers. Arnold T. Peters and Malcolm S. Wild made an extensive investigation of dyes **9** with their synthesis, structure-color and structure-dyeing properties [29–31].

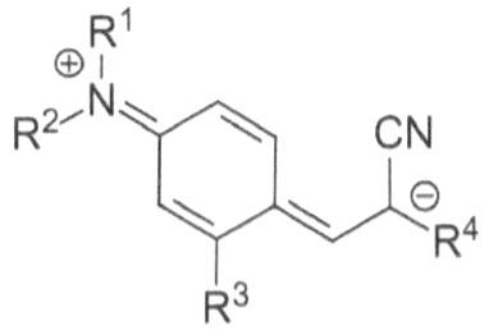

R^1, R^2: H, CH_3, C_2H_5, C_2H_4OH, $C_2H_4OOC_2H_5$, C_2H_4CN
R^3: H, CH_3, R^4: CN, $COOC_2H_5$

9

The lightfastness of the dyes **9** is considerably better if malonodinitrile ($R^4 = CN$) is used instead of cyanoacetate ($R^4 = COOC_2H_5$) [4, 29–31]. With the substituents R^1 and R^2 application properties like e. g. color shade, lightfastness and sublimation properties are optimized.

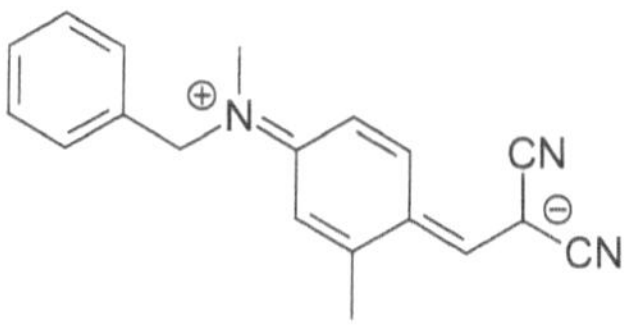

10

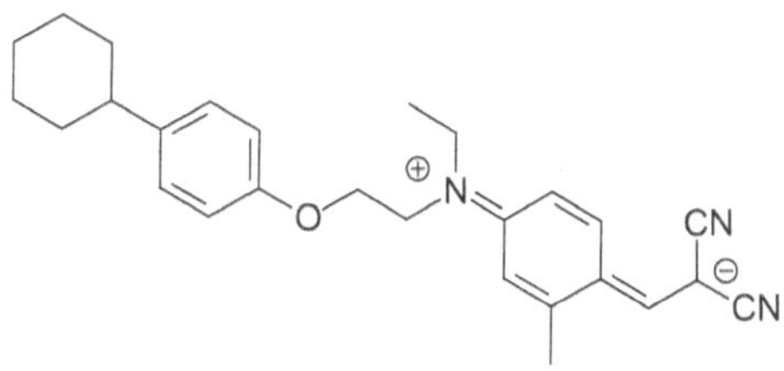

11

Dye **10** colors polyester fibers a red-violet shade and **11** (CI Disperse Yellow 201, CI Solvent Yellow 179) is used for coloration of thermoplastics.

Due to very wide possibilities for structural variations the application properties of streptomerocyanine dyes can be optimized in a wide range. A hi-tech application is as imaging dyes for *dye diffusion thermal transfer printing* (D2T2) [32]. Here the dyes need to be bright and stable to light and heat and, once transferred, the dyes should not show tendency to migrate. These requirements are met with dye **10**. In addition, the storage stability of the printed image is improved by *para* substitution of a phenoxy group with cyclohexyl **11** [32].

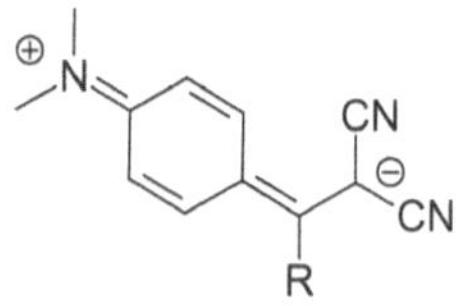

12a: R = H
12b: R = CN

According to Dewar´s rules [5, 6, 21–23] a third cyano substituent on the central methine group leads to a strong bathochromic shift from **12a** (λ_{max} = 430 nm; EtOH) [3] to **12b** (λ_{max} = 515 nm; EtOH) [3]. This effect is used in commercial dyes to manufacture red dyes for different applications.

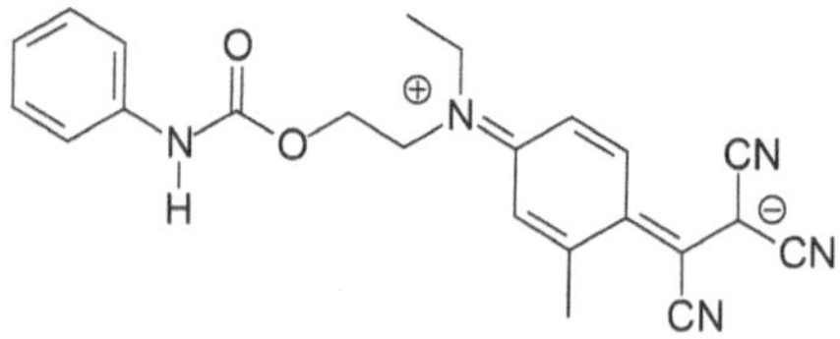

13

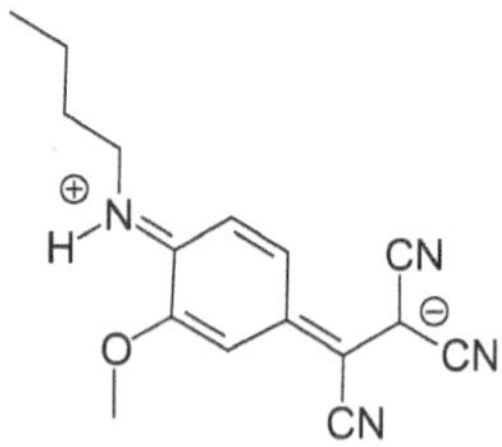

14

So, **13** dyes polyester fibers in a red-violet shade [4] and **14** is used as a bright magenta dye in D2T2 [32].

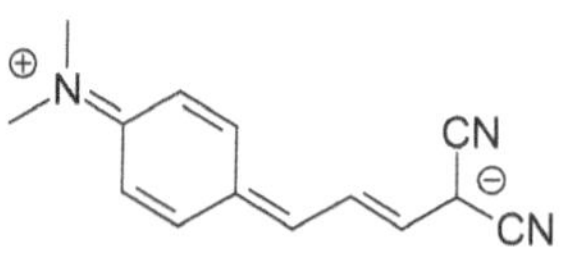

15

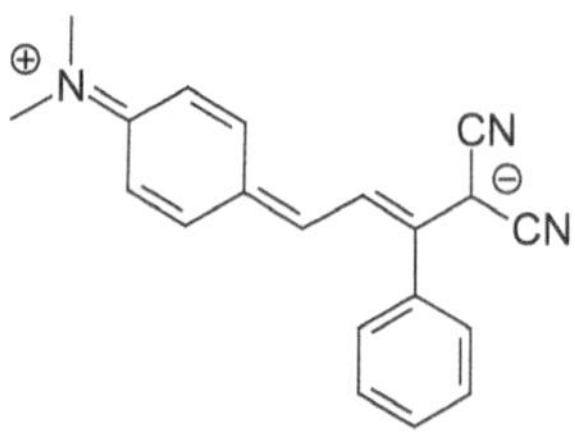

16

Extending the polymethine chain by a vinylene group gives red dyes like e. g. **15** (λ_{max} = 498 nm; EtOH) [33] and **16**, (λ_{max} = 502 nm) [34].

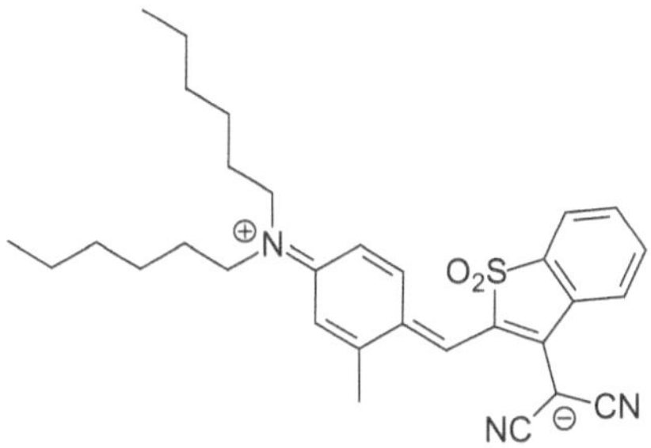

17

In mid-1977 the Central Research Group of the Swiss company Sandoz began a project to find new disperse dyes for polyester [35]. The result of systematic industrial R&D was the brilliant and stable blue dye **17** (Foron Brilliant Blue S–R, CI Disperse Blue 354) [36] – one of the brightest blue disperse dyes commercially available for polyester, giving a brilliant turquoise shade. The red dye **16** is the acyclic analogue of the blue dye **17** (λ_{max} = 610 nm; EtOH) [37] and the (formal) cyclization to **17** is an amazing effect regarding bathochromic shift of λ_{max} and stability.

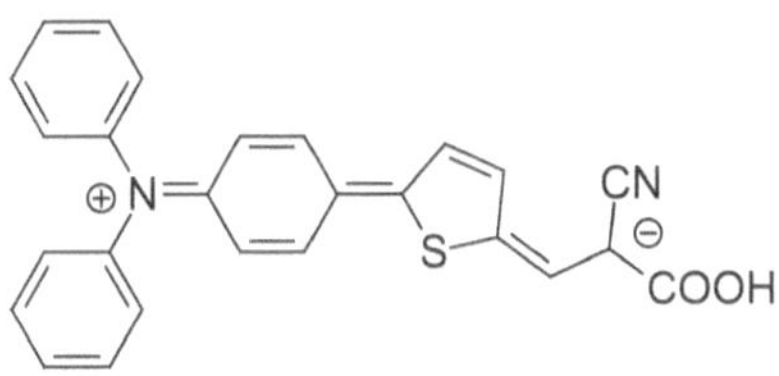

18

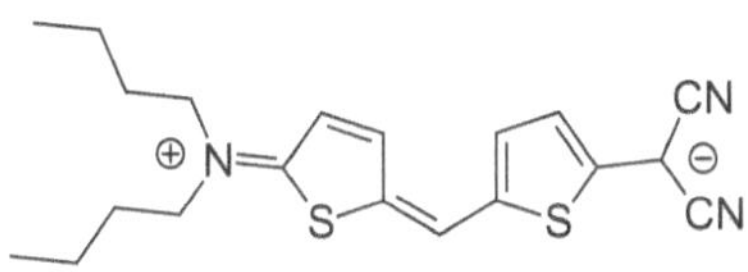

19

In addition, streptomerocyanine dyes have been investigated for their potential as sensitizer dyes in dye sensitized solar cells (DSSC) [38] like e. g. **18** (λ_{max} = 410 nm; *t*-BuOH, ACN) and for photorefractive applications [1] like e. g. **19** (λ_{max} = 647 nm; dioxane) [2]. Despite all the effort being expended on these applications it is not clear which, if any, of these potential applications will become commercial reality.

References

1. Würthner F, Wortmann R, Meerholz K. Chromophore design for photorefractive organic materials. Chem Phys Chem. 2002;3:17.
2. Würthner F, Thalacker C, Matschiner R, Lukaszuk K, Wortmann R. Optimization of neutrocyanine chromophores based on five-membered heterocycles for photorefractive applications. Chem Commun. 1998;1739.
3. Irick G, Straley JM. Yellow to red methine dyes derived from 1,3- indandione. Text Chem Color. 1969;1:178.
4. Berneth H. Methine dyes and pigments. In: Elvers B, editor. Ullmann´s encyclopedia of industrial chemistry. Weinheim: Wiley-VCH, 2012.
5. Tyutyulkov N, Fabian J, Mehlhorn A, Dietz F, Tadjer A. Polymethine dyes – structure and properties. Sofia: St Kliment Ohridski University Press, 1991.
6. Mustroph H. Polymethine dyes. Phys Sci Rev. 2020;5. DOI:10.1515/psr-2019-0084.
7. Pauling L. A theory of the colors of dyes. Proc Natl Acad Sci. 1939;25:577.
8. Förster T. Quantenmechanische Rechnungen zur Theorie der organischen Farbstoffe. Z Phys Chem B. 1940;47:245.
9. Herzfeld KF, Sklar AL. Colour and constitution of polymethine dyes. Rev Mod Phys. 1942;14:294.
10. Gregory P. Classification of dyes by chemical structure. In: Waring DR, Hallas G, editors. The chemistry and application of dyes. New York: Plenum Press; 1990.
11. Hamer FM, Winton BS, Trimethincyanine A. a trimethinoxonol and some compounds related to trimethinoxonols; some di- and tetra-methinmerocyanines. J Chem Soc. 1949;1126.
12. Mustroph H. Merocyanine dyes. Phys Sci Rev 2021;6:accepted. DOI:10.1515/psr-2020-0145.
13. Radeglia R. ^{13}C-NMR-Spektroskopische Untersuchungen der Elektronenstruktur von einfachen Polymethinen. ^{13}C-Chemische Verschiebungen, ^{13}C-H- Kopplungskonstanten und Eu(DPM)$_3$-Verschiebungseffekte. J Prakt Chem. 1973;315:1121.
14. Scheibe G, Seiffert W, Hohlneicher G, Jutz C, Springer HJ. Trans-Protonenkopplungskonstanten und C–C-Abstände bei Cyaninen, Merocyaninen und Olefinen. Tetrahedron Lett. 1966;5053.
15. Radeglia R, Dähne S. Lösungsmittelinduzierte Elektronenverschiebungen in einfachen Merocyanin-Farbstoffen. J Mol Struct. 1970;5:399.
16. Scheibe P, Schneider S, Dörr F, Daltrozzo E. Spektroskopische Untersuchungen an Merocyaninen I. UV- und NMR-Spektren von malodinitril- substituierten vinylogen Säureamiden. Ber Bunsenges Phys Chem. 1976;80:630.
17. Blanchard-Desce M, Alain V, Bedworth PV, Marder SR, Fort A, Runser C, et al. Large quadratic hyperpolarizabilities with donor-acceptor polyenes exhibiting optimum bond length alternation: correlation between structure and hyperpolarizability. Chem Eur J. 1997;3:1091.
18. Mustroph H, Reiner K, Senns B, Mistol J, Ernst S, Keil D, et al. The effects of substituents and solvents on the ground-state π-electronic structure and electronic absorption spectra of a series of model merocyanine dyes and their theoretical interpretation. Chem Eur J. 2012;18:8140.
19. Brooker LG. Some recent developments in the chemistry of photographic sensitizing dyes. Experientia Suppl. 1955;2:229.
20. König W. Über den Begriff der „Polymethinfarbstoffe" und eine davon ableitbare allgemeine Farbstoff-Formel als Grundlage einer neuen Systematik der Farbenchemie. J Prakt Chem. [2] 1926;112: 1.
21. Hallas G. Effects of extended conjugation in some basic dyes. J Soc Dyers Colourists. 1970;86:237.

22. Griffiths J. Colour and constitution of organic molecules. London: Academic Press, 1976.
23. Fabian J, Hartmann H. Light absorption of organic colorants. Berlin: Springer, 1980.
24. Mustroph H, Ernst S, Senns B, Towns AD. Molecular electronic spectroscopy: from often neglected fundamental principles to limitations of state- of-the-art computational methods. Color Technol. 2015;131:9.
25. Mustroph H. Dyes: quantum chemical calculation of electronic spectra. Phys Sci Rev. 2019;4. DOI:10.1515/psr-2019-0040.
26. Mustroph H, Towns A. Fine structure in electronic spectra of cyanine dyes: are sub-bands largely determined by a dominant vibration or a collection of singly- excited vibrations? Chem Phys Chem. 2018;19:1016.
27. Malhotra SS, Whiting MC. Researches on polyenes. Part VII. The preparation and electronic absorption spectra of homologous series of simple cyanines, merocyanines, and oxonols. J Chem Soc. 1960;3812.
28. König W, Seifert K. Über das Vinylen-Homologe des „Michlerschen Hydrolblaus". Ber Dtsch Chem Ges. 1934;67:2112.
29. Peters AT, Wild MS. Styryl dyes for synthetic-polymer fibres. Part I – synthesis and characterization of some 4-amino-β,β-dicyanonstyrenes and ethyl- α-cyano-4-aminocinnamates. J Soc Dyers Colour. 1977;93:126.
30. Peters AT, Wild MS. Styryl dyes for synthetic-polymer fibres. Part II – colour, dyeing and fastness properties of some 4-amino-β,β-dicyanonstyrenes and ethyl- α-cyano-4-aminocinnamates. J Soc Dyers Colour. 1977;93:133.
31. Peters AT, Wild MS. Styryl dyes for synthetic-polymer fibres. Part III – derivatives of 2.methyl-4-(N-ethyl-N-β-hydroxyethyl) amino β,β-dicyanostyrenes. J Soc Dyers Colour. 1978;94:106.
32. Bradbury R. Dyes for dye diffusion thermal transfer (D2T2) printing. In: Freeman HS, Peters AT, editors. Colorants for non-textile applications. Amsterdam: Elsevier, 2000.
33. Hertel E, Hoffmann KA. Über die Wirkung von Substituenten an olefinischen Kohlenstoffatomen auf die physikalisch-chemischen Eigenschaften der chromophoren Gruppen Vinylen und Divinylen. Z Phys Chem B. 1941;50:382.
34. Gokhale UV, Seshadri S. Synthesis of 1,1-dicyano-2,4-diaryl-1,3-butadienes – a novel chromophore. Dyes Pigm. 1986;7:389.
35. Baumann W. The world´s most brilliant blue. A new dyestuff for polyester fibres. Foron brilliant blue S-R. Int Gazette (Sandoz). 1983;48:4.
36. Baumann W. CH 636 116. 22 Dec 1978.
37. Hu J-Z, Skrabal P, Zollinger H. A comparison of the absorption spectra of a series of blue disperse dyes with the calorimetric evaluation of their dyeings. Dyes Pigm. 1987;8:189.
38. Mishra A, Fischer MK, Bäuerle P. Metal-free organic dyes for dye-sensitized solar cells: from structure : property relationships to design rules. Angew Chem Int Ed. 2009;48:2474.

Gerhard Pfaff

66 Titanium dioxide pigments

Abstract: Titanium dioxide (TiO_2) is by far the most important white pigment. The extraordinary scattering behavior of TiO_2 together with the lack of absorption in the visible range and the stability in different application systems is the reason why TiO_2 is the most commonly used pigment. Of the three TiO_2 modifications, only rutile and anatase play a role for pigment uses, whereas brookite has no technical importance. Two main processes are used for the production of TiO_2 pigments, the sulfate process and the chloride process. Many TiO_2 pigments are equipped with an additional surface treatment (aftertreatment) to improve weather resistance and lightfastness of the pigment itself and of the pigmented organic matrix. A second aim of the surface treatment is the improvement of the dispersibility of the pigments in the application system.

Keywords: titanium dioxide pigments, anatase, rutile, chloride process, sulfate process, surface treatment, aftertreatment

66.1 Fundamentals and properties

Titanium dioxide (TiO_2) exists in the three modifications rutile, anatase, and brookite. All these modifications occur in nature. Rutile and anatase are manufactured industrially in large quantities, whereas brookite is difficult to produce and has no technical importance. A significant proportion of the TiO_2 produced is used for pigment purposes (C.I. Pigment White 6). Further industrial applications of TiO_2 are found in catalysts and ceramic materials TiO_2 [1–4].

Rutile is the thermodynamically most stable TiO_2 modification. Anatase and brookite have, however, similar lattice energies as rutile. They are therefore also stable under normal conditions. A monotropic phase conversion from anatase to rutile takes place at about 915 °C, for brookite at about 750 °C. Rutile is characterized by a very high stability up to its melting range at about 1830 to 1850 °C.

All three TiO_2 modifications have the same structural units, titanium atoms, which are surrounded octahedrally by six oxygen atoms. TiO_6 octahedrons are therefore the dominant assembly units in the three crystal lattices. The arrangement of these octahedrons is, however, different in the three modifications. Linking of the TiO_6 octahedrons among each other occurs in different ways via corners and edges. There are two common edges each between the TiO_6 octahedrons in rutile. The octahedrons in anatase

This article has previously been published in the journal Physical Sciences Reviews. Please cite as: G. Pfaff Titanium Dioxide Pigments *Physical Sciences Reviews* [Online] 2021, 6. DOI: 10.1515/psr-2020-0199

https://doi.org/10.1515/9783110587128-066

are linked via four and in brookite via three common edges. The oxygen atoms in all three crystal lattices are surrounded by three titanium atoms in a trigonal arrangement. Rutile and anatase crystallize in a tetragonal lattice, brookite has a rhombic crystal structure.

Thermal treatment of TiO_2 at temperatures above 1000 °C leads to the liberation of oxygen in tiny quantities. Lower oxides of titanium are formed under these conditions (TiO_{2-x}). As a consequence, changes in color and electrical conductivity are observed. Above 400 °C, a significant yellowing of TiO_2 takes place caused by the thermal expansion of the lattice. The effect is reversible after lowering of the temperature.

Rutile has the highest density of the three TiO_2 modifications based on the most compact structure. It has a Mohs hardness of 6.0 to 6.5 and is thus the hardest form of TiO_2 (anatase and brookite 5.5 to 6.0). The lower hardness of anatase is beneficial in applications where abrasiveness is of relevance, e.g. in the synthetic fiber, the plastics, and the paper industry.

The high refractive index of TiO_2 (2.7 for rutile and 2.55 for anatase) together with the absent absorption in the visible spectral range between 380 and 700 nm is the basis for its broad use as white pigment.

Figure 66.1 shows the UV-VIS spectra of rutile and anatase pigments with their characteristic differences [2]. Noticeable is the difference in the spectra concerning the absorption edge in the near UV. The absorption of rutile in the range of about 385 nm (the absorption of anatase starts at lower wavelengths and not in the visible range) leads to the situation that rutile pigments have an extremely small yellow tinge in the white. Anatase pigments, on the other hand, exhibit a slight blue tint and appear for the human eye really white. Anatase is therefore regarded as the whiter white compared with rutile.

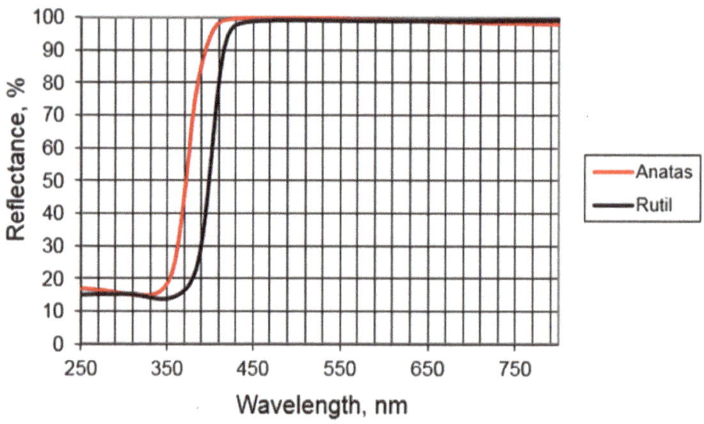

Figure 66.1: UV/VIS spectra of rutile and anatase pigment pellets [2].

TiO_2 is photoactive and belongs to the light-sensitive semiconductors. It is characterized by absorption of electromagnetic radiation in the near-UV region. This absorption is the reason why electrons from the valence band are raised to the conduction band (band model). The energy difference between the valence and the conductivity bands in the solid state is 3.05 eV for rutile and 3.29 eV for anatase. These energies correspond to an absorption band at <415 nm for rutile and <385 nm for anatase. The raise of electrons from the valence band to the conduction band explains the formation of separated electron/hole pairs, which are called "excitons" and thus the photo-semiconductor properties of TiO_2. The photoactivity of TiO_2 pigments is a disadvantage for nearly all applications because the excitons can cause undesired redox reactions associated with a damage of the binder system. Reduction of the photoactivity of TiO_2 pigments is therefore an important ongoing topic for the development of these materials. An application where the photoactivity of TiO_2 is desired is the photocatalysis.

TiO_2 is the white pigment with by far the greatest technical importance. The extraordinary scattering behavior of TiO_2 together with the lack of absorption in the visible range and the stability in different media and binder systems is the reason why TiO_2 is the most widely used pigment.

66.2 Production of titanium dioxide pigments

Two main processes, referred to as sulfate process and chloride process, are industrially used for the manufacture of TiO_2 pigments with controlled properties, such as mean particle size, particle size distribution, particle shape, crystal structure (anatase, rutile), and composition. The surface treatment of TiO_2 pigments based on inorganic and organic compositions plays an important role with respect to the stabilization of the pigments and the compatibility with the application medium.

The raw materials for the TiO_2 production include natural products such as ilmenite, leucoxene, and rutile as well as synthetic materials such as titanium slag and synthetic rutile. Important titanium minerals are summarized in Table 66. 1 [3]. From these minerals, only ilmenite, rutile and leucoxene, a weathering product of ilmenite, have gained economic importance for the production of TiO_2 pigments. Mined ilmenite and rutile are mainly used for the manufacture of TiO_2 pigments. A smaller part of these mined minerals is used for the production of titanium metal and for welding electrodes.

Ilmenite ($FeTiO_3$) is found worldwide in primary massive ore deposits or as secondary alluvial deposits (sands). Important deposits exist in Canada, USA, Brazil, Russia, Norway, Australia, South Africa, and India. A characteristic of all ilmenite sources is the presence of other metals besides titanium and iron. The use of ilmenite from beach sands in existing or fossil coastlines for the TiO_2 production is likewise possible because the action of surf, currents and wind has led to the enrichment of ilmenite and other heavy minerals in certain dunes or beaches.

Table 66.1: Titanium-containing minerals [3].

Mineral	Formula	TiO_2 content (wt %)
Rutile	TiO_2	92–98
Anatase	TiO_2	90–95
Brookite	TiO_2	90–100
Ilmenite	$FeTiO_3$	35–60
Leucoxenes	$Fe_2O_3TiO_2$	60–90
Perovskite	$CaTiO_3$	40–60
Sphene (titanite)	$CaTiSiO_5$	30–42
Titanomagnetite	$Fe(Ti)Fe_2O_4$	2–20

Leucoxenes are mixtures of minerals with TiO_2 contents of about 90%, which are typically formed in corroded ilmenite. They are mined and treated in some deposits separately. The leucoxene quantities used for the production of TiO_2 are small in comparison to those of ilmenite.

Most rutile ores are magmatic rocks with high titanium and low iron contents. Rutile ores can also be formed by the metamorphosis of titanium-bearing sediments or magmatites. The rutile content in primary rocks is low in most cases. Only sands in which rutile is accompanied by zircon, ilmenite and other heavy minerals can therefore be regarded as suitable deposits. The overall quantities of mineable rutile are limited and too low to meet the volume demands for titanium-containing raw materials. Synthetic rutile is therefore used to substitute the natural form in part [3].

The preparation of the titanium-containing ores for the production of TiO_2 pigments starts mostly from heavy mineral sands. The usually 3 to 10% heavy minerals containing raw sands are typically obtained by wet dredging. Next steps are sieving of the sands and gravity concentration in several stages with Reichert cones and/or spirals to get a product containing 90 to 98% heavy minerals. Heavy and lighter minerals are separated by this procedure from each other [5].

Ilmenite and other magnetic components are then separated from the nonmagnetic parts, especially from rutile, zircon and leucoxenes, by dry or wet magnetic separation. The weakly magnetic weathered ilmenites and leucoxenes are recovered by high-intensity magnetic separation in the following step. Rutile is separated from zircon electrostatically utilizing the different conductivity of both materials [3].

Synthetic raw materials used for the production of TiO_2 pigments are characterized by a high TiO_2 content. Titanium slag is suitable for the sulfate process, synthetic rutile for the chloride process. Both materials are produced starting from ilmenite and leucoxene ores. The production of titanium slag proceeds mainly by a metallurgical process in which the iron from ilmenite is reduced by anthracite or coke to iron metal at 1200 to 1600 °C in an electric arc furnace followed by separation (Sorel process). Titanium slags with a TiO_2 content of 70 to 85% are formed together with titanium-free pig iron.

66.2.1 Sulfate process

A scheme for the production of TiO_2 pigments with the sulfate process is shown in Figure 66.2. Ilmenite or titanium slag from the Sorel process, which are used as the raw materials, are dried at the beginning to a moisture content of less than 0.1% and ground in ball mills (a) to give a mean particle size of less than 40 µm. The material obtained in this way passes a screen (b), a magnetic separator (c) and a cyclone (d) before it reaches a silo (e) for a temporary storage.

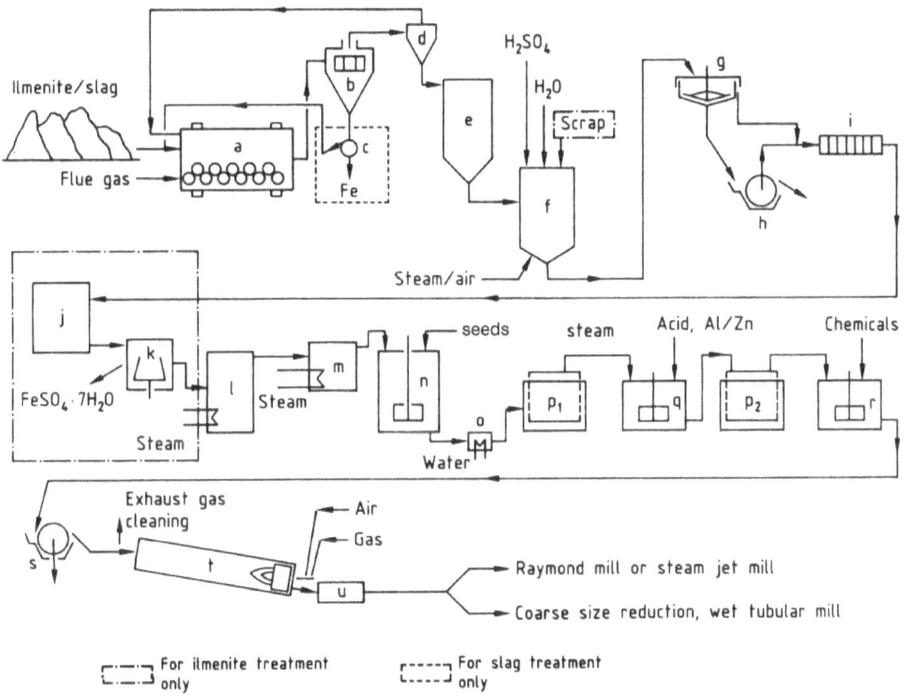

Figure 66.2: Flow diagram of TiO_2 production by the sulfate process. (a) ball mill/dryer; (b) screen; (c) magnetic separator (optionally); (d) cyclone; (e) silo; (f) digestion vessel; (g) thickener; (h) rotary filter; (i) filter press; (j) crystallizer; (k) centrifuge; (l) vacuum evaporator; (m) preheater; (n) stirred tank for hydrolysis; (o) cooler; (p) Moore filters; (q) stirred tank for bleaching; (r) stirred tank for doping; (s) rotary filter for dewatering; (t) rotary kiln; (u) cooler [3].

The following digestion is usually applied batch-wise in suitable vessels (f). The ground raw material is mixed with 80 to 98% H_2SO_4. Another process variant uses 80% sulfuric acid and starts the reaction by cautious addition of oleum. Mixing of the reactants is sufficient to initiate the process. A vigorous digestion reaction starts reaching temperatures of about 170 to 220 °C. Oxysulfates and sulfates of titanium, iron and other metals are formed under these conditions.

$$FeTiO_3 + 2\,H_2SO_4 \rightarrow TiOSO_4 + FeSO_4 + 2\,H_2O \tag{66.1}$$

Sulfuric acid and titanium-containing raw materials are used in a ratio that the weight proportion of free H_2SO_4 to TiO_2 in the suspension produced by the hydrolysis, the so-called "acid number" is between 1.8 and 2.2. The reaction mixture is kept in the vessel to mature for up to 12 hours, after the maximum temperature has been reached. The duration of the maturing period depends on the raw material. It is important at this point that the titanium-containing components become as soluble as possible. The digestion of the raw materials can be accelerated by blowing air through the reaction mixture during the heating process and also during the maturing period.

In the following step, the cake obtained after cooling of the digested reaction mixture is dissolved in cold water or in diluted sulfuric acid recycled from the process. Premature hydrolysis is avoided by keeping the temperature below 85 °C. Air is blown into the suspension to keep it in motion during the dissolution. The TiO_2 concentration of the resulting solution is in the range from 8 to 12 wt % (ilmenite derived product) or 13 to 18 wt % (slag derived product) TiO_2.

Solved Fe^{3+} hydrolyzes together with Ti^{4+} and accumulates on the surface of formed TiO_2 hydrate. It has therefore to be removed prior to the subsequent hydrolysis step. A suitable way to do it is the reduction of Fe^{3+} to Fe^{2+} by adding of scrap iron during the dissolution or immediately afterwards. The addition of small amounts of Ti^{3+} prevents the re-oxidation of Fe^{2+} ions during the subsequent process steps. The deeply violet colored Ti^{3+} ions lead to a black appearance of the digestion solution, which has a high viscosity. It is therefore called "black liquor".

The aim of the following clarification step is the removal of all undissolved solid material from the solution. A preliminary settling in a thickener (g) takes place followed by filtration of the sediment with a rotary filter (h). The filtrate and the supernatant from the thickener are put through a filter press (i) with the objective to remove fine particles.

The digestion solutions contain after the reduction of the Fe^{3+} ions 5 to 6 wt % $FeSO_4$ (from titanium slag) or 16 to 20 wt % $FeSO_4$ (from ilmenite). The ilmenite derived solution is cooled now under vacuum in a crystallizer (j). $FeSO_4 \cdot 7\,H_2O$ crystallizes under these conditions and can be separated. The concentration of TiO_2 in the solution corresponds to approximately 25 wt %. The crystallized $FeSO_4 \cdot 7\,H_2O$, which is called copperas salt, is separated by filtration or centrifugation (k). It can be used as raw material for the production of iron oxide pigments and for several other technical purposes.

The hydrolysis step is the second central part of the sulfate process after the digestion. It consists mainly in the precipitation of titanium dioxide hydrate ($TiO_2 \cdot x\,H_2O$) by diluting of the previously received solution with water in a temperature range of 94 to 110 °C.

$$TiOSO_4 + (x+1)\ H_2O \rightarrow TiO_2 \cdot x\,H_2O + H_2SO_4 \tag{66.2}$$

The precipitate contains also other components of the raw material, which are soluble in sulfuric acid and not yet crystallized together with the iron(II) sulfate heptahydrate, mainly niobium oxide hydrate. The hydrolysis reaction takes place in brick-lined, stirred tanks (n). Steam is injected in these tanks during the hydrolysis reaction. The formed precipitated hydrolysate does not show yet pigment properties. The mean primary size of the precipitated particles is in the range of about 5 nm, far from the particle sizes of 200 to 300 nm for TiO_2 pigments.

The hydrolysis of concentrated solutions of titanium sulfate in absence of accelerators proceeds sluggishly and incompletely. Suitable nuclei are therefore added or formed to accelerate the reaction. The nuclei are produced using the Mecklenburg method or the Blumenfeld method. The Mecklenburg method leads to colloidal TiO_2 hydrate, which is precipitated with sodium hydroxide at 100 °C from titanium oxysulfate solutions, whereby only 1% of the precipitate is sufficient for nucleation. The Blumenfeld method uses a small part of the sulfate solution for a separate hydrolysis reaction in boiling water. A precipitate is formed, which is added to the bulk solution. The size of the hydrolyzate particles depends on the number of nuclei added.

Several parameters have an influence on the hydrolysis process and on the properties of the formed hydrolysate. These include the agitation (in particular during the initial stage of the hydrolysis), the titanium concentration (is adjusted, if necessary by vacuum evaporation, to give a converted TiO_2 content of 170–230 g/l during hydrolysis; lower concentrations result in a coarser particle size), the “acid number” (considerable effect on the TiO_2 yield and on the particle size of the precipitate, typical hydrolysis periods of 3 to 6 hours have a TiO_2 yield of 93 to 96%), the concentration of other salts present (especially $FeSO_4$, high concentrations lead to finely divided hydrolyzates), and the temperature regime (affects mainly the volume–time yield and the purity of the hydrolyzate).

The liquid phase of the TiO_2 hydrate suspension contains 20 to 28% H_2SO_4 and various amounts of dissolved sulfates. The hydrate is filtered using Moore filters (p_1) and washed with water and diluted acid. Washing, however, is still not sufficient to remove all interfering ions, which are partly adsorbed on the surface of the hydrate particles. A bleaching step is suitable to remove most of the remaining impurities by chemical reduction. For that purpose, the filter cake is slurried with 3 to 10% diluted acid at 50 to 90 °C and mixed with zinc or aluminum powder in a stirred tank (q). The reduction leads to lower valence states for the impurities associated with a better solubility. A second filtration and washing step follows (p_2), after which the hydrate has only very low quantities of disturbing impurities, but still chemisorbed 5 to 10% H_2SO_4. This acid is removed by the following calcination step [3].

TiO_2 of highest purity is obtained when the hydrate is now calcined without any further additions (dopants). A relatively coarse TiO_2 is formed, which is mostly not yet suitable for pigment purposes. The rutile content of this TiO_2 depends on the heating temperature. High-quality pigment grades are obtained when the hydrate is treated in a separate tank (r) with alkali metal compounds and phosphoric acid as mineralizers (<1%) prior to calcination. Rutile pigments are formed with less phosphoric acid than anatase pigments. The formation of rutile pigments needs the addition of rutile nuclei (<10%). The addition of zinc oxide, aluminum oxide, and/or antimony(III) oxide (<3%) stabilizes the crystal structure of certain rutile pigments (intrinsic stabilization). The ions of the added metals are incorporated in the rutile lattice (preferably on Ti^{4+} sites, only in higher concentration on interstitial sites) during the following calcination process and contribute from there to the improvement of the rutile pigments concerning their weather resistance and UV stability.

The aim of the calcination step is the conversion of the TiO_2 hydrate into a final TiO_2 pigment. The doped hydrate is filtered with rotary vacuum filters (s). Water is removed during filtration up to a TiO_2 content of 30 to 40%. The use of pressure rotary filters or automatic filter presses is likewise possible. TiO_2 contents of about 50% can be achieved in this case. A low solubility of the dopants in water can lead to a certain loss during filtering. This loss can be complemented by adding further dopant amounts to the filter cake before the start of the high-temperature reaction.

Calcination is carried out in rotary kilns (t) using the countercurrent principle. The residence time of the material in the kiln is typically 7 to 20 hours. The first phase of the thermal treatment, approximately two-thirds of the residence time, is a drying step. Sulfur trioxide is released from the sulfate in the second phase above 500 °C. It is decomposed partially to sulfur dioxide and oxygen at higher temperatures. The temperature is increased in the third phase to a maximum of 800 to 1100 °C depending on the required pigment quality, the desired throughput, and the temperature profile of the kiln. Water is removed from the hydrate during the main reaction and TiO_2 is formed.

$$TiO_2 \cdot x\,H_2O \rightarrow TiO_2 + x\,H_2O \qquad (66.3)$$

The operating program for the kiln is decisive for the rutile content, respectively, the rutile/anatase ratio, the mean particle size, the particle size distribution, and the formation of aggregates. The product formed in the kiln is a TiO_2 clinker, which is cooled in drum coolers (u). The temperature of the exhaust gas must be above 300 °C at the exit of the kiln. In this way, condensation of sulfuric acid in the ducting is avoided.

The final step of the sulfate process for the production of TiO_2 pigments is a suitable grinding of the calcined material. Wet or dry grinding leads to the destruction of agglomerates and aggregates in the clinker. The required size of the pigment particles is adjusted effectively at the same time. Clinker lumps are crushed in hammer mills prior to wet grinding in tube mills. The addition of dispersion agents is common in order to optimize the grinding process. Coarse particles from the suspension are separated by centrifugation and return to the mills. Roller milling is another technology, which is used in combination with subsequent de-agglomeration by wet milling. Aggregates used for dry grinding are hammer mills, cross-beater mills, roller mills, pendular mills, and steam-jet mills. Suitable additives used in the mills support wetting of the particles during the grinding procedure and improve the dispersibility of final untreated pigments.

66.2.2 Chloride process

The chloride process for the production of TiO_2 pigments is illustrated in Figure 66.3. The first step of the process is the chlorination of natural and/or synthetic rutile. These raw materials are converted to titanium tetrachloride in a reducing atmosphere. Calcined petroleum coke is used as the reducing agent. The ash and volatile content of the coke is very low. The reaction of rutile with chlorine gas and coke, also referred to as carbochlorination, is strongly exothermic.

$$TiO_2 + 2\,Cl_2 + C \rightarrow TiCl_4 + CO_2 \tag{66.4}$$

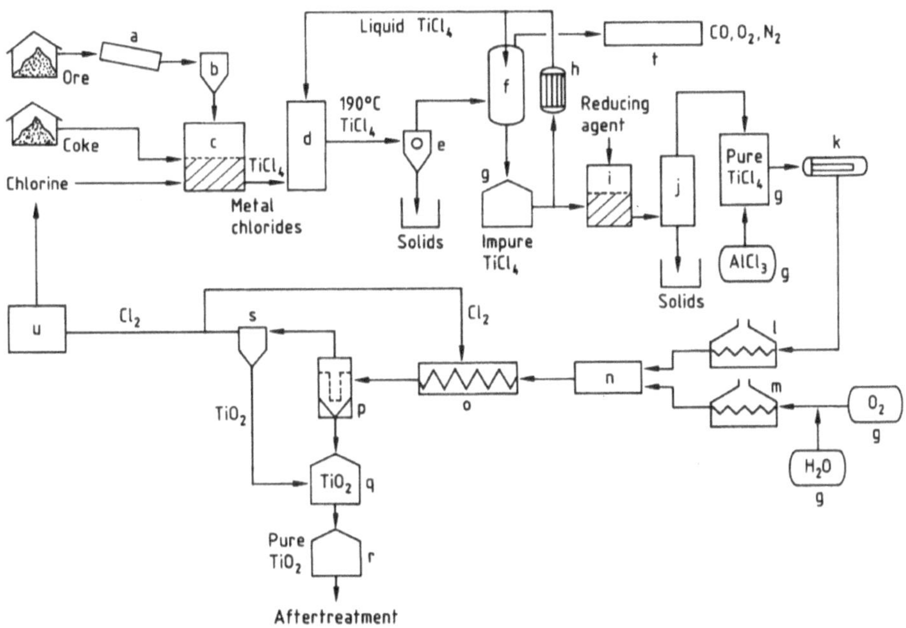

Figure 66.3: Flow diagram of TiO_2 production by the chloride process. (a) mill; (b) silo; (c) fluidized-bed reactor; (d) cooling tower; (e) separation of metal chlorides; (f) $TiCl_4$ condensation; (g) tank; (h) cooler; (i) vanadium reduction; (j) distillation; (k) evaporator; (l) $TiCl_4$ superheater; (m) O_2 superheater; (n) combustion reactor; (o) cooling coil; (p) filter; (q) TiO_2 purification; (r) silo; (s) gas purification; (t) waste-gas cleaning; (u) Cl_2 liquefaction unit [3].

There is also an endothermic reaction in the process, whereby carbon reacts with the generated carbon dioxide under formation of carbon monoxide at rising temperatures

($C + CO_2 \rightarrow 2\,CO$). The reaction process during the carbochlorination is therefore more complex. A partial reaction can be described as follows:

$$TiO_2 + 2\,Cl_2 + 3\,C + CO_2 \rightarrow TiCl_4 + 4\,CO \qquad (66.5)$$

The formed carbon monoxide contributes to the reaction as a further reducing agent:

$$TiO_2 + 2\,Cl_2 + 2\,CO \rightarrow TiCl_4 + 2\,CO_2 \qquad (66.6)$$

Oxygen is blown into the reactor in addition to the chlorine to maintain the reaction temperature in the range of 800 to 1200 °C. The consumption of coke per ton of TiO_2 is 250 to 300 kg in case that pure chlorine is used.

The chemical reaction of rutile with chlorine gas and coke is carried out using a fluidized-bed reactor. The particle size of the titanium–containing raw material is comparable with the grains of sand, the size of the coke particles is five times larger. The adjustment of the particle size is done by a suitable milling procedure (a). The ground material is stored temporarily in a silo (b) from where it is conveyed to a brick-lined fluidized-bed reactor (c), which is heated to temperatures of 800 to 1200 °C. The titanium in the raw material and the chlorine react to nearly 100% to $TiCl_4$. A part of the other metals contained in the raw material besides titanium is also converted into volatile chlorides. These metal chlorides are also found in the reaction gases.

Cooling of the reaction gases is done with liquid $TiCl_4$ either indirectly or directly (d). Crystallization of metal chlorides during cooling is problematic because the solid chlorides tend to deposit on cold surfaces of the equipment. $FeCl_2$ and $FeCl_3$ formed during the chlorination of iron-containing raw materials are separated. The reaction gases are cooled down primarily only to temperatures slightly below 300 °C. The accompanying chlorides can be separated under these conditions from the volatile $TiCl_4$ by condensation or sublimation (e). $TiCl_4$ is the main component of the reaction gas at that time. The gas is cooled down to a temperature below 0 °C to bring the $TiCl_4$ to condensation (f). Small amounts of $TiCl_4$ and chlorine in the exhaust gas, mainly consisting of carbon monoxide, carbon dioxide and nitrogen, are removed by wet scrubbing with alkaline compounds (t).

Purification of the $TiCl_4$ is done by distillation. Remaining metal chlorides can be removed from the $TiCl_4$ by evaporation (j). Only the volatile $TiCl_4$ evaporates at this point of the process. It can be collected after condensation as clear and colorless liquid. Dissolved chlorine can be removed by reduction with iron, copper or tin powder [3]. The removal of residual vanadium tetrachloride (VCl_4) and vanadium oxychloride ($VOCl_3$) from the $TiCl_4$ is more complicated. The two vanadium compounds are separated by reduction to form solid vanadium chlorides, mainly VCl_3 (i). Convenient reducing agents for this step are copper, titanium(III) chloride, hydrogen sulfide, and several hydrocarbons. After subsequent evaporation (j), the $TiCl_4$ contains typically less than 5 ppm vanadium. The separation of vanadium is necessary because larger quantities of this metal in the TiO_2 would lead to a yellow tinge.

The purified titanium tetrachloride reacts in the following step with oxygen at 900 to 1400 °C in a combustion reactor to form TiO_2 and chlorine (n). The purified $TiCl_4$ is vaporized in an evaporating unit (k). The vapor is indirectly heated in a superheater (l) to temperatures of 500 to 1000 °C. Oxygen is heated separately in another superheater to more than 1000 °C (m). The reaction of $TiCl_4$ with oxygen can be described according to the following equation:

$$TiCl_4 + O_2 \rightarrow TiO_2 + 2\,Cl_2 \qquad (66.7)$$

The thus formed TiO_2 has already useful pigment properties. The reaction for the TiO_2 formation is moderately exothermic. A high reaction temperature is necessary and titanium tetrachloride and oxygen are therefore heated up prior to the reaction. The high temperature can be achieved in the case of oxygen by using an electric plasma flame or by indirect heating. Hot $TiCl_4$ and oxygen are blown through separate nozzles into the reaction chamber and are mixed there as fast and complete as possible. In this way, a high transformation rate is assured. A subsequent cooling unit (o) is constructed in such a way that the pigment is cooled down very quickly to temperatures below 600 °C.

The TiO_2 formed during the process is cooled further during the separation of the TiO_2 either indirectly or directly by solid particles, mostly by sand. Filtration (p) is used to separate the TiO_2 particles from the gas mixture and from the sand, which is much coarser than the pigment. The gas is returned to the cooling zone (o) of the combustion reactor. It can also be conveyed back to the carbo-chlorination process as oxygen-containing chlorine via the liquefaction unit (u). Chlorine adsorbed on the TiO_2 particles can be removed by flushing with nitrogen and oxygen or by heating.

Most important parameters with an influence on the properties of the final TiO_2 pigments are the reaction temperature, the excess of oxygen, and the rheological conditions in the combustion reactor. The reactor design is a decisive factor for the achievement of selected pigment qualities. Caking of TiO_2 on the reactor walls must be avoided because it leads in most cases to a loss in quality.

The combustion of vaporized $TiCl_4$ with oxygen using the described conditions typically generates TiO_2 pigments in the rutile modification. The calcination of TiO_2 hydrates as practiced in the sulfate process leads either to anatase or to rutile pigments. The TiO_2 modification depends in this case strongly on the lattice stabilization by doping. Many of the TiO_2 pigments produced in this way contain anatase together with rutile. The ratio of the two modifications in a specific pigment can decisively be adjusted by the reaction parameters, mainly by the temperature used for the TiO_2 formation. As a rule: the higher the calcination temperature the more rutile is formed.

TiO_2 pigment particles are often coated with a layer consisting of selected metal oxides, metal hydroxides or organic compounds. Such so-called surface treatments are necessary depending on the requirements of a specific application. The variety of TiO_2 pigments ranges from anatase to rutile types and from untreated to surface treated pigments. The diversity of TiO_2 pigments is based also on the many different requirements for the granulometric properties (mean particle size, particle size distribution, particle shape, specific surface area).

66.2.3 Inorganic and organic surface treatment

Surface treatment, also referred to as aftertreatment is an additional process step after the synthesis of a pigment. It is carried out with the aim to improve weather resistance and lightfastness of the pigment itself and of the pigmented organic matrix. Another aspect is the improvement of the dispersibility of the pigments in the application system. Surface treatment includes the coating of the individual pigment particles with thin and colorless inorganic and organic layers of low solubility.

The appearance of surface treated pigments compared to untreated pigments should be thereby not or only little affected. Suitable treatments prevent the direct contact between the application medium and the reactive surface of the pigment particles. The properties of TiO_2 and other pigments can be controlled mainly by the chemical composition and by the reaction parameters used for the treatment. The adjustment of suitable densities and porosities of the coating is of high importance for the pigment characteristics.

Inorganic surface treatments can be produced using different chemical processes. A widely used route is the precipitation of the components of the surface treatment from aqueous solutions onto the suspended TiO_2 particles. Metal oxides and oxide hydrates are deposited in most cases using specific precipitation parameters. The deposition of the components of the treatment is typically proceeded in a batch process in stirred vessels, either successively or jointly. Common compounds used for the surface coating besides metal oxides and oxide hydrates are silicates, and phosphates of titanium, zirconium, and aluminum. Cerium, tin, vanadium, boron, zinc, manganese, or antimony are used as components of the treatment in some special cases [6–8].

A second route for the aftertreatment of TiO_2 pigments is the adsorption of metal oxides, oxide hydrates or other suitable compounds on the surface of the TiO_2 particles during grinding. The surface treatment obtained using this method leads in most cases only to a partial coating of the pigment surface.

The deposition of the components of the surface treatment from the gas phase by hydrolysis with water vapor or by decomposition of volatile substances, e.g., of chlorides or suitable organometallic compounds is another technique used for the aftertreatment of TiO_2 pigments. This route can be considered mainly for TiO_2 pigments produced by the chloride process.

TiO_2 pigments with an inorganic surface treatment can be divided in qualities with dense surface coatings for paints and plastic applications (prepared by homogeneous precipitation of SiO_2 in aqueous suspension [9], by successive precipitation of the treatment components in aqueous suspension combined with calcination at temperatures of 500 to 800 °C [10], by aftertreatment with zirconium, titanium, aluminum, and silicon compounds [11, 12], by aftertreatment with merely 1 to 3% of alumina), in pigments with porous surface coatings for the use in emulsion paints (prepared by treatment with titanium, aluminum, and silicon compounds), and in pigments with dense surface coatings and excellent lightfastness for the paper laminate industry (prepared by surface coating of highly lattice-stabilized TiO_2 types with silicates and phosphates of zirconium, titanium, and aluminum [2–4].

66.3 Pigment properties and uses

TiO_2 is chemically very stable and insoluble in water and organic reagents. It is dissolved in hot concentrated sulfuric acid, hydrofluoric acid and hot alkaline solutions and can be digested by potassium hydrogen sulfate (acid digestion). Solved alkali metal titanates and free titanic acids are unstable in aqueous systems and amorphous TiO_2 hydrates are formed by the reaction with water.

TiO_2 reacts at high temperatures with reducing agents, in particular with hydrogen or ammonia, under formation of titanium suboxides (TiO, Ti_2O_3, Ti_3O_5, Ti_4O_7). The reaction of TiO_2 with chlorine in the presence of coke takes place above 700 °C. Titanium tetrachloride is formed under these conditions (chloride process).

The specific surface area of commercially available TiO_2 powders range from 0.5 to more than 300 m^2/g depending on the manufacturing parameters. Coordinatively bonded water is located at the surface of the TiO_2 particles. This water is bound in the form of hydroxyl groups. These groups have acidic or basic properties depending on the nature of the bonding of the hydroxyl species to the titanium atoms [13, 14]. In consequence of this, the surface of TiO_2 particles has always a polar character. Hydroxyl groups at the surface of the particles have a decisive influence on the properties of TiO_2 pigments, mainly on the dispersibility and the weather resistance. The known photochemically induced reactions of TiO_2 are strongly related with the presence of hydroxyl groups at the surface of the particles. An example of this is the decomposition of water into hydrogen and oxygen and the reduction of nitrogen to ammonia and hydrazine in the presence of TiO_2 [15].

The most important properties of TiO_2 pigments for the application are scattering power, hiding power (tinting strength), whiteness, dispersibility, lightfastness, and weather resistance. These characteristics depend strongly on the chemical purity, the lattice stabilization, the particle size, the size distribution, and the surface treatment. There is, on the other hand, also a significant dependency on the composition of the application system.

The optical behavior of rutile and anatase pigments is related to their exceptionally high refractive indices. These are with 2.70 and 2.55 even higher than the refractive index of diamond (2.42). The pigmentation of a binder or another application medium (refractive indices mostly between 1.5 and 1.6) with TiO_2 leads to a relative refraction coefficient (refraction index of the pigment/refraction index of the binder) of 1.5 to 2.0.

The scattering power, respectively, the hiding power depends on the refractive index of a TiO_2 powder and on its particle size. The maximum for the scattering of rutile pigments is according to Mie's theory at a particle size of 0.19 µm and for anatase pigments at 0.24 µm (calculated for a wavelength of 550 nm) [16]. The optimal particle sizes for rutile and anatase powders correspond consistently with these values. The scattering power of a TiO_2 pigment is also dependent on the wavelength of the incident light. The pigments with smaller particle sizes scatter light of shorter wavelengths more strongly and exhibit therefore a slightly bluish shade, while larger particles show a more yellowish tone.

Whiteness, brightness, and mass tone of TiO_2 pigments depend mainly on the crystal modification, the particle size, and the purity of the TiO_2. The shift of the absorption band of anatase pigments into the UV region is the reason why they have a less yellow undertone compared with rutile pigments. Pigments produced by the chloride process are characterized by higher color purity (neutral hue) and higher brightness values than pigments from the sulfate process. This observation can be explained by the effective distillation step in the chloride process, which leads to very pure $TiCl_4$ and after oxidation to very white and bright TiO_2 pigments. High-quality TiO_2 pigments with sufficiently good whiteness and brightness properties are also produced using the sulfate process, where purification steps are part of the synthesis.

TiO_2 pigments manufactured by the sulfate process and by the chloride process are exchangeable for a wide range of applications. However, sulfate-derived pigments offer advantages for some uses, e.g., in printing inks, because they are less abrasive than chloride-derived products.

Good dispersion of the TiO_2 particles in the application medium is a prerequisite for high gloss and low haze in the pigmented surface or product. This can normally be achieved by intensive grinding and by the use of pigments with a suitable organic surface treatment.

Lightfastness and weather resistance of TiO_2 pigments in the application system is of highest importance. A problem of TiO_2 containing paints and coatings, which often appears during weathering, is pigment chalking [17]. Binder systems with low permeability to oxygen have an advantage with regard to chalking. The absence of oxygen during weathering contributes also to the prevention of chalking. On the other hand, a graying reaction can happen under these conditions, which is less pronounced during the exposure to air. Water has an influence on the graying of binder systems, too. Graying is greatly reduced in the absence of water. Both chalking and graying are less severe with rutile pigments because rutile has more stable crystal structures. A suitable inorganic surface treatment improves the stability for both pigment types significantly. Doping and lattice stabilization are used in addition to stabilize TiO_2 pigments. In particular, rutile pigments are stabilized in this way.

The problems of anatase and rutile pigments in various binder systems with regard to lightfastness and weather resistance can be described using the so-called chalking cycle [2, 18]. In a first step, water molecules interact with the TiO_2 particles. Hydroxyl groups are formed thereby and are bound on the TiO_2 surface.

The absorption of UV light leads to a charge separation in the crystal lattice of the TiO_2. Bond electrons are raised to a higher energy level (from valence band to the conduction band corresponding to the band model), where they move freely. Electrons in the conduction band and positively charged "holes" in the valence band are formed.

$$TiO_2 + h\nu \rightarrow TiO_2\,(e^- + p^+) \qquad (66.8)$$

The holes are mobile and use "hole conduction" for their motion. This means that charge exchange with neighboring lattice sites occurs. Separated electron/hole pairs, so-called excitons, are formed. These can be understood as chemically very active states of the TiO_2 particles. Excitons that migrate to the surface of the TiO_2 particles interact with the components of the organic binder system. The organic material is degraded initially accompanied by a loss of gloss and finally by chalking. Chalking means in this context that the pigment particles in suchlike weathered surfaces are no longer firmly embedded. The particles can be removed from such surfaces by simply wiping.

OH^- ions from the surface of the pigment particles are oxidized in a next step by the positively charged holes to $OH^\cdot$ radicals on the TiO_2 surface.

$$p^+ + OH^- \rightarrow OH^\cdot \qquad (66.9)$$

The so-formed $OH^\cdot$ radicals are very reactive and able to attack polymer molecules of the binder system. A separation of hydrogen by these radicals under formation of water and polymer radicals $R^\cdot$ can take place now. The polymer radicals react further, with oxygen to form peroxyl radicals R-O-$O^\cdot$. Their disintegration to other polymer radicals of the general type $R_1^\cdot$ or $R_2^\cdot$ is likewise possible. The erasure of holes initiated by the reaction with OH^- ions leads to an excess of electrons in the conduction band of the TiO_2. These electrons are able to reduce Ti^{4+} in the TiO_2 lattice to Ti^{3+}.

$$Ti^{4+} + e^- \rightarrow Ti^{3+} \qquad (66.10)$$

The Ti^{3+} ions can be oxidized again by present or newly adsorbed oxygen. Ti^{4+} ions and oxygen radical anions are formed during this reaction.

$$Ti^{3+} + O_2 \rightarrow Ti^{4+} + {}^\cdot O_2{}^- \qquad (66.11)$$

Oxygen radical anions generated in this way are adsorbed on the TiO_2 surface. These radical ions react with moisture under formation of hydroperoxyl radicals.

$$Ti^{4+} + {}^\cdot O_2{}^- + H_2O \rightarrow Ti^{4+} + OH^- + {}^\cdot O_2H \qquad (66.12)$$

Hydroperoxyl radicals can desorb from the TiO_2 surface and attack the polymer molecules of the binder. The OH^- anions react with the TiO_2 to OH groups bound to the surface.

In summary, the chalking cycle can be regarded as the reaction of water and oxygen to form $OH^{\cdot}$ and $HO_2^{\cdot}$ radicals under the influence of UV radiation and the catalytic activity of the TiO_2 surface.

$$H_2O + O_2 + h\nu \rightarrow OH^{\cdot} + HO_2^{\cdot} \tag{66.13}$$

The exclusion of air and/or water interrupts the cycle after the reduction of Ti^{4+} to Ti^{3+} (eq. (66.10)). The concentration of Ti^{3+} in the TiO_2 lattice is increased when oxygen is excluded or a binder with a very low diffusion rate for oxygen is used. In this case, graying due to the inherent color of the Ti^{3+} ions can be observed. Graying is reversible and decreases with gradual exposure to oxygen. The formation of surface hydroxyl groups and hydroperoxyl radicals (eq. (66.12)) does not take place when water can be excluded. Moisture extension is therefore a way to stop the degradation of the binder. A suitable surface treatment, doping, and lattice stabilization of rutile pigments are in most cases suitable to prevent the photochemical degradation of the binder. Application media without pigmentation such as coatings are degraded by exposure to sun light and weathering. Stabilized TiO_2 pigments prevent light penetration into deeper areas of the coating film and thus inhibit the degradation and finally the breakdown of the binder.

TiO_2 pigments are used universally in nearly all application systems such as paints and coatings, printing inks, plastics, fibers, papers, cosmetics and ceramics. Paints and coatings constitute the largest part of these TiO_2 pigment applications. The pigment volume concentration of TiO_2 is usually 10 to 35% in gloss paints and coatings. In exceptional cases, e.g., in matt emulsion paints, the concentration can exceed 80%. TiO_2 pigments concentrations in printing inks are normally in the range from 8 to 15%.

TiO_2 pigments are used in plastics in a broad manner. White coloration of durable and nondurable goods are important applications for TiO_2 pigments. Brightening of colored plastics by addition of TiO_2 is another application purpose. The ability of TiO_2 pigments to absorb UV radiation in the wavelength range of smaller than 415 nm is a valuable property usable for the protection against these harmful rays. The use of TiO_2 pigments in synthetic fibers creates a solid appearance.

Another application field for TiO_2 pigments is the paper industry. Fillers like kaolin, chalk, or talc are preferably used as brightening materials and opacifiers for paper applications. TiO_2 pigments, on the other hand, are used for very white paper that has to be opaque even when it is very thin. The pigments can be incorporated directly into the paper during paper manufacturing or applied as a coating to give a superior quality ("art" paper).

TiO_2 pigments, particularly anatase pigments, are used in many cosmetic formulations. TiO_2 in cosmetic products generates white effects and covers the skin. It has the best coverage behavior amongst all white pigments due to the very high refractive index. TiO_2 pigments are used in all varieties of cosmetic emulsions, pastes, pencils, and suspensions. They are also applied in soaps and toothpastes. Day creams and eye shadows belong to the typical decorative applications of TiO_2 pigments in the cosmetics industry. The pigments are also used, based on their spectral properties, as UV absorbers in sunscreen products. Nano-sized rutile pigments with narrow particle size distribution are preferably used for sun protection.

TiO_2 pigments are also applied as food additives. They are nontoxic and approved for food without any quantity restriction. The pigments are particularly used for food coatings, chewing gum, and coloring of medicinal products, e.g. coating of tablets.

TiO_2 pigments are applied also in the enamel and ceramic industries, for the manufacture of white cement, and for coloring of rubber and linoleum. The pigments fulfill the requirements regarding the high temperatures used in some of these industries.

TiO_2 white pigments can be converted to electrically conductive pigments by an additional coating with conductive oxidic compositions, such as indium tin oxide (ITO) or antimony tin oxide (ATO) [19]. Such pigments are characterized by a slightly or more intensive gray color depending on the composition of the conductive coating. They are used in photosensitive papers for electro-photography or for the antistatic treatment of coatings and plastics.

TiO_2 pigments are nontoxic unless they are contaminated with heavy metal compounds. They do not exhibit acute toxicity (LD_{50} value rat oral: >5000 mg/kg, inhalational LC_{50} value rat >6.82 mg/l, 24 h). TiO_2 pigments are not irritating to skin or mucous membranes. Mild irritation to the eyes and the respiratory tract due to mechanical abrasion initiated by TiO_2 particles is possible. It is important to avoid the inhalation of pigment dust, no matter whether it is TiO_2 or another pigment. The inhalation of smaller particles into the lung is more detrimental compared to larger particles Specific effects of TiO_2 resulting in inflammation of living beings are not known. Nevertheless, there is a classification of TiO_2 under the EU CLP regulation from February 18, 2020 classifying certain TiO_2 powders and powder mixtures containing TiO_2 as suspected carcinogens (category 2) via inhalation [20].

Chronic effects observed during the manufacture and application of TiO_2 powders over many years and physical hazards caused by the use of TiO_2 have not been reported.

Special risks of nano-sized TiO_2 pigments (transparent pigments) in manufacture and application must be discussed in conjunction with the general evaluation of nanopowders (powders with a substantial proportion of particles with a size smaller than 100 nm). It should be pointed out here that TiO_2 nanopigments, as well as other nano-sized pigments, are typically agglomerated and aggregated into larger units, which are not nanomaterials per se. The pigment particles are dispersed in the application in a liquid matrix, which is typically transferred into cured products such as paints, plastics, or printed materials. There is no potential for the inhalation of nanoparticles starting with their transfer into a dispersion.

References

1. Elfenthal L, Schmelzer J. In Kittel - Lehrbuch der Lacke und Beschichtungen. Spille J, editor. vol. 5. 2nd ed. Stuttgart/Leipzig: S. Hirzel Verlag, 2003:24.
2. Winkler J. Titanium Dioxide. Hannover: Vincentz Verlag; 2003.
3. Auer G. In industrial inorganic pigments. Buxbaum G, Pfaff G, editors. 3rd ed. Weinheim: Wiley-VCH Verlag, 2005:51.
4. Pfaff G. Inorganic pigments. Berlin/Boston: Walter de Gruyter GmbH, 2017:48.
5. Leutz R. Gewinnung und Aufbereitung von Schwermineralsanden in Camaratuba/Brasilien. Erzmetall. 1989;42:383.
6. Patent DE. 23 42 889 (DuPont). 1973.
7. Patent US. 4,781,761 (DuPont). 1987.
8. Gesenhues U. Coprecipitation of hydrous alumina and silica with TiO_2 pigments as substrate. J Colloid Interface Sci. 1994;168:428.
9. Patent US. 2,885,366 (DuPont). 1956.
10. Patent GB. 1008652 (Tioxide Group Ltd.). 1961.
11. Patent DE. 12 08 438 (Titangesellschaft mbH). 1960.
12. Patent DE. 146 712 (British Titanium Products). 1965.
13. Boehm HP. Funktionelle Gruppen an Festkörperoberflächen. Chem Ing Tech. 1974;46:716.
14. Primet M, Pichat P, Mathieu MV. Infrared studies of the surface of titanium dioxides. I. Hydroxyl groups. J Phys Chem. 1971;75:1216.
15. Schrauzer GN, Guth TD. Photolysis of water and photoreduction of nitrogen on titanium dioxide. Am Chem Soc Div Org. 1977;99:7189.
16. Kubelka P, Munk FZ. Ein Beitrag zur Optik der Farbanstriche. Tech Phys. 1931;12:539.
17. Kämpf G, Papenroth W, Holm R. Systematische Untersuchungen der mikromorphologischen Abbauvorgänge an bewitterten Beschichtungssystemen mit unterschiedlichem Aufbau. Farbe + Lack. 1973;79:9.
18. Völz HG, Kämpf G, Klaeren A. Die photochemischen Abbaureaktionen bei der Bewitterung TiO_2-pigmentierter Bindemittel. Farbe + Lack. 1976;82:805.
19. Patent JP. 58025-363 (K. K. Ricoh). 1981.
20. EU Regulation No 1272/2008 on classification. labelling and packing (CLP) of substances and mixtures. 2008.

Gerhard Pfaff

67 Transparent pigments

Abstract: Transparent pigments are characterized by very small particles with sizes in the range below 100 nm and large specific surface areas. Most of the technically relevant pigments consist of inorganic compounds. Transparent pigments are classified as nanomaterials. Nanosized primary particles do not or only slightly scatter visible light, which is the reason for their transparency. Pigmentation with these pigments leads to a transparent appearance of the application systems. There are colored and colorless transparent pigments. If colored pigments such as α-FeOOH, α-Fe2O3, and $CoAl_2O_4$ are used, the application medium takes on color while remaining transparent. Colorless transparent pigments are nearly invisible in the application system. Only in combination with effect pigments or conventional colored pigments, effects with a special appearance are possible, e.g., frost effects. There are two industrially relevant colorless transparent pigments, TiO_2 and ZnO. Both pigments find their main applications as functional powders. They are characterized by strong UV absorption and are therefore applied as UV absorbers. As such, they are used for the protection of organic materials such as plastics and coatings. They are used in addition in a broad diversity of sun screen products where they protect human skin. The manufacture of transparent pigments takes place mostly using wet-chemical or gas-phase reactions.

Keywords: transparent pigments, cobalt blue, iron blue, iron oxide, titanium dioxide, zinc oxide, UV absorption

67.1 Fundamentals and properties

Transparent pigments are colorants showing transparent properties in application systems. They consist typically of inorganic oxidic materials. Organic transparent pigments have nearly no technical importance. Typical particle sizes of transparent pigments are in the range below 100 nm. The smaller the particles, the more transparent the pigments are. Transparent inorganic pigments can be classified into colored and colorless types. Colorless transparent pigments such as nano-sized titanium dioxide and zinc oxide scatter the light of the visible spectral range complete without any absorption. To a certain degree, these pigments may be regarded as functional nanomaterials with special properties. Nano-sized colored pigment types such as transparent

This article has previously been published in the journal Physical Sciences Reviews. Please cite as: G. Pfaff Transparent Pigments *Physical Sciences Reviews* [Online] 2021, 6. DOI: 10.1515/psr-2020-0200

https://doi.org/10.1515/9783110587128-067

iron oxides are mainly characterized by their specific color properties. Some of these colorants have also functional properties [1, 2].

Transparency and hiding power of pigments depend mainly on the light scattering behavior of the particles in the respective application system. The size of the particles is the decisive factor for the scattering behavior. Scattering of systems pigmented with transparent colorless particles depends in addition on the difference between the refractive indices of the pigment and the binder, respectively. The scattering power grows with increasing particle diameters and decreases with declining sizes. Maximum transparency of a medium containing a transparent pigment is achieved when the dispersion of the particles is done in the optimal manner. The destruction of aggregates and agglomerates is of high importance for the effect to be achieved. Powders consisting of very small particles have typically a large surface area and need more effort to destroy aggregates and agglomerates. The specific surface area of transparent pigments exceeds in most cases the value of 100 m^2/g.

Dispersion with the highest possible number of primary particles in the application medium is for transparent pigments just as important as for all other pigments. The dependency of transparency, scattering, absorption, and opaqueness (hiding power) of a colored pigment can be summarized as follows:

Very small particles (30–100 nm): high transparency, very low scattering, high absorption.

Particles in the range from 100 to 800 nm: increasing scattering up to a maximum in the range from 300 to 700 nm, medium absorption, high opaqueness (hiding power).

Particles in the range from 800 nm to 2000 nm: decreasing scattering, low absorption, medium transparency, respectively, opaqueness.

Transparent pigments with a high degree of agglomeration and/or aggregation act comparably to pigments consisting of larger particles. The appearance is similar in these cases to a conventional pigment.

Technically important transparent pigments are titanium dioxide (micronized TiO_2), zinc oxide, the iron oxides α-FeOOH and α-Fe_2O_3, cobalt blue, and iron blue.

Main application segments for transparent pigments are decorative paints (effect lacquers, e.g., frost effects), printing inks, plastics, and wood protection varnishes. In all these cases, the high UV absorption and the light fastness of the pigments are used for the protection of the binder matrix or the subsurface. The production processes for transparent pigments correspond to those of conventional inorganic pigments. There is one decisive difference in manufacturing processes, which is to avoid excessive particle growth.

Transparent pigments are nanomaterials with all the advantages and disadvantages of this substance class. A special risk management for the manufacture and application of transparent pigments is therefore necessary. From the occupational health and safety perspective, especially the operation with nanoscale dust requires attention. Measures must be taken therefore to minimize the exposure to nanomaterials for employees affected [2]. Information on the toxicology of the transparent pigments can be found in related publications for the single pigments.

67.2 Transparent titanium dioxide

Titanium dioxide powders with primary particle sizes of 5–30 nm are characterized by transparency in the visible spectral range. Rutile and anatase are of application-technical interest in form of their transparent pigment forms. The specific surface areas of transparent titanium dioxides are in the range from 80 to 200 m^2/g and correspond with the very large specific surface areas.

Transparent TiO_2 pigments with their very small primary particles do not or only slightly scatter visible light. They have, however, a strong UV absorption ability. This property is used in some of the main application fields for transparent TiO_2 pigments. They act as UV absorbers in various systems, where they protect the organic materials. Typical examples for this application are plastics and coatings. The protection function is also used for the human skin where transparent TiO_2 pigments are applied in a broad diversity of sun screen products. TiO_2 nanoparticles are also characterized by a relatively high surface reactivity that need to be considered for several applications. This reactivity is particularly important for photocatalytic processes.

Transparent TiO_2 pigments are suitable to achieve so-called frost effects and of hue shifts in coatings pigmented with colored pigments [3]. Frost effects are generated by the optical interaction of transparent rutile pigments and metal effect pigments in base coats of metallic coatings [4]. The human eye is able to recognize a certain subdued color travel effect if the observation angle on a coated surface is changed. A different interaction of the wavelengths of the visible spectrum with single TiO_2 nanoparticles is the reason for this appearance. The blue parts of the spectrum are scattered strongly whereas the red and the green parts are scattered only slightly by the TiO_2 particles. The scattered blue parts of the light leave the coated surface at a flat angle. Hue shifts in colored coatings modified by the addition of a transparent titanium dioxide pigment are not dependent on the observation angle [5]. The interaction of the TiO_2 particles in the coating with the incident light is responsible for the visible hue shift, e.g., from red to magenta.

Nanosized TiO_2 powders are also used for several functional purposes, e.g., for catalysis (DENOX catalysts), gas purification (gas adsorption at the large surface), and heat stabilization of silicone rubber.

The synthesis of transparent titanium dioxide pigments is carried out by using various technical processes. The ratio of rutile and anatase in the final product can be affected by the chosen reaction parameters. In some processes, almost exclusively rutile is formed. Untreated transparent TiO_2 types show a very high reactivity in regard to photochemical processes, similar to conventional white titanium dioxide pigments. A suitable surface treatment is the best method to stabilize the nanosized powders. The treatment consists in nearly all cases of various combinations of inorganic oxides (e.g., SiO_2, Al_2O_3, ZrO_2) [1].

The following wet-chemical routes can be used for the production of transparent titanium dioxide pigments [1, 2]:

$$TiOSO_4 + H_2O \rightarrow TiO_2 + H_2SO_4 \quad \text{(mainly anatase)} \tag{67.1}$$

$$TiOCl_2 + H_2O \rightarrow TiO_2 + 2\ HCl \quad \text{(mainly rutile)} \tag{67.2}$$

$$TiCl_4 + 4\ NaOH \rightarrow TiO_2 + 4\ NaCl + 2\ H_2O \quad \text{(mainly rutile)} \tag{67.3}$$

$$Na_2TiO_3 + 2\ HCl \rightarrow TiO_2 + 2\ NaCl + H_2O \quad \text{(mainly rutile)} \tag{67.4}$$

$$Ti(OC_3H_7)_4 + 2\ H_2O \rightarrow TiO_2 + 4\ C_3H_7OH \quad \text{(amorphous } TiO_2\text{)} \tag{67.5}$$

Precipitation, neutralization, filtration, washing, drying, and micronizing are the typical process steps for all these routes.

It is basically possible to produce nanosized TiO_2 powders by a suitable gas-phase process. One of these processes is the combustion of titanium tetrachloride in a mixture of hydrogen and oxygen at temperatures below 700 °C (flame pyrolysis) [6].

67.3 Transparent zinc Oxide

Zinc oxide powders with primary particle sizes in the lower nanometer range are characterized by transparent properties in the visible spectral range, comparable with the related titanium dioxides. ZnO has distinct UV absorption properties, comparable with TiO_2. Transparent zinc oxide types are mainly applied for sun protection. They can therefore be classified as functional pigments.

The industrial manufacture of transparent ZnO pigments is carried out mainly by the reaction of vaporized high-purity zinc metal with oxygen in a plasma chamber [1].

$$2\ Zn + O_2 \rightarrow 2\ ZnO \tag{67.6}$$

ZnO powders with primary particle sizes in the range from 20 to 30 nm are formed in this combustion process.

The hydrolysis of diluted zinc organic compounds can also be used to produce nanosized ZnO powders [7].

$$Zn(OR)_2 + H_2O \rightarrow ZnO + 2\ ROH \tag{67.7}$$

The primary particles obtained under these conditions have a size in the range of about 15 nm.

Further possibilities for the production of transparent zinc oxide pigments are sol-gel routes or precipitation reactions in the presence of protective colloids. The colloids are used in this case to limit the particle growth [8]. Typical starting materials for both processes are purified zinc sulfate and zinc chloride solutions.

67.4 Transparent iron oxides

Two compositions of transparent iron oxides are of technical importance: α-FeOOH and α-Fe_2O_3. The particle sizes of both transparent oxides are with 1 to 50 nm in the typical range of nanoparticles. The two pigments cover the color range from yellow to red, as expected. Main applications of transparent iron oxide red and yellow are effect coatings (together with effect pigments), pure yellow and red shades (in some cases together with titanium dioxide pigments), and wood protection coatings. Suitable application systems for the two pigments are baking finishes, water-based binders, acrylic-isocyanate systems, acid-curing systems, amine-curing systems, and air-drying binders [1, 2]. Transparent iron oxide red is additionally used for coloring of transparent plastics. The pigments are preferably introduced as user-friendly preparations because they are much more difficult to disperse in form of powders.

Transparent iron oxide pigments are often manufactured using a precipitation process. The resulting α-FeOOH or α-Fe_2O_3 particles are mostly needle-shaped. Transparent iron oxide red can also be prepared in form of spherical particles (reaction of iron pentacarbonyl with oxygen).

The production of transparent iron oxide yellow starts with the precipitation of extremely fine iron(II) hydroxide or iron(II) carbonate crystallites from very diluted aqueous iron(II) salt solutions:

$$FeSO_4 + 2\ NaOH \rightarrow Fe(OH)_2 + Na_2SO_4 \tag{67.8}$$

Subsequent oxidation of the washed and dried precipitate with atmospheric oxygen leads to the formation of nanosized α-FeOOH particles.

$$2\ Fe(OH)_2 + 1/2\ O_2 \rightarrow 2\ \alpha\text{-}FeOOH + H_2O \tag{67.9}$$

Technical parameters determining the quality of the final pigment are the concentrations of the solutions, the temperature, the oxidation time, the pH value, and the maturing time of the precipitated particles in the suspension. The resulting precipitate is washed by decanting to remove remaining salt, filtered, dried, and ground. Needle-shaped yellow particles with lengths of 50 to 150 nm and thicknesses of 2 to 5 nm are finally obtained [9].

Transparent iron oxide red can be formed by thermal treatment of transparent iron oxide yellow at temperatures in the range from 300 to 500 °C. The resulting α-Fe_2O_3 is needle-shaped as well as the α-FeOOH used as starting material.

$$2\ \alpha\text{-}FeOOH \rightarrow \alpha\text{-}Fe_2O_3 + H_2O \tag{67.10}$$

The iron oxide red obtained by this reaction is ground to destroy agglomerates and aggregates and to achieve the desired particle size distribution.

Another process for the preparation of transparent iron oxide red pigments with a Fe_2O_3 content of about 85% is the precipitation of iron(II) hydroxide or iron(II) carbonate from iron(II) salt solutions followed by the direct oxidation with air at 30 °C in the presence of $MgCl_2$, $CaCl_2$, or $AlCl_3$ [10].

A specific route for the production of semitransparent α-Fe_2O_3 pigments of very high chemical purity is the combustion of iron pentacarbonyl in an excess of air at temperatures of 580 to 800 °C. Carbon dioxide is formed as a by-product of this reaction [1].

$$2\ Fe(CO)_5 + 6\ 1/2\,O_2 \rightarrow \alpha\text{-}Fe_2O_3 + 10\ CO_2 \qquad (67.11)$$

The color of the resulting pigments may range from orange to red. X-ray amorphous particles with sizes from 10 to 20 nm are possible to produce in this way. The particles are characterized by an isometric shape [11, 12]. Transparent α-Fe_2O_3 pigments derived from the carbonyl process show a favorable dispersion behavior in comparison with the products obtained from the precipitation route [1].

The fastness properties of transparent iron oxides are sufficient for the desired applications and comparable with those of conventional iron oxide types. The UV absorption properties of transparent iron oxide pigments are used in a favorable way when plastic bottles and food packaging films are colored with these colorants. Transparent iron oxides are also applied effectively in wood protection coatings.

67.5 Transparent cobalt blue

Transparent cobalt blue pigments consist of very small and thin flakes of mainly hexagonal shape. They crystallize in the spinel lattice like conventional $CoAl_2O_4$ pigments. Transparent $CoAl_2O_4$ pigments are characterized by diameters in the range from 20 to 100 nm, thicknesses of about 5 nm, and specific surface areas of about 100 m^2/g. They have a very good chemical stability combined with an excellent light and weather fastness. Their low color strength derived from the relatively low refractive index is the reason why the pigments find only limited consideration in common application systems.

Transparent cobalt blue powders are synthesized by precipitation of cobalt(II) hydroxide and aluminum hydroxide from diluted solutions followed by filtration, washing and drying. The dried mixture of the two hydroxides in the stoichiometric ratio of 1 to 2 is heated up to temperatures of about 1000 °C. $CoAl_2O_4$ is formed by a solid-state reaction under release of water.

$$2\ Al(OH)_3 + Co(OH)_2 \rightarrow CoAl_2O_4 + 4\ H_2O \qquad (67.12)$$

The resulting powder is ground to destroy aggregates and agglomerates and to achieve the required pigment quality, especially the suitable particle size distribution [13].

67.6 Transparent iron blue

Transparent iron blue pigments are niche products even more than transparent cobalt blue pigments. Main reasons for their low utilization are the weak thermal and the poor alkali stability. Iron blue pigments with particle sizes in the range of 20 to 30 nm are synthesized like conventional iron blue pigments with the difference that solutions of higher dilution are used [14]. Specific grinding and de-agglomeration steps are necessary to adjust the final pigment quality.

References

1. Etzrodt G. In industrial inorganic pigments. Buxbaum G, Pfaff G, editors. 3rd ed. Weinheim: Wiley-VCH Verlag, 2005:261.
2. Pfaff G. Inorganic pigments. Berlin/Boston: Walter de Gruyter GmbH, 2017:197.
3. Winkler J. Titandioxid. Hannover: Vincentz Verlag, 2003:108.
4. Patent US. 4,753,829 (BASF). 1986.
5. Patent DE. 43 23 372 (Sachtleben Chemie). 1993.
6. Patent EP. 0 609 533 (Degussa). 1994.
7. Patent DE. 199 07 704 (Bayer). 2000.
8. Patent JP. 07 232 919 (Nippon Shokubai Co.). 1994.
9. Etzrodt G. In Kittel - Lehrbuch der Lacke und Beschichtungen. Spille J, editor. vol. 5. 2nd ed. Stuttgart/Leipzig: S. Hirzel Verlag, 2003:86.
10. Patent DE. 25 08 932 (Bayer). 1975.
11. Patent DE. 22 10 279 (BASF). 1972.
12. Patent DE. 23 44 196 (BASF). 1973.
13. Patent DE. 28 40 870 (BASF). 1978.
14. Pfaff G. Inorganic pigments. Berlin/Boston: Walter de Gruyter GmbH, 2017:144.

Gerhard Pfaff

68 Ultramarine pigments

Abstract: Ultramarine pigments are representatives of the inorganic blue pigments. They are characterized by intensive shades of blue and relatively weak violet and pink color tones. The composition of the pigments can be summarized by the formula $Na_7Al_6Si_6O_{24}S_3$. The color of ultramarine is generated by incorporation of free polysulfide radicals in the structure of a sodalite-derived crystal lattice. Ultramarine pigments are synthesized using a series of complex solid-state reactions. The pigments are mainly used in plastics, paints and powder coatings and printing inks, but also in paper and paper coatings.

Keywords: ultramarine pigments, polysulfide radicals, solid-state reaction

68.1 Fundamentals and properties

Ultramarine pigments belong to the most important inorganic blue pigments. Some variations of ultramarine with minor technical importance are characterized by violet or red colors. Synthetic ultramarine is the synthetic equivalent to the naturally occurring mineral lapis lazuli, which is found in high quality in Afghanistan and Chile. Lapis lazuli is a representative of the semi-precious stones. It is used since prehistoric times for decorative applications. A blue pigment was obtained out of the mineral already in early days by grinding and separation of the lighter rock using a kneading process. Blue pigment powders produced in this way were named "ultramarines", because they were brought to the main users "across the sea".

Ultramarine pigments were soon an indispensable colorant for artists' colors. The deep blue color, the excellent light fastness and the color purity were decisive for the use in art. High transportation costs, depletion of mines and subsequent processing were the reasons, why the pigments were more expensive than gold at the end of the eighteenth century. Scientists in Europe started activities in the 1820's to find a synthetic and economic process for the manufacture of ultramarine blue [1–4]. It is supposed that Guimet in France and Gmelin in Germany developed independently similar processes for the synthetic preparation of ultramarine in 1828 [5].

Synthetic ultramarines pigments are commercially available in three colors:

This article has previously been published in the journal Physical Sciences Reviews. Please cite as: G. Pfaff Ultramarine Pigments *Physical Sciences Reviews* [Online] 2021, 6. DOI: 10.1515/psr-2020-0201

https://doi.org/10.1515/9783110587128-068

Reddish blue, C.I. Pigment Blue 29.
Violet, C.I. Pigment Violet 15.
Pink, C.I. Pigment Red 259.

Ultramarines are based on sulfur-containing sodium aluminosilicate compositions. They have a sodalite crystal lattice corresponding to the formula $Na_6Al_6Si_6O_{24}(NaS_n)$ with n-values from 2 to 4. Modern spectroscopic methods contributed finally to the clarification of the nature of the color giving components in ultramarines [6–9]. The attendance of free polysulfide radicals in the crystal structure, responsible for the color, was detected by electron paramagnetic resonance spectroscopy. Such radicals are normally very unstable in the solid state, but incorporated as negatively charged polysulfide ions, e.g., as S_3^- in the aluminosilicate lattice together with sodium counterions, they are stabilized. The crystal structure of ultramarine pigments is shown in Figure 68.1.

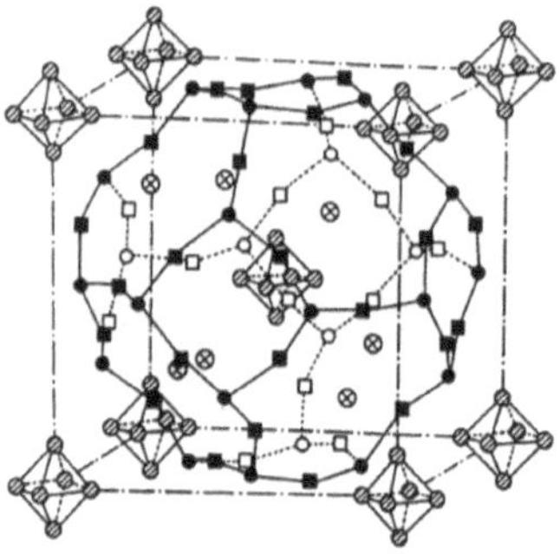

Figure 68.1: Crystal structure of ultramarine pigments [4].

Equal numbers of silicon and aluminum ions are present in the simplest case of the three-dimensional basic structure of ultramarine. The basic lattice unit is $Na_6Al_6Si_6O_{24}$ or $(Na^+)_6(Al^{3+})_6(Si^{4+})_6(O^{2-})_{24}$. The specific nature of the incorporated polysulfide groups is responsible for the color of the pigments derived from this structure [10–15]. The basic lattice $(Na^+)_6(Al^{3+})_6(Si^{4+})_6(O^{2-})_{24}$ can be derived from SiO_2 respectively $Si_{12}O_{24}$ by substituting six of the silicon ions by aluminum ions. Every Al^{3+} ion must be incorporated together with a Na^+ ion to keep the overall ionic charge for the structure at zero. Six of the available eight sodium sites in the crystal structure are always occupied by sodium ions that are required for the stability of the lattice. The remaining two sodium sites are filled with sodium associated with polysulfide groups. Only one S_3^{2-} polysulfide ion can therefore be inserted into the sodalite lattice (as Na_2S_3). The subsequent oxidation to S_3^- leads to a loss of one of the accompanying sodium ions. This explains a basic ultramarine lattice formula of $Na_7Al_6Si_6O_{24}S_3$ [3].

There are two types of anionic sulfur groups in ultramarine blue, S_3^- and S_2^-. Both species are free radicals that are stabilized in the basic crystal lattice. S_3^- radical ions absorb a broad energy band in the visible green–yellow–orange region centered at 600 nm. S_2^- radical ions, on the other hand, absorb in the violet–ultraviolet region at 380 nm [3].

The color quality of ultramarine pigments is adjusted by the sulfur content in the crystal lattice. Higher sulfur amounts in the structure leading to more intense colors can be obtained using high silicon felspar in the production process. The number of sodium ions needed for lattice stabilization is reduced by this approach and more polysulfide groups are incorporated. A product resulting from this manufacturing procedure is $Na_{6.9}Al_{5.6}Si_{6.4}O_{24}S_{4.2}$. It is characterized by a stronger, redder shade of blue compared with the simpler composition.

Ultramarine violet and ultramarine pink have a slightly modified lattice structure. The polysulfide color centers are further oxidized in this case, possibly to S_4, or S_4^- [3].

Synthetic as well as natural ultramarines with sodalite-related crystal structures belong to the class of clathrates. They are characterized by a system of very small cavities (cages). Their lattice paths are up to 0.4 nm corresponding with the diameter of the cavities in the structure. It is possible to exchange the sodium ions by other metal ions, e.g., potassium, lithium, silver or copper. Slightly redder shades of ultramarine blue can be generated by the substitution of sodium by potassium ions [16].

68.2 Production of ultramarine pigments

The manufacture of ultramarine pigments is based on a complex solid-state chemistry. The process consists of the steps clay activation, blending and heating of the raw materials, oxidation and finally purification and refinement. China clay (kaolinitic clay), feldspar, anhydrous sodium carbonate, and sulfur are used as raw materials. In addition, a reducing agent, e.g. oil or coal, is necessary.

The clay activation is done continuously in a rotary or a tunnel kiln or in a batch process in crucibles in a tiled stove. Kaolinite ($Al_4(OH)_8Si_4O_{10}$) is converted to metakaolinite ($Al_4(OH)_{8-2x}OSi_4O_{10}$) with the release of water at temperatures of about 700 °C.

$$Al_4(OH)_8Si_4O_{10} \rightarrow Al_4(OH)_{8-2x}OSi_4O_{10} + x\ H_2O \quad (68.1)$$

The activated clay is then blended with the other raw materials followed by dry-grinding. This process step is carried out in batch or continuous ball mills. Mean particle sizes of about 15 µm are obtained. Table 68.1 contains typical recipes for a green shade and a red shade type of ultramarine blue pigments.

The ground mixture is fired at temperatures of 750 to 800 °C under reducing conditions, mostly in a batch process. Directly fired kilns with the blend in lidded

Table 68.1: Typical recipes for ultramarine blue pigments (in wt %) [3].

Component	Green shade type	Red shade type
Calcined clay	32.0	30.0
Feldspar		7.0
Sodium carbonate	29.0	27.0
Sulfur	34.5	33.0
Reducing agent	4.5	3.0

crucibles of controlled porosity or muffle kilns are used for this step. The powder mixture is often compacted to form bricks with the aim to improve the throughput. The bricks are stacked in a defined pattern. They are indirectly heated up in chambers using gas-fired burners [3]. Sodium carbonate reacts in a very complex process under reducing conditions (sulfur dioxide atmosphere) with the sulfur and the reducing agent at 300 °C to form sodium polysulfide [16].

$$Na_2CO_3 + x\ S + SO_2 \rightarrow Na_2S_x + CO_2 + SO_3 \tag{68.2}$$

The metakaolinite reacts with sodium carbonate at higher temperatures of 700 to 800 °C and over a period of several days. It is transformed in this time into a framework with the three-dimensional sodalite structure:

$$3\ Al_4(OH)_{8-2x}OSi_4O_{10} + 6\ Na_2CO_3 \rightarrow 2\ Na_6Al_6Si_6O_{24} + 6\ CO_2 + (12-3x)\ H_2O \tag{68.3}$$

Na_2S_x formed according to eq. (68.2) reacts with the generated $Na_6Al_6Si_6O_{24}$ to pre-ultramarine, which is still colorless.

$$Na_6Al_6Si_6O_{24} + Na_2S_x \rightarrow Na_6Al_6Si_6O_{24} \cdot Na_2S_x \tag{68.4}$$

The pre-ultramarine formed in this reaction contains entrapped sodium and polysulfide ions in the crystal structure.

The next step is the oxidation of the polysulfide ions at temperatures of about 500 °C, whereby air is admitted in controlled amounts. Oxygen from the air reacts slowly in up to 20 days with excess sulfur to form sulfur dioxide. In an exothermic reaction, the di- and triatomic polysulfide ions are oxidized to S_2^- and S_3^- free radicals. Sodium-sulfur compounds and sulfur are formed as by-products.

$$2\ Na_6Al_6Si_6 \cdot O_{24}Na_2S_x + O_2 + SO_2 \rightarrow 2\ Na_7Al_6Si_6O_{24} \cdot S_x + Na_2SO_4 \tag{68.5}$$

The furnace is cooled down after the oxidation of the polysulfide ions. A full kiln cycle needs a period of 3 to 4 weeks. The formed product is called raw ultramarine. It contains typically 75 wt % blue ultramarine, 23 wt % sodium sulfoxides, and 2 wt % sulfur [3].

Purification and refinement procedures are done in a next step. They can take place batch-wise or continuously. The raw ultramarine blue is crushed and ground, transferred with warm water into a slurry, filtered and washed to remove the sulfur compounds. The sulfurous impurities are released by repeated elutriating and wet grinding. At the same time, the particle size of the ultramarine is reduced to a range of 0.1 to 10.0 µm.

Specific particle size distributions are obtained by gravity or centrifugal separation processes. Flocculation and filtration are typical techniques used for the removal of residual fine particles. The separated fractions are dried and ground to achieve pigment grades of different particle size distribution [3].

The production of violet ultramarine pigments takes place by heating a mid-range blue grade with ammonium chloride at about 240 °C in the presence of air. The treatment of the violet pigment with hydrogen chloride gas at 140 °C leads to pink ultramarine.

68.3 Pigment properties and uses

Ultramarine pigments are characterized by intensive shades of blue and relatively weak violet and pink color tones. The basic blue color of ultramarine is a rich, bright reddish blue. The blue color tone of ultramarine pigments is adjusted by varying the chemical composition and the particle size distribution. The violet and pink pigment types have significantly weaker and less saturated colors.

The final color quality of ultramarine pigments is adjusted by grinding to reduce the particle size and to achieve a specific size distribution. The mean particle size for most of the pigments is in the range from 0.7 to 5.0 µm. Scanning electron micrographs of representative ultramarine pigments show that the particles are characterized by an irregular shape [4]. The specific surface area of the pigments varies with the particle size distribution and has values in the range of 1 to 3 m^2/g.

Ultramarine pigments create a transparent blue in gloss paints and clear plastics. The reason for this is the refractive index of the pigments of 1.5. This value is similar to that of paints and plastics. Opacity of systems containing ultramarine pigments can be achieved by adding small quantities of white pigments [3].

Ultramarine blue is thermally stable up to around 400 °C in various applications. Violet pigments are stable up to 280 °C and pink pigments up to 220 °C, respectively. All three pigments are characterized by an excellent light fastness. A certain disadvantage of ultramarines is their instability against acids. The pigments decompose under the action of acids. A short-term stability against acids can be achieved by protecting the pigment particles with a silica surface treatment. Blue and violet ultramarine pigments are stable in diluted alkaline solutions, whereas pink grades undergo a color change to violet.

The stability and safety of ultramarine pigments are the basis for their use in many applications. The pigments are used among others in plastics, paints and powder coatings, printing inks, paper and paper coatings, rubber and thermoplastic elastomers, cosmetics, and leather finishes. Ultramarine pigments do not bleed or migrate from paints or polymers. They are therefore approved for a wide range of food contact applications.

Ultramarine pigments do not exhibit acute toxicity (LD_{50} value rat oral: >5000 mg/kg). They are not irritating to skin or mucous membranes. The pigments have a long-term and widespread human and environmental exposure without any reported ill effects. The worldwide application of ultramarine as an additive for sugar, to produce a whitening effect, and as a whitening agent for the use in household washing did not have adverse health effects.

A hazard of ultramarines is the liberation of hydrogen sulfide on contact with acids. It must therefore be ensured that no contamination with acids is possible during storage and disposal. Ultramarine pigments are not combustible, but in case of fire, sulfur dioxide can be formed and released. During the production process, sulfur dioxide and water-soluble sodium-sulfur compounds are generated. These must be disposed in accordance with the applicable environmental regulations [3].

References

1. Endriss H. Aktuelle anorganische Buntpigmente. Zorll U, editor. Hannover: Vincentz Verlag, 1997:85.
2. Wehner M. In Kittel - Lehrbuch der Lacke und Beschichtungen. Spille J, editor. vol. 5. 2nd ed. Stuttgart/Leipzig: S. Hirzel Verlag, 2003:125.
3. Calvert D. In industrial inorganic pigments. Buxbaum G, Pfaff G, editors. 3rd ed. Weinheim: Wiley-VCH Verlag, 2005:136.
4. Pfaff G. Inorganic pigments. Berlin/Boston: Walter de Gruyter GmbH; 2017:137.
5. Ball P. Bright earth - the invention of color. Sydney: Penguin Random House, 2001.
6. Podschus E, Hofmann U, Leschewski K. Z. Röntgenographische Strukturuntersuchung von Ultramarinblau und seinen Reaktionsprodukten. Anorg Allg Chem. 1936;228:305.
7. Tarling SE, Barnes P, Mackay AL. Simulation of industrial furnacing with powder X-ray diffraction. J Appl Cryst. 1984;17:96.
8. Tarling SE, Barnes P, Klinowsky J. The structure and Si, Al distribution of the ultramarines. Acta Cryst B. 1988;44:128.
9. Booth DG, Dann SE, Weller MT. The effect of the cation composition on the synthesis and properties of ultramarine blue. Dyes Pigm. 2003;58:73.
10. Clark RJ, Dines TJ, Kurmoo M. On the nature of the sulfur chromophores in ultramarine blue, green, violet, and pink and of the selenium chromophore in ultramarine selenium: characterization of radical anions by electronic and resonance Raman spectroscopy and the determination of their excited-state geometries. Inorg Chem. 1983;22:2766.
11. Clark RJ, Franks ML. The resonance Raman spectrum of ultramarine blue. Chem Phys Lett. 1975;34:69.

12. Clark RJH, Cobbold DG. Characterization of sulfur radical anions in solutions of alkali polysulfides in dimethylformamide and hexamethylphosphoramide and in the solid state in ultramarine blue, green, and red. Inorg Chem. 1978;17:3169.
13. Reinen D, Lindner GG. The nature of the chalcogen colour centres in ultramarine-type solids. Chem Soc Rev. 1999;28:75.
14. Gobeltz N, Demortier A, Lelieur JP, Duhayon C. Encapsulation of the chromophores into the sodalite structure during the synthesis of the blue ultramarine pigment. J Chem Soc Faraday Trans. 1998;94:2257.
15. Gobeltz-Hautecoeur N, Demortier A, Lede B, Lelieur JP, Duhayon C. Occupancy of the sodalite cages in the blue ultramarine pigments. Inorg Chem. 2002;41:2848.
16. Gobeltz N, Demortier A, Lelieur JP, Duhayon C. Identification of the products of the reaction between sulfur and sodium carbonate. Inorg Chem. 1998;37:136.

Gerhard Pfaff

69 White pigments

Abstract: White pigments are inorganic pigments, whose optical action is mainly based on non-selective light scattering. White pigments do not show absorption in the range of visible light, but have high scattering power leading to high hiding power in their applications. The greater the difference between the refractive index of the white pigment and that of the surrounding medium, the higher the scattering power.

Keywords: fillers, lithopone, titanium dioxide pigments, white pigments, zinc oxide pigments, zinc sulfide pigments

White pigments are inorganic pigments, whose optical action is mainly based on non-selective light scattering. White pigments do not show absorption in the range of visible light, but have high scattering power leading to high hiding power in their applications. The greater the difference between the refractive index of the white pigment and that of the surrounding medium, the higher the scattering power [1, 2].

White fillers, which are also used as pigments in emulsion paints and paper are distinguished from white pigments by the refractive index and by the intended use. Organic white pigments are unknown [1].

The first synthetically produced white pigment was white lead (basic lead carbonate). It was already known in Roman times. At 1800, zinc white (zinc oxide) was developed, followed by antimony white (antimony oxide) and zinc sulfide. Lithopones (zinc sulfide barium sulfate mixtures) were manufactured for the first time in the second half of the nineteenth century. During the twentieth century, the main processes for the production of titanium dioxide pigments were developed (sulfate and chloride process). Titanium dioxide in the anatase and the rutile modification is the most important white pigment today. Lithopones, zinc sulfide and zinc white have also a significant importance besides titanium dioxide pigments, whereas white lead and antimony white are not larger applied due to toxicological reasons [1, 2].

White pigments are used for coloration of coatings, emulsion paints, printing inks and plastics but also for brightening of different colored systems. They show good stability against chemical substances and high weather resistance.

Table 69.1 contains a summary of white pigments and fillers. Titanium dioxide, zinc oxide and zinc sulfide pigments are described in own chapters.

This article has previously been published in the journal Physical Sciences Reviews. Please cite as: G. Pfaff White Pigments *Physical Sciences Reviews* [Online] 2021, 6. DOI: 10.1515/psr-2020-0202

https://doi.org/10.1515/9783110587128-069

Table 69.1: Overview on white pigments and fillers [1].

Name	Formula	Refractive index n_D	C.I.
White pigments			
titanium dioxide	TiO_2		PW 6
anatase		2.55	
rutile		2.75	
zinc sulfide	ZnS	2.37	PW 7
zinc oxide (zinc white)	ZnO	2.01	PW 4
lithopones	$ZnS + BaSO_4$	1.84–2.08	PW 5
basic lead carbonate (white lead)	$2\,PbCO_3 \cdot Pb(OH)_2$	1.94–2.09	PW 1
antimony(III) oxide (antimony white)	Sb_2O_3	2.00–2.09	PW 11
Fillers			
calcium carbonate	$CaCO_3$	1.48–1.65	PW 18
barium sulfate	$BaSO_4$	1.64	PW 21
talc	$3\,MgO \cdot 4\,SiO_2 \cdot H_2O$	1.54–1.59	PW 26

References

1. Pfaff G. Inorganic pigments. Berlin/Boston: Walter de Gruyter GmbH, 2017:47.
2. Buxbaum G, Pfaff G, editors. Industrial inorganic pigments. Weinheim: Wiley-VCH Verlag, 2005:51.

Gerhard Pfaff

70 Zinc oxide pigments

Abstract: Zinc oxide pigments belong besides titanium dioxide and zinc sulfide pigments to the most important inorganic whites. The pigments are mostly used in paints, coatings, artists' colors, and cosmetics. Zinc oxide pigments are synthesized using various processes. The direct process (American process), the indirect process (French process), and the wet process have the greatest industrial importance for the production of ZnO pigments.

Keywords: direct process, indirect process, wet process, zinc oxide, zinc oxide pigments

70.1 Fundamentals and properties

Zinc oxide powders with the composition ZnO belong to the most important white pigments. ZnO pigments are also known as zinc white (C.I. Pigment White 4). Other terms for zinc oxide pigments are Chinese white and flowers of zinc. Zinc oxide occurs in nature in form of the mineral zincite and crystallizes in the hexagonal wurtzite structure [1–3].

Zinc oxide in its pure form is white, but changes the appearance to yellow when heated to temperatures above 300 °C. The change from white to yellow is reversible after cooling (thermochromism). There is a weak luminescence in the dark after heating up and cooling down. The thermochromism of ZnO is caused by a small excess of zinc (ca. 0.03%) generated by an oxygen release during the thermal treatment. The zinc excess respectively the concentration of oxygen defects in the ZnO lattice is dependent on the route used for the synthesis.

Zinc oxide is a direct semiconductor with by a band gap of 3.2 to 3.4 eV. It absorbs UV radiation (at wavelengths below 366 nm) and is transparent for visible light. The conductivity of ZnO is increased by doping with suitable elements. Aluminum and boron are preferably used for n-doping of zinc oxide. Other elements for n-doping are gallium and indium. The synthesis of p-doped ZnO is likewise possible, but more complicate. Zinc oxide interacts at high temperatures in a solid-state reaction with other oxides, e. g., with iron or manganese oxides to form ternary oxides such as zinc ferrites [1–3].

This article has previously been published in the journal Physical Sciences Reviews. Please cite as: G. Pfaff Zinc Oxide Pigments *Physical Sciences Reviews* [Online] 2021, 6. DOI: 10.1515/psr-2020-0203

https://doi.org/10.1515/9783110587128-070

70.2 Production of zinc oxide pigments

Zinc oxide pigments can be synthesized using various processes [1–8]. The direct process (American process), the indirect process (French process), and the wet process have attained the greatest industrial importance for the production of ZnO pigments.

Raw materials for the manufacture of zinc oxide pigments where formerly mainly zinc ores and concentrates (direct process) or metal from zinc producers (indirect process). Today, zinc oxide producers use preferably zinc residues and secondary zinc (zinc ashes from hot dip galvanization, zinc concentrates, metallization residues, skimming from casting furnaces, oxide residues from the indirect process).

Increasing demands for chemically pure and user-friendly white pigments have led to the modification of the traditional production processes. Sophisticated purification steps and a specific surface treatment are very important for the properties of the final pigments.

70.2.1 Direct process (American process)

The direct process for the manufacture of zinc oxide pigments consists of two steps, the high-temperature reduction of a zinc-containing raw material to zinc metal and the oxidation of the formed zinc to zinc oxide by oxygen. The raw material with a typical zinc oxide content of 60% to 75% is reduced with coal in rotary kilns at temperatures of 1000 to 1200 °C. The products of this high-temperature reaction are zinc vapor and carbon monoxide.

$$ZnO + C \rightarrow Zn + CO \tag{70.1}$$

The formed carbon monoxide acts as an additional reducing agent and carbon dioxide is generated as a further reaction gas besides zinc vapor.

$$ZnO + CO \rightarrow Zn + CO_2 \tag{70.2}$$

The formation of ZnO is already possible in this step, when a sufficient amount of oxygen is present in the kiln. The gas mixture, which leaves the kiln, contains zinc vapor, CO, and CO_2. The gases are passed into a reactor where the zinc and the carbon monoxide are oxidized to ZnO and CO_2, respectively.

$$Zn + CO + O_2 \rightarrow ZnO + CO_2 \tag{70.3}$$

The oxidation conditions are controlled carefully to adjust particle shape and size and to obtain the required yield. The formed zinc oxide is extracted and collected in filters. A direct use of the so-obtained ZnO powder as a pigment is possible. An improvement of the pigment properties is possible by thermal treatment at temperatures of up to 1000 °C and by a suitable surface treatment. The thermal treatment is executed particularly for the adjustment of the properties of high-purity zinc oxide, which is used in photocopiers. The surface treatment of ZnO particles improves their weather resistance, lightfastness and dispersibility in the most relevant application systems. The surface of the particles is coated typically with an organophilic coating generated from oil and propionic acid.

The direct process is considered, on the whole, as the most simple, cost-effective, and energy-efficient route for the manufacture of zinc oxide pigments.

70.2.2 Indirect process (French process)

The indirect process consists also of two steps, the vaporization of zinc by boiling and the oxidative combustion to ZnO in air. The adjustment of the combustion conditions is decisive for the achievement of the desired physical and chemical properties of the resulting zinc oxide powders. The chemical purity of the formed ZnO depends primarily on the composition of the zinc vapor.

The furnace type used for the necessary production of zinc vapor in the first process step is selected according to the raw materials and the highest possible yield. Pure metallic zinc, zinc-containing metal residues or secondary zinc are used as raw materials. The separation of various interfering metals, mainly of aluminum, cadmium, iron, or lead, is important. It is done by liquid or vapor phase techniques before the zinc is oxidized. Muffle furnaces or retorts of graphite or silicon carbide are used to vaporize the zinc respectively the zinc-containing metal. Undesired nonvolatile residues such as solid aluminum, iron and lead are collected in the muffle furnace or in the retort. They must be removed within certain time periods. Amounts of aluminum, cadmium, copper, iron and lead in the vapor have to be removed before the zinc oxidation. A suitable technique for the purification of the vapor is the fractional distillation in columns with silicon carbide plates.

After leaving the columns, the purified zinc vapor is oxidized with atmospheric oxygen in a combustion chamber connected with the distillation unit.

$$2\,Zn + O_2 \rightarrow 2\,ZnO \qquad (70.4)$$

The zinc oxide formed in this process step is extracted after cooling and collected in filters. It is possible to use the so-obtained ZnO powder as a pigment, but the properties can be improved further by a suitable surface treatment.

A variation of the indirect process uses a furnace with two separate chambers. Pieces of the metallic raw material are put into the first chamber of the furnace where the metal is heated and melted. The molten metal runs in the second chamber, which is connected with the first one. Distillation of the melt in the absence of air takes place in this electrically heated chamber. Residues of non-metals are removed from the surface before the distillation starts. Metallic impurities in the melt are collected in the residue from the distillation and are removed from there at certain intervals. The molten zinc is vaporized and converted to zinc oxide by the reaction with oxygen.

In a third variant, zinc oxide is synthesized in a rotary kiln starting also from zinc or zinc-containing metal. Melting, distillation and a part of the oxidation take place in the same zone of the kiln. Shape and size of the resulting zinc oxide particles as well as their impurity content are controlled and adjusted by the temperature and the gas atmosphere in the kiln [2].

70.2.3 Wet process

Zinc oxide pigments are also produced using a wet-chemical route. The process starts with the precipitation of zinc carbonate from purified solutions of zinc sulfate or zinc chloride by addition of Na_2CO_3 solution.

$$ZnSO_4 + Na_2CO_3 \rightarrow 2\,ZnCO_3 + Na_2SO_4 \tag{70.5}$$

The precipitation of zinc carbonate can also be performed with $KHCO_3$ instead of Na_2CO_3.

$$ZnSO_4 + 2\,KHCO_3 \rightarrow ZnCO_3 + K_2SO_4 + CO_2 + H_2O \tag{70.6}$$

Basic zinc carbonate can be precipitated as well, depending on the reaction conditions.

$$5\,ZnSO_4 + 5\,Na_2CO_3 + 3\,H_2O \rightarrow 2\,ZnCO_3 \cdot 3\,Zn(OH)_2 + 5\,Na_2SO_4 + 3\,CO_2 \tag{70.7}$$

Concentration of the reactants, choice of the raw materials, pH value, and temperature are the most decisive factors for the chemical composition and the physical properties of the formed precipitate, which is filtered, washed, and finally calcined to form zinc oxide.

$$ZnCO_3 \rightarrow ZnO + CO_2 \tag{70.8}$$

The wet process is particularly suitable for the production of zinc oxide pigments with a very fine particle size and a high specific surface area.

70.3 Pigment properties and uses

Zinc oxide pigments produced at high temperatures are crystalline and have particle sizes in the range from 0.1 to 5 µm. Depending on the synthesis conditions, the particles can have an acicular shape with sizes from 0.5 to 10 µm. Wet-chemically generated ZnO pigments are amorphous, have a sponge form, and have particle sizes in the range of up to 50 µm. The pigments are stable in all relevant application systems, but are soluble in organic and inorganic acids.

Zinc oxide pigments are typically offered as powders and to a lesser extent as pellets and granules. Dosage forms like pellets and granules have benefits for the pigment handling. They are often easier to disperse and show a significant dust reduction during processing. The pigments are mainly used in paints, coatings, and artists' colors. Other applications of zinc oxide pigments are exterior paints for wood preservation, antifouling and anticorrosion paints [9]. The use of ZnO improves film formation and the durability of cured films. Zinc oxide used in paints and coatings contributes to the avoidance of mildew.

The most important technical use of zinc oxide powders is natural and synthetic rubber. ZnO acts in this case as an activator for the vulcanization accelerators. The specific surface area of the zinc oxide correlates with its reactivity. The larger the surface area, the more reactive the ZnO powder. Moreover, zinc oxide contributes to a good durability of the vulcanized rubber and improves its thermal conductivity. Typical ZnO contents in rubbers are in the range of 2–5%.

Zinc oxide is also used as a component of glasses, ceramics, and enamels. ZnO has the advantage in these applications to reduce the thermal expansion, to lower the melting point, and to increase the resistance against chemicals. It should be noted that zinc oxide has the ability to modify the gloss and to improve the opacity of these materials.

ZnO is characterized by beneficial antibacterial properties. It is therefore used in form of powders and medical ointments in the pharmaceutical and cosmetic industries. The reaction of ZnO with eugenol is relevant for the formation of dental cements (zinc oxide eugenol cement).

Varistors based on zinc oxide are very important functional ceramic components. They are produced by calcination of highly purified ZnO with additives such as Bi_2O_3 [10]. The photoconductive properties of ZnO are used for photo-reproduction purposes. Doped ZnO is of importance for the fabrication of master papers for the offset reproduction [11].

Zinc oxide pigments are nontoxic and do not exhibit acute toxicity (LD_{50} value rat oral: >5000 mg/kg). ZnO is soluble in acids and can therefore be resorbed from living organisms. Zinc is an essential element playing a role for the growth of cells, skin, and hair and for the liver function. There is no indication for skin or eye irritating properties of ZnO powders. Zinc oxide pigments can be used in skin ointments in concentrations of up to 40%. Resorption of ZnO through the intact skin takes place to only a very limited extent. There is no evidence of carcinogenicity, genotoxicity and reproduction toxicity in human beings. Whereas zinc is essential for mammals, even very small quantities can be unfavorable for aquatic organisms. The presence of too much zinc in the water inhibits growth and photosynthesis and leads finally to the death of many organisms. Zinc ions have therefore to be removed from the waste water by precipitation or flocculation [3].

References

1. Jahn B. In Kittel-Lehrbuch der Lacke und Beschichtungen. Spille J, editor. vol. 5. 2nd ed. Stuttgart/Leipzig: S. Hirzel Verlag, 2003:179.
2. Jahn B. In industrial inorganic pigments. Buxbaum G, Pfaff G, editors. 3rd ed.Weinheim: Wiley-VCH Verlag, 2005:88.
3. Pfaff G. Inorganic pigments. Berlin/Boston: Walter de Gruyter GmbH, 2017:80.
4. Brown H. Zinc oxide rediscovered. New York: New Jersey Zinc Co, 1957.
5. Klingshirn C. ZnO: Material, physics and applications. ChemPhysChem. 2007;8:782.
6. Moezzi A, McDonagh AM, Cortie MB. Zinc oxide particles: synthesis, properties and applications. Chem Eng J. 2012;185–186;1.
7. Hänig G, Ulbrich KH. ZnO – Produkt zwischen Pigmentchemie und Hüttenwesen. Erzmetall. 1979;32:140.
8. Heiland G, Mollwo E, Stöckmann F. Electronic processes in zinc oxide. Solid State Phys. 1959;8:191.
9. Meyer G. New application for zinc. New York: Zinc Institute Inc., 1986.
10. Ziegler E, Helbig R. Der Metall-Oxid-Varistor (MOV). Phys unserer Zeit. 1986;17:171.
11. Patent DE. 3045591 (Matsushita). 1980.

Gerhard Pfaff

71 Zinc sulfide pigments

Abstract: Zinc sulfide pigments belong besides titanium dioxide and zinc oxide pigments to the most important inorganic whites. The pigments are mostly used in primers, plastics, fillers, putties, artists' colors, and emulsion paints. Zinc sulfide pigments are divided into pure zinc sulfide (ZnS) and lithopone (mixtures of ZnS with $BaSO_4$) pigments. They are synthesized using precipitation processes in aqueous medium. Compared with pure zinc sulfide pigments, lithopone pigments have a much larger technical importance.

Keywords: zinc sulfide pigments, lithopone pigments, precipitation, co-precipitation

71.1 Fundamentals and properties

Zinc sulfide pigments belong together with titanium dioxide and zinc oxide pigments to the most important white pigments. Besides pure ZnS pigments (C.I. Pigment White 7), also pigment mixtures consisting of zinc sulfide and barium sulfate (C.I. Pigment White 5) are available. Such mixtures are known under the name lithopone. Compared with pure zinc sulfide pigments, lithopone pigments have a much larger sales volume [1–4].

Zinc sulfide pigments find their application especially where good light scattering behavior is required, but also properties such as low abrasion, low oil number, or low Mohs hardness are necessary. The production of zinc sulfide pigments can start from industrial zinc-containing effluents. Thus, an environmentally friendly reuse of zinc from wastes is possible.

ZnS pigments excellently fulfill the main requirements for white pigments in regard to spectral properties, those are no absorption and optimal scattering of the incident light in the visible range. There are also zinc sulfide pigments available that are doped with cobalt to generate a bluish-white tinge, which goes back to the specific absorption edge in the near-UV range based on the doping.

ZnS crystallizes in the sphalerite (zinc blende) or wurtzite lattice depending on the specific production parameters. The refractive index, which is decisive for the scattering properties, is with 2.37 significantly higher than that of binders or other application media (1.5 to 1.6). Spherical ZnS particles have the maximum scattering

This article has previously been published in the journal Physical Sciences Reviews. Please cite as: G. Pfaff Zinc Sulfide Pigments *Physical Sciences Reviews* [Online] 2021, 6. DOI: 10.1515/psr-2020-0204

https://doi.org/10.1515/9783110587128-071

power when their diameters are in the range of 290–300 nm. The contribution of barium sulfate to the light scattering phenomena in the case of lithopone pigments is only small due to the relatively low refractive index of $BaSO_4$ (1.64). Barium sulfate acts as a filler (extender) in these pigments and contributes to the scattering efficiency of the zinc sulfide to a certain degree.

71.2 Production of zinc sulfide pigments

Raw materials for the manufacture of zinc sulfide pigments are zinc oxide and zinc slags from smelters, ammonium chloride slags from hot dip galvanizing, and liquid zinc-containing waste such as pickle liquors from galvanizing plants.

Barium sulfate powders are produced starting with the synthesis of barium sulfide of sufficient purity. BaS is produced by the reaction of an intimate mixture of crushed natural barium sulfate (barite), with petroleum coke at temperatures of 1000 to 1300 °C in a rotary kiln.

$$BaSO_4 + 2\ C \rightarrow BaS + 2\ CO_2 \tag{71.1}$$

BaS is soluble in water and can therefore be dissolved in warm water. Insoluble reaction products can be separated by filtration and collected in the filter cake. The obtained barium sulfide solution is adjusted to a concentration of ca. 200 g/l BaS and pumped into the lithopone precipitation vessel. The almost clear barium sulfide solution can be stored only for a short period of time in the vessel to avoid undesired formation of polysulfides.

High-purity barium sulfide can be obtained in smaller quantities by thermal treatment of barium carbonate in a stream of hydrogen sulphide at 1000 °C [1,2,3,4].

$$BaCO_3 + H_2S \rightarrow BaS + H_2O + CO_2 \tag{71.2}$$

71.2.1 Pure zinc sulfide pigments

The pigment synthesis starts with dissolving of the raw materials in an acid. Typically, sulfuric acid is used to form a zinc-containing solution. Impurities in the acidic solution thus obtained are separated by filtration or precipitation. Iron, nickel, chromium, manganese, silver, cadmium, and other heavy metals have to be removed before the next reaction step. The purification starts with the precipitation of iron and manganese as oxide hydroxides and of cobalt, nickel, and cadmium as hydroxides. In the following step, zinc dust is added at 80 °C to the solution. All the metals that are situated at a higher position in the electrochemical voltage series than zinc (e. g., chromium, cadmium, cobalt, nickel, lead, iron, copper, silver) are reduced and completely precipitated. Zinc is the only metal, which remains dissolved under these conditions. The metal slime formed after the zinc addition is filtered off and recycled. A small amount of a soluble cobalt salt is now added to the purified zinc salt solution. Its task is to stabilize the ZnS that should be formed in the next step by incorporation of cobalt ions (0.02 to 0.5%) in the zinc sulfide crystal lattice. This incorporation leads to a significant stabilization of the ZnS against UV light. Without such a treatment, zinc sulfide pigments become gray when exposed to sunlight.

Next step of the pigment synthesis is the precipitation of zinc sulfide. The purified cobalt-containing zinc salt solution reacts with a highly purified sodium sulfide solution. Finely precipitated ZnS is formed during this reaction. The conditions, especially pH value, temperature and concentration, play an important role for the achievement of the desired pigment quality.

$$ZnSO_4 + Na_2S \rightarrow ZnS + Na_2SO_4 \qquad (71.3)$$

The white precipitate is separated from the suspension by filtration, washed until it is free of sulfate and dried. It is then calcined at 700 °C to get the optimal narrow particle size distribution of about 290–300 nm and thus the best scattering power for the pigment. The cobalt ions from the cobalt salt solution, which was added to the zinc salt solution before the precipitation, are incorporated in the ZnS crystal lattice during the annealing.

There is also the possibility to produce zinc sulfide pigments using a hydrothermal route. ZnS is precipitated in this case with a slight excess of sulfide at a pH value of 8.5. The pH value is increased in the following to 12 to 13 and sodium carbonate is added. The suspension is autoclaved for 15 to 20 min at 250 to 300 °C. Zinc sulfide formed by the hydrothermal process crystallizes in the sphalerite structure in contrast to the product formed by precipitation and calcination, which has the wurtzite structure and about 10% less scattering power. ZnS pigments synthesized by the hydrothermal process have the higher quality. The process for their production is, however, less economic due to the higher production costs [3].

71.2.2 Lithopone

Lithopone pigments are manufactured by the reaction of a soluble zinc salt with barium sulfide in aqueous solution. $ZnSO_4$ is the preferred zinc salt for the synthesis. It is obtained by the reaction of the raw materials described above with sulfuric acid. The reaction of $ZnSO_4$ and BaS in the ratio 1:1 is a co-precipitation from a chemical point of view and leads to a white co-precipitate with the theoretical composition 29.4 wt % ZnS and 70.6 wt % $BaSO_4$.

$$ZnSO_4 + BaS \rightarrow ZnS + BaSO_4 \qquad (71.4)$$

Different molar ratios of ZnS and $BaSO_4$ are likewise possible. A starting solution containing for example $ZnSO_4$, $ZnCl_2$ and BaS in the molar ratio 1:3:4 leads to a co-precipitate with the composition 62.5 wt % ZnS and 37.5 wt % $BaSO_4$.

$$ZnSO_4 + 3\ ZnCl_2 + 4\ BaS \rightarrow 4\ ZnS + BaSO_4 + 3\ BaCl_2 \qquad (71.5)$$

Purified solutions of the zinc salt respectively of the zinc salts and of BaS are a necessity to obtain white pigments of high quality. These solutions are mixed thoroughly in a precipitation reactor. Temperature, pH value, salt concentration, stirring speed reactor geometry and other parameters have to be controlled carefully. A "raw lithopone" precipitate is formed consisting of very fine ZnS and $BaSO_4$ particles. It is filtered, washed, and dried using a suitable dryer, e. g., a turbo dryer. The filter cake is crushed to lumps of ca. 2 cm size. These lumps are directly heated up in a rotary kiln with natural gas to temperatures of 650 to 700 °C. Traces of sodium, potassium, and magnesium salts are added to take influence on the crystal growth. Temperature and residence time in the kiln are the two main parameters for the control of the particle sizes of ZnS and $BaSO_4$. The ZnS particles should have an optimal size of about 300 nm, whereas the size of the $BaSO_4$ particles should be in the range of up to 1 μm.

The hot ZnS/$BaSO_4$ mixture is quenched in water after the high-temperature treatment. The product thus obtained typically passes classifiers and hydro-separators, a thickener unit, and rotary filters. It is washed on the filter until it is free of salts, dried in a suitable dryer and ground in a high-intensity mill. The product should now have the desired pigment properties. It can be equipped additionally with an organic surface treatment depending on the intended use [3].

71.3 Pigment properties and uses

Pure zinc sulfide as well as lithopone pigments are thermally stable in air up to temperatures of about 550 °C. Their texture is soft compared with other white pigments. Zinc sulfide is stable in aqueous systems in the pH range of 4 to 10 and in most organic solvents. Slow oxidative decomposition of ZnS under the influence of UV radiation takes place when oxygen and water are present. Barium sulfate is very stable and insoluble in acids, bases, and organic media.

The most common types of lithopone pigments have a ZnS content of 30% (red seal) or 60% ZnS (silver seal). Other compositions are of minor importance. They contain 15% (yellow seal), 40% (green seal), or 50% (bronze seal) zinc sulfide. A specialty is the product Sachtolith with a ZnS content above 97%. Pure zinc sulfide pigments belong to the Sachtolith type products. Zinc sulfide pigments are often equipped with a surface treatment. Such a treatment can consist of inorganic and organic compounds and can have a hydrophilic or hydrophobic character.

Lithopone pigments are mainly used in primers, plastics, fillers, putties, artists' colors, and emulsion paints. Only low binder amounts are needed for the pigments. Lithopone pigments have good wetting and dispersion properties. Lithopone and TiO_2 pigments can be used together in various application media in a favorable way. The addition of TiO_2 to a lithopone pigment increases the refractive index of the latter one and improves its scattering behavior in the application.

Lithopone pigments are often used in UV-cured paint systems. They are responsible for a strong shift of the absorption band towards the blue caused by the ZnS.

Pure zinc sulfide and lithopone pigments as well as other zinc compounds have fungicidal and algicidal effects. Their application is beneficial in paint formulations for exterior use, where they prevent the growth of algae or fungi on the hardened paint surface.

The use of lithopone pigments in plastics is based on their good lightfastness and the clear bluish-white shade. The pigments show an advantageous behavior in liquid plastic materials during the extrusion process, mainly high throughput rates and economic operation of the extruder equipment. Lithopone pigments find also application as flame retardants. They can substitute in fire-resistant systems up to 50% of the toxic antimony trioxide without any adverse effect [3, 4].

Zinc sulfide pigments cannot keep up with titanium dioxide in many applications. A replacement of lithopone by TiO_2 pigments may be beneficial especially in coating systems. However, ZnS types of high quality will maintain their importance for the market., especially in applications where special technical properties in addition to a specific light scattering are needed.

Zinc sulfide pigments do not exhibit acute toxicity (LD_{50} value rat oral: >5000 mg/kg). Studies on rats, which had zinc sulfide pigments administered, only showed an extremely low toxicity. Inhalation of pigment dust may cause mechanical respiratory tract irritation and contact with skin or eyes can lead to irritation by mechanical friction.

The effects of zinc ions on living organisms are very complex. The human body contains 2 g of zinc and requires daily quantities of 10 to 20 mg for metabolic processes. On the other hand, soluble zinc in large quantities is toxic. The low solubility of ZnS is the reason why it does not cause problems for the human body. The insufficient concentration of stomach acid and the relatively low solubility of zinc sulfide contribute to the fact that there are no physiologically significant amounts of soluble zinc detectable. The application of both zinc sulfide and barium sulfate in contact with foods is permitted by the most authorities responsible for toxicology and occupational health [4].

References

1. Clausen H. Zinc based pigments. In Pigments Handbook. Lewis PA, editor. vol. 1. 2nd ed. New York: John Wiley & Sons, 1988:43.
2. Küppers H-J. In Kittel-Lehrbuch der Lacke und Beschichtungen. Spille J, editor. vol. 5. 2nd ed. Stuttgart/Leipzig: S. Hirzel Verlag, 2003:105.
3. Griebler W-D. In industrial inorganic pigments. Buxbaum G, Pfaff G, editors. 3rd ed. Weinheim: Wiley-VCH Verlag, 2005:81.
4. Pfaff G. Inorganic pigments. Berlin/Boston: Walter de Gruyter GmbH, 2017:74.

Index

https://doi.org/10.1515/9783110587128-205